Environmental Hydraulics, Turbulence and Sediment Transport, 2nd edition

Environmental Hydraulics, Turbulence and Sediment Transport, 2nd edition

Editors

Jaan H. Pu
Prashanth Reddy Hanmaiahgari
Manish Pandey
Mohammad Amir Khan

MDPI

Basel • Beijing • Wuhan • Barcelona • Belgrade • Novi Sad • Cluj • Manchester

Editors

Jaan H. Pu
Faculty of Engineering and Digital Technologies
University of Bradford
Bradford
UK

Prashanth Reddy Hanmaiahgari
Civil Engineering
Indian Institute of Technology Kharagpur
Kharagpur
India

Manish Pandey
Civil Engineering
Indian Institute of Technology Kharagpur
Kharagpur
India

Mohammad Amir Khan
Galgotias College of Engineering and Technology
Uttar Pradesh
India

Editorial Office
MDPI
St. Alban-Anlage 66
4052 Basel, Switzerland

This is a reprint of articles from the Special Issue published online in the open access journal *Fluids* (ISSN 2311-5521) (available at: https://www.mdpi.com/journal/fluids/special_issues/hydraulics_bedforming_II).

For citation purposes, cite each article independently as indicated on the article page online and as indicated below:

Lastname, A.A.; Lastname, B.B. Article Title. *Journal Name* **Year**, *Volume Number*, Page Range.

ISBN 978-3-7258-1427-5 (Hbk)
ISBN 978-3-7258-1428-2 (PDF)
doi.org/10.3390/books978-3-7258-1428-2

Contents

About the Editors

Jaan H. Pu

Jaan H. Pu: Dr Pu has been an Associate Professor at the University of Bradford, UK, since 2017. Dr Pu's research concentrates on numerical and laboratory approaches to represent various water engineering applications, which include the naturally compound riverine flow, sediment transport, scouring, water quality, and vegetated flow. His research outputs have led to several high-quality journal articles (70+), conference proceedings (10+), edited books (6), and book chapters (2). He has supervised/is supervising eight PhD students at Bradford (six as their principal supervisor) to investigate river hydrodynamics, vegetation, and sediment transport-challenging applications. He has been appointed as an Associate Editor for *Frontiers in Environmental Science* and *Frontiers in Built Environment* and as a Guest Editor for several Special Issues in *Water*, *Fluids*, and *Frontiers in Environmental Science.*

Prashanth Reddy Hanmaiahgari

Prashanth Reddy Hanmaiahgari: Dr. Hanmaiahgari is an Associate Professor in the Department of Civil Engineering at the Indian Institute of Technology Kharagpur. His research interests include the experimental and numerical modelling of free surface flows, pressurized flow in closed conduits and turbulence, sediment transport, and river morphology. He has taught unsteady flows, open channel flow hydraulics, turbulence, and sediment transport. He has also published several peer-reviewed journal papers on turbulence, unsteady flows, and sediment transport.

Manish Pandey

Manish Pandey: Dr Pandey is currently an Assistant Professor in the Department of Civil Engineering at the Indian Institute of Technology Kharagpur. His research interests include experimental hydraulics, sediment transport, river training works, hydraulics structures, and flow through open channels. He has more than six years of teaching and research experience in experimental hydraulics and water resources engineering. He has authored more than 60 peer reviewed journal papers, book chapters, and conferences. He has supervised two PhD and ten MTech students. Presently, he is supervising three Ph.D. and three MTech students to investigate river hydrodynamics and sediment transport-challenging applications.

He has been a member of the Editorial Board of several high-ranked journals. He has collaborated with a number of leading universities and research organizations, including the University of Glasgow, the University of Bradford, Southeast University China, NCHU Taiwan, etc.

Mohammad Amir Khan

Mohammad Amir Khan: Dr Khan has completed P.hd. from IIT Roorkee. Currently he is working on the flow turbulence in the vicinity of mid-channel braided bar. Besides, he is working on the flow turbulence in the tetrahedron field.

fluids

Editorial

Environmental Hydraulics, Turbulence, and Sediment Transport, Second Edition

Jaan H. Pu [1,*], Manish Pandey [2] and Prashanth Reddy Hanmaiahgari [2]

1 Faculty of Engineering and Digital Technologies, University of Bradford, Bradford BD7 1DP, UK
2 Indian Institute of Technology, Kharagpur 721302, India; manishpandey3aug@gmail.com (M.P.); hpr@civil.iitkgp.ac.in (P.R.H.)
* Correspondence: j.h.pu1@bradford.ac.uk

Within river systems, the process of bed-forming is intricate, dynamic and is shaped by different factors. Hydraulic forces exerted by water flow play a crucial role, forming the bed substrate over time. Additionally, the presence of vegetation within the riverbed and along its banks introduces further complexity, as the interaction between plants and hydrodynamics can alter sediment transport patterns and riverbed morphology. The movement of both suspended particles and bedload materials within the water column contributes to the ongoing riverbed landscape evolution. The primary aim of this editorial collection is to assemble an extensive range of research methodologies aimed to inform engineering practices pertinent to river management. Through an exhaustive exploration of various topics, including water quality indexing, erosion and sedimentation patterns, influence of vegetation, hydrological modelling for understanding flow dynamics, and identification of critical hydraulic parameters with the utilisation of both analytical and experimental modelling techniques, this paper endeavours to provide valuable insights derived from rigorous research efforts. By synthesising and presenting these findings, we offer a resource that can effectively guide future endeavours in river engineering and related disciplines.

The purpose of this Research Topic (RT) is to examine the latest advances and developments in addressing the challenges in the fluvial systems as well as to discuss the opportunities they create for sustainable city solutions. This RT comprises eleven research articles contributed by 49 authors, organised into several distinct themes. These themes encompass studies on managing scour phenomenon, mitigating water pollution, understanding behavioural aspects of fluvial dynamics, and exploring efforts towards river restoration through representative modelling. The key insights are as follows:

- Water quality and ecological impact on fluvial systems;
- Drainage network and sediment management;
- Hydrological and hydrodynamic modelling and analysis;
- Experimental approaches of the scour phenomenon.

Several proposed models were used to predict the Water Quality Index (WQI) and to investigate how industrialisation has affected the quality of surface water in the Kelani River basin, Sri Lanka [1]. From 2005 to 2012, nine criteria were assessed, and the results showed that the water quality around industrial zones, specifically at Raggahawatte Ela, was impaired. The results highlight the necessity of taking proactive steps in water management to address the declining quality of surface water from industrial activity. The impact of finite-length vegetation on flow structure during inland tsunami events was assessed using numerical simulations incorporating a Volume of Fluid (VOF) technique combined with the Reynolds Stress Model (RSM) to accurately track the free surface level [2]. The study outcomes indicate that the wider vegetation patches generate large vortices and low-velocity areas downstream, while longer patches reduce velocities within the vegetation

Citation: Pu, J.H.; Pandey, M.; Hanmaiahgari, P.R. Environmental Hydraulics, Turbulence, and Sediment Transport, Second Edition. *Fluids* **2024**, 9, 120. https://doi.org/10.3390/fluids9060120

Received: 7 May 2024
Accepted: 10 May 2024
Published: 22 May 2024

but create high-velocity zones downstream. This study underscores the importance of considering vegetation layout based on Aspect Ratio (AR) for effective tsunami mitigation.

Drainage network and sediment management can effectively indicate the erosion in rural and urban drainage channels on tidal lowland [3]. However, the study suggests the lack of suitable erosion assessment models. Using the MIKE-11 2D program, different erosion scenarios were simulated, with scenario of 75% O&M model identified as being the most suitable. Numerical simulation methods were applied for hydraulic flushing in reservoir sediment management, covering drawdown flushing, pressure flushing, and turbidity current venting [4]. The study discusses empirical/analytical methods and 1D to 3D computational fluid dynamic models. Case studies illustrated model selection, domain determination, inputs, performance, and results comparison. The key findings include recommending 1D models for narrow reservoirs, 2D models for wide reservoirs or significant lateral changes, and 3D models for pressure flushing or near-field processes.

The performance of hydrological and hydrodynamic study was compared using SWAT and HEC-HMS models [5]. Both models were calibrated and validated for the period 2007–2014. The results showed satisfactory performance for both models, with SWAT capturing high flows more accurately and HEC-HMS performing better for medium flows. Both models accurately simulated low flows. SWAT outperformed HEC-HMS in seasonal flow simulations, making it accurate for wet and dry seasons. The critical hydraulic parameters (CHPs) downstream of the hydraulic structure were investigated using FLOW-3D numerical models [6]. The study identified significant parameters like velocity profiles, Froude number, and turbulent kinetic energy. The results show discrepancies from previous studies, with differences in hydraulic jumps (HJs) and sequent depths. The study suggests that the barrage efficiently holds HJs at investigated tailwater levels. The numerical model has been proven to be able to assess the behaviour of round buoyancy-driven particle clouds in still ambient conditions [7]. The results of the numerical models showed that the flow behaviour varied depending on particle buoyancy, indicating that the Boussinesq assumption is not fully suitable for such flow representation.

As observed by [3,8], the suspended sediment distribution can be represented using the Rouse model and Kundu and Ghoshal model for hyper-concentrated and diluted flows. Machine learning techniques, including XGBoost Classifier, Linear Regressor (Ridge), Linear Regressor (Bayesian), K Nearest Neighbours, Decision Tree Regressor, and Support Vector Machines (Regressor), were further applied to determine the relationship between different sediment and flow parameters.

Using cable flow–altering experimental approaches of the scour phenomenon, one can evaluate scour countermeasures [9]. It was observed that scour reduction increased with larger cable diameters and finer sediment sizes. Cables attenuated the flow within the scour holes, reducing the downflow and horseshoe vortices. Increasing the cable–pier diameter ratio further reduced the scour depth, especially for finer sediments. It was also found that cables mitigated downstream scour action and reduced the vortex shedding frequency. Experimental investigations demonstrated the effectiveness of riprap, submerged vanes, and the combination of the two in reducing scour around vertical walls and spill-through abutments under clear-water conditions [10]. Riprap proved more effective than submerged vanes alone, but the combination yielded the highest reduction in scour depth (up to 54% for vertical walls and 39% for spill-through abutments). Properly scaled riprap aprons also reduced the required riprap volume significantly. Submerged vanes improved the riprap stability and reduced edge failure, particularly with square-shaped riprap layers [11]. The effectiveness of roughened elements for mitigating bridge abutment scour were tested on rectangular abutments, with the optimal installation depth found to be 0.6–0.8 times the abutment length. Elements with a thickness and protrusion of 0.2 times the length exhibited the most effective protection, reducing the maximum scour depth by up to 32.8%.

Finally, this collective research sheds light on the complexities and possibilities inherent in managing and mitigating impacts of fluvial systems. Through a diverse array of research papers and methodologies, this RT explores current and emerging approaches aimed at enhancing the resilience of water infrastructures. Each selected contribution offers insights into innovative theories, cutting-edge technologies, and practical applications within their specific areas of study. Lastly, we extend our gratitude to all reviewers and authors whose invaluable contributions have enriched this RT.

Author Contributions: Conceptualisation, J.H.P. and M.P.; writing—original draft preparation, M.P.; writing—review and editing, J.H.P. and P.R.H. All authors have read and agreed to the published version of the manuscript.

Data Availability Statement: All data of this editorial paper are allocated in this paper itself.

Conflicts of Interest: The authors declare no conflicts of interest.

References

1. Makubura, R.; Meddage, D.P.P.; Azamathulla, H.M.; Pandey, M.; Rathnayake, U. A Simplified Mathematical Formulation for Water Quality Index (WQI): A Case Study in the Kelani River Basin, Sri Lanka. *Fluids* **2022**, *7*, 147. [CrossRef]
2. Amina; Tanaka, N. Numerical Investigation of 3D Flow Properties around Finite Emergent Vegetation by Using the Two-Phase Volume of Fluid (VOF) Modeling Technique. *Fluids* **2022**, *7*, 175. [CrossRef]
3. Syarifudin, A.; Satyanaga, A.; Wijaya, M.; Moon, S.-W.; Kim, J. Sediment Transport Patterns of Channels on Tidal Lowland. *Fluids* **2022**, *7*, 277. [CrossRef]
4. Lai, Y.G.; Huang, J.; Greimann, B.P. Hydraulic Flushing of Sediment in Reservoirs: Best Practices of Numerical Modeling. *Fluids* **2024**, *9*, 38. [CrossRef]
5. Chathuranika, I.M.; Gunathilake, M.B.; Baddewela, P.K.; Sachinthanie, E.; Babel, M.S.; Shrestha, S.; Jha, M.K.; Rathnayake, U.S. Comparison of Two Hydrological Models, HEC-HMS and SWAT in Runoff Estimation: Application to Huai Bang Sai Tropical Watershed, Thailand. *Fluids* **2022**, *7*, 267. [CrossRef]
6. Zaffar, M.W.; Haasan, I.; Ghumman, A.R. Numerical Investigation of Critical Hydraulic Parameters Using FLOW-3D: A Case Study of Taunsa Barrage, Pakistan. *Fluids* **2023**, *8*, 310. [CrossRef]
7. Alnahit, A.O.; Kaye, N.B.; Khan, A.A. Understanding the Influence of the Buoyancy Sign on Buoyancy-Driven Particle Clouds. *Fluids* **2024**, *9*, 101. [CrossRef]
8. Kumar, S.; Singh, H.P.; Balaji, S.; Hanmaiahgari, P.R.; Pu, J.H. Inclusive Hyper- to Dilute-Concentrated Suspended Sediment Transport Study Using Modified Rouse Model: Parametrized Power-Linear Coupled Approach Using Machine Learning. *Fluids* **2022**, *7*, 261. [CrossRef]
9. Muhawenimana, V.; Foad, N.; Ouro, P.; Wilson, C.A.M.E. Local Scour Patterns around a Bridge Pier with Cable-Wrapping. *Fluids* **2023**, *8*, 3. [CrossRef]
10. Fathi, A.; Zomorodian, S.M.A.; Zolghadr, M.; Chadee, A.; Chiew, Y.-M.; Kumar, B.; Martin, H. Combination of Riprap and Submerged Vane as an Abutment Scour Countermeasure. *Fluids* **2023**, *8*, 41. [CrossRef]
11. Zolghadr, M.; Zomorodian, S.M.A.; Fathi, A.; Tripathi, R.P.; Jafari, N.; Mehta, D.; Sihag, P.; Azamathulla, H.M. Experimental Study on the Optimum Installation Depth and Dimensions of Roughening Elements on Abutment as Scour Countermeasures. *Fluids* **2023**, *8*, 175. [CrossRef]

fluids

MDPI

Case Report

A Simplified Mathematical Formulation for Water Quality Index (WQI): A Case Study in the Kelani River Basin, Sri Lanka

Randika Makubura [1], D. P. P. Meddage [2], Hazi Md. Azamathulla [3], Manish Pandey [4] and Upaka Rathnayake [1,*]

1 Department of Civil Engineering, Faculty of Engineering, Sri Lanka Institute of Information Technology, New Kandy Road, Malabe, Colombo 10115, Sri Lanka; randikamk.96@gmail.com
2 Department of Civil and Environmental Engineering, Faculty of Engineering, University of Ruhuna, Hapugala 80000, Sri Lanka; meddage.p@cee.ruh.ac.lk
3 Department of Civil and Environmental Engineering, The Faculty of Engineering, The University of West Indies, St. Augustine 32080, Trinidad and Tobago; hazi.azamathulla@sta.uwi.edu
4 Department of Civil Engineering, Water Resources and Environmental Division, National Institute of Technology Warangal, Warangal 506004, India; mpandey@nitw.ac.in
* Correspondence: upaka.r@sliit.lk; Tel.: +94-719883318

Abstract: Surface water quality is degraded due to industrialization; however, it is one of the widely used sources for water supply systems worldwide. Thus, the polluted water creates significant issues for the health of the end users. However, poor attention and concern can be identified on this important issue in most developing countries, including Sri Lanka. The Kelani River in Sri Lanka is the heart of the water supply of the whole Colombo area and has the water intake for drinking purposes near an industrialized zone (Biyagama). Therefore, this study intends to analyze the effect of industrialization on surface water quality variation of the Kelani River basin in Sri Lanka in terms of the water quality index (WQI). We proposed a regression model to predict the WQI using the water quality parameters. Nine water quality parameters, including pH, total phosphate, electric conductivity, biochemical oxygen demand, temperature, nitrates, dissolved oxygen, chemical oxygen demand, and chlorine evaluated the Kelani River water quality. The proposed regression model was used to examine the water quality of samples obtained at twelve locations from January 2005 to December 2012. The highest WQI values were found in Raggahawatte Ela throughout the 8 years, located near the Biyagama industrial zone. The relationship of industries to water quality in the Kelani River is stated. The surface water quality gradually decreased as a result of development and industrialized activities. Therefore, this work showcases and recommends the importance of introducing necessary actions and considerations for future water management systems.

Keywords: Kelani River; simplified mathematical model; spatiotemporal variation; water quality index (WQI)

Citation: Makubura, R.; Meddage, D.P.P.; Azamathulla, H.M.; Pandey, M.; Rathnayake, U. A Simplified Mathematical Formulation for Water Quality Index (WQI): A Case Study in the Kelani River Basin, Sri Lanka. *Fluids* **2022**, *7*, 147. https://doi.org/10.3390/fluids7050147

Academic Editor: Mahmoud Mamou

Received: 6 April 2022
Accepted: 22 April 2022
Published: 23 April 2022

1. Introduction

Water is a primary necessity for all humans and natural systems and the most critical resource for society's sustainable development [1]. Surface water has a significant contribution to world drinking water volumes. Therefore, surface water can be treated as one of the most important ecological factors, which balance the hydrologic cycle. The river basins are an essential component of the surface water and thus of hydrology. They have unique geographical characteristics that guide various flow situations. Each river basin is a self-contained drainage system in which different types of water are cycled through constant flow and mutual transformation [2]. Increased anthropogenic activities, such as sand mining, transportation, irrigation, small-scale fishing, manufacturing, and disposal of municipal and agricultural waste including toxic chemicals, have increased the amount of waste and pollutants discharged directly into rivers. As a result, natural river systems become contaminated [3–5]. Additionally, numerous studies have revealed that the impact of industries on surface water has increased considerably in recent years while highlighting the critical parameters affecting the quality of surface water [5–7]. Statistics showcase that around 10% of industries release treated wastewater directly into streams, while the remainder dumps untreated effluent directly into rivers and other bodies of water [8]. However, the dumping percentage for untreated wastewater can be higher in developing countries.

Surface water deterioration due to increased pollution is one of the significant issues facing the growing populations worldwide [9]. This is a considerable threat to the drinking water supply in most countries and impacts economic development due to various health issues. According to United Nations International Children's Emergency Fund (UNICEF) [10], especially in Asia, improving the quality of water supplies is a key priority because poor water quality significantly threatens community health, primarily through water-borne diseases. To minimize these problems, many countries executed their water quality protection measures through continuous water quality monitoring [11]. It is imperative to evaluate the quality of water and the key parameters, which usually vary spatially and temporally, to have a clearer understanding of water resource conditions for human consumption and other activities.

The quality of surface water and the quality of groundwater are at a threat due to anthropogenic activities all over the world [12–14]. The situation is similar in Sri Lanka, and both surface and groundwater quality have deteriorated dramatically during the last few decades. Rapid industrialization in Sri Lanka is one of the main reasons for the water quality deterioration [15–17]. Major industries and factories such as textile industries, raw rubber factories, rubber latex factories, beverage factories, milk and food industries, plywood factories, steel manufacturing factories, chemical industries, fertilizer manufacturing factories, petroleum refineries, and other commercial industries directly contribute to water pollution-induced problems in Sri Lanka [18,19]. The Kelani River has been identified as one of the most polluted rivers in Sri Lanka. However, the Kelani River is the primary source of water supply for Colombo, the commercial heart of Sri Lanka. Therefore, the catchment area of the Kelani River basin from Glencorse (6.9643° N, 80.1877° E) to the river mouth (6.9787° N, 79.8700° E) has been identified as an environmentally sensitive area considering the importance of maintaining acceptable water quality concentration limits.

However, most of the manufacturing industries (both government and private sector) are concentrated in the western province of Sri Lanka and occupy a significant proportion of the area of the Kelani River basin. These include the largest export processing zones and industrial estates, such as the Biyagama and Katunayake export processing zones and the Seethawaka Industrial estate. Some industries have direct effluents to the river, while some other industries release the treated wastewater. Therefore, the acceptable water quality concentrations are exceeded at several sites along the river.

Therefore, the drinking water supply to Colombo is under threat, and solid research and implementation of the outcomes are required to provide safe drinking water to more than 1 million people. However, the study is a minimum to provide a holistic view of the water quality along the Kelani River as a single index. The water quality index (WQI) combines a set of water quality parameters into a single index and is extensively used in surface water quality assessments [20–23]. The index provides the overall water quality at an identified location and time. Therefore, to overcome the identified research gap, the measured water quality parameters along the Kelani River were presented as a single index, the WQI. The variations in water quality concentrations were analyzed, and spatial representations of the affected areas were presented.

2. Study Area and Methodology

2.1. Study Area

The Kelani River is one of the longest rivers in Sri Lanka. It is 144 km long and drains an area of 2230 km^2 [24]. It starts in the middle of the central hills in Sri Lanka (Nallatanniya area) and ends at Mattakuliya in Colombo, Sri Lanka's most prominent commercial city. As was already stated, the Kelani River is one of the most important rivers in Sri Lanka as it covers more than 80% of the water demand of Colombo. The other main uses of Kelani River's water are transportation, irrigation, fisheries, and hydropower generation. The flow in the Kelani River varies between 800 and 1500 m^3/s during monsoon seasons, whereas it falls to 20–25 m^3/s in the dry season, depending on the operation of 3 reservoirs in the catchment [25]. The land use patterns of the Kelani River basin are showcased in Figure 1. The upstream is somewhat greenish, whereas the downstream of the river basin has significant built-up land.

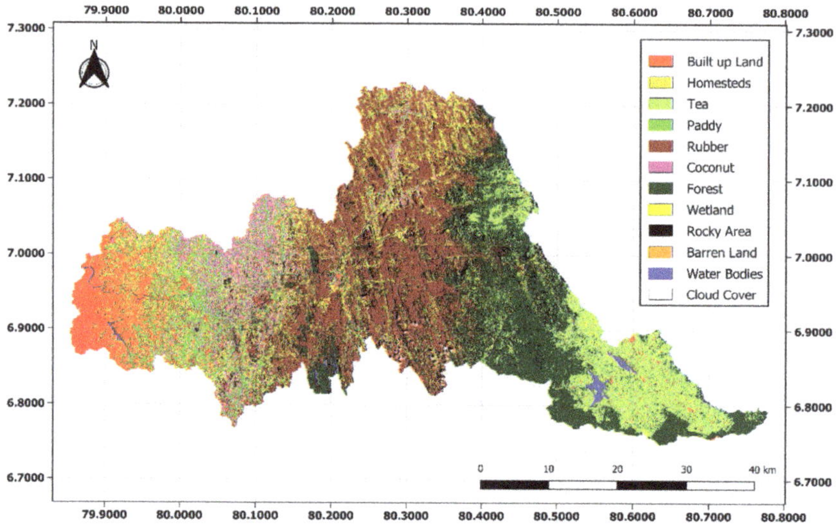

Figure 1. Land use pattern in the Kelani River basin.

Cabinet Paper No 93/340/166 [26] identified and designated the Kelani River basin as an environmentally sensitive area, including the divisional secretariats of Kolonnawa, Kelaniya, Hanwella, Biyagama, Ruwanwella, Homagama, and Dehiovita [27]. Therefore, the downstream of the Kelani River basin was selected for this study.

Openly available water quality data were collected from The Central Environmental Authority (CEA) for twelve sampling sites: (1) Thalduwa Bridge, (2) Seethawaka Ferry, (3) Pugoda Ferry, (4) Hanwella Bridge, (5) Kaduwela Bridge, (6) Welivita Bridge, (7) Victoria Bridge, (8) Pugoda Ela, (9) Wak Oya, (10) Pusseli Oya, (11) Maha Oya, and (12) Raggahawatte Ela over eight years (i.e., from January 2005 to December 2012). The water quality data can be freely obtained from The Central Environmental Authority website (http://203.115.26.11:8881/environmentalreport/dl, accessed on 21 February 2022). These sampling locations are given in Figure 2.

Nine commonly used water quality parameters including pH, chemical oxygen demand (COD), biochemical oxygen demand (BOD), total nitrate (NO_3^-), total phosphate (PO_4^{2-}), dissolved oxygen (DO), temperature, electric conductivity (EC), and chloride (Cl^-) were obtained for this analysis. Based on the literature [5–7,25,28–31] and the availability of the data, these water quality parameters were selected.

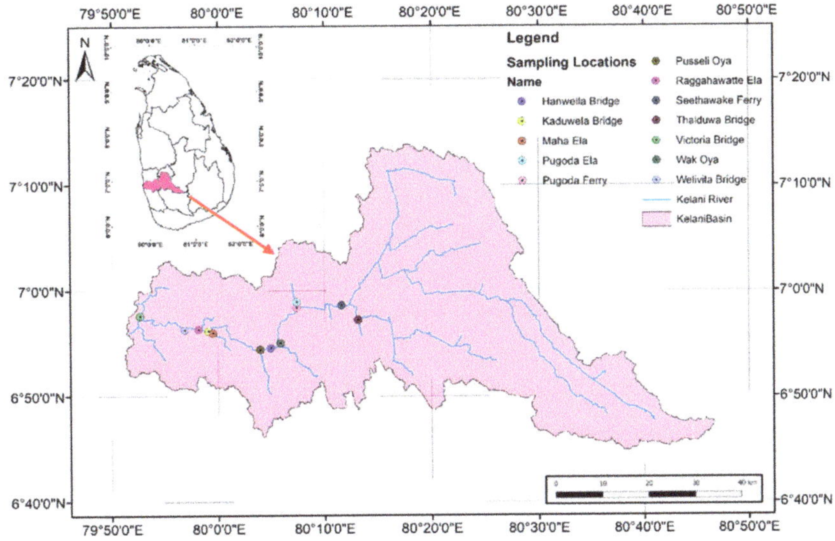

Figure 2. Sampling locations along the Kelani River basin.

2.2. WQI Model Development

The water quality index (*WQI*) was developed using the weighted arithmetic index method [32] and is given in Equation (1).

$$WQI = \frac{\sum Q_i W_i}{\sum W_i} \tag{1}$$

where Q_i is the quality rating of ith water quality parameter, and W_i ($\sum W_i = 1$) is the unit weight of ith water quality parameter. The quality rating, Q_i can be calculated using Equation (2).

$$Q_i = \frac{100(V_i - V_o)}{(S_i - V_o)} \tag{2}$$

where V_i is the actual amount of ith parameter, V_o represents the ideal value of the parameter [$V_o = 0$, except for pH ($V_o = 7$) and DO ($V_o = 14.6$ mg/L)], and S_i is the standard allowable value for the ith parameter. The unit weight (W_i) is calculated using Equation (3).

$$W_i = \frac{K}{S_i} \tag{3}$$

The term K is a proportional constant and calculated as per Equation (4).

$$K = \frac{1}{\sum \left(\frac{1}{S_i}\right)} \tag{4}$$

Water quality status (WQS) was determined based on the WQI ratings and is presented in Table 1 [32].

The water quality concentrations obtained for the 12 sampling locations were used to calculate the WQI. The results were interpreted to have discussion points in this research. In addition, the water quality standards and guidelines under Sri Lanka Standards (SLS 614) for each water quality constituent are given in Table A1 in Appendix A.

Table 1. Recommended range of WQI for intended usage: [32].

WQI	Status of Water Quality (WQS)	Intended Usage		
		Drinking	Irrigation	Industrial
0–25	Excellent	✓	✓	✓
26–50	Good	✓	✓	✓
51–75	Poor	X	✓	✓
76–100	Very poor	X	✓	X
Above 100	Not suitable for drinking or fish culture	Proper treatment is required		

3. Results and Discussion

3.1. Water Quality Parameter Presentation

Nine water quality parameters were analyzed for the development of the WQI. Highly fluctuating locations were selected from the spatiotemporal plots of each water quality constituent. Dissolved oxygen concentration is an important water quality parameter for the equilibrium of the aquatic ecosystem as it is a standard indicator for assessing water resource quality. DO levels in natural water paths can be varied mainly due to wastewater discharges. It was found that DO was negatively correlated to the temperature and ammonium concentrations (NH_4^+) [33]. Kumari et al. [34] reported that an increase in temperature decreases the dissolution of ambient DO in river water. In addition, higher activities of microorganisms cause low DO values in summer as they require a substantial amount of oxygen for metabolizing activities and for organic matter degradation [35].

Raggahawatte Ela, Seethawaka ferry, and Maha Ela showed a distinct pattern of variation of DO levels, whereas the other locations showcased DO levels within acceptable dissolve oxygen standards. These illustrations for 2005 and 2012 are shown in Figure 3 (*Illustrations for the years 2006–2011 are given in Appendix A*).

These variations showcase lower DO values along the Kelani River (less than 4 mg/L). Therefore, aquatic life may be facing a concerning situation. The lowest DO values over the years are presented in Table 2. Sampling locations in Maha Oya and Raggahawatte Ela showcase continuous lower DO values.

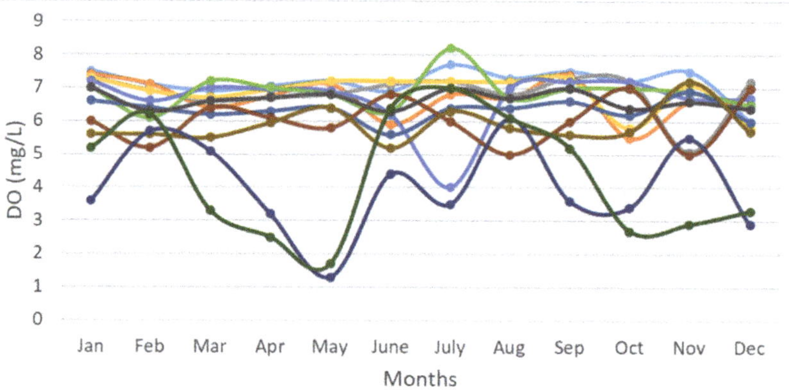

(**a**)

Figure 3. *Cont.*

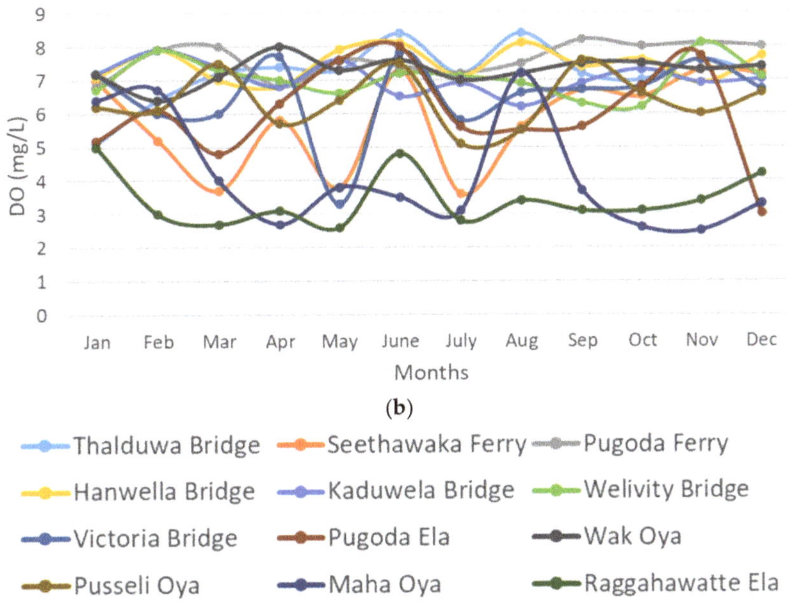

(**b**)

Figure 3. DO variation along the Kelani River basin in (**a**) 2005 (**b**) 2012.

Table 2. Minimum DO values along the Kelani River.

Month/Year	Location	Minimum DO Value (mg/L)
May/2005	Maha Oya	1.3
May/2006	Raggahawatte Ela	2.6
February/2007	Raggahawatte Ela	2.5
December/2008	Maha Oya	2.3
August/2009	Pugoda Ela	2.1
January/2010	Raggahawatte Ela	1.9
June/2011	Wak Oya	1.0
November/2012	Maha Oya	2.5

The spatiotemporal variations of the other water quality constituents (COD, BOD, and NO_3^-) were also analyzed; however, they are not presented in this paper (as they are not the results from this study but the visualization of the measurements). Nevertheless, the most essential features are stated here for information. (Interested readers may request these details from the corresponding author.)

Substantial variation in COD can be found at the Pugoda Ela, Victoria Bridge, Raggahawatte Ela, and Seethawaka ferry sampling stations. In addition, Maha Ela and Raggahawatte Ela had the most significant BOD fluctuations over the years. Furthermore, Raggahawatte Ela showcased the highest variations of nitrates. These observations can interpret a draft view of the water quality in the Raggahawatte Ela area.

3.2. WQI Analysis

The weighted arithmetic method was employed to obtain the WQI along the Kelani River. Figure 4 showcases the spatiotemporal variation of the calculated WQI.

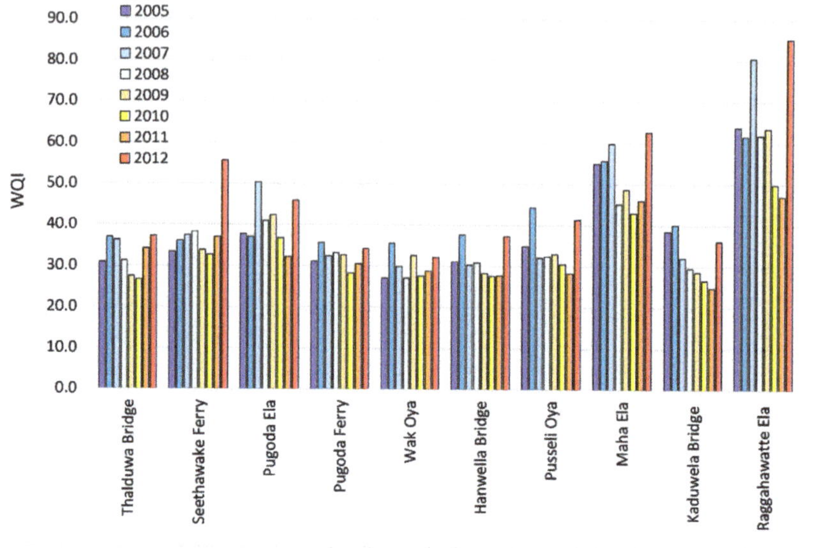

Figure 4. Spatiotemporal variation of calculated WQI along the Kelani River.

The WQI analysis showcases that it has higher values for the Raggahawatte Ela sampling location in all years. In addition, Maha Ela illustrates higher WQI values. The minimum and the maximum WQI for the years from 2005 to 2012 are given in Table 3. The highest WQI values in Raggahawatte Ela imply that the area had the poorest water quality. However, Wak Oya had comparatively good water quality from 2005 to 2008 and 2012, while Thalduwa and Kaduwela had the best water quality in the remaining years.

Table 3. Maximum and minimum values of WQI along the Kelani River.

Year	Max Value	Location	Min Value	Location
2005	63.842	Raggahawatte Ela	27.091	Wak Oya
2006	61.670	Raggahawatte Ela	35.563	Wak Oya
2007	80.623	Raggahawatte Ela	29.570	Wak Oya
2008	61.903	Raggahawatte Ela	27.105	Wak Oya
2009	63.541	Raggahawatte Ela	27.480	Thalduwa Bridge
2010	49.905	Raggahawatte Ela	26.615	Kaduwela Bridge
2011	47.091	Raggahawatte Ela	24.813	Kaduwela Bridge
2012	85.309	Raggahawatte Ela	32.167	Wak Oya

Raggahawatte Ela, the closest site to the Biyagama industrial zone, surpassed the COD, BOD, and DO values, resulting in the poorest water quality compared to other test locations, indicating a clear picture of the influence of industrial effluents on surface water quality.

Similarly, as illustrated in Figure 5, the water quality state at the Seethawaka ferry site has deteriorated significantly over the previous three years from 2010 to 2012, primarily as a result of receiving industrial wastewater from the Seethawaka industrial zone. Therefore, industrialization directly influences the water quality in the Kelani River.

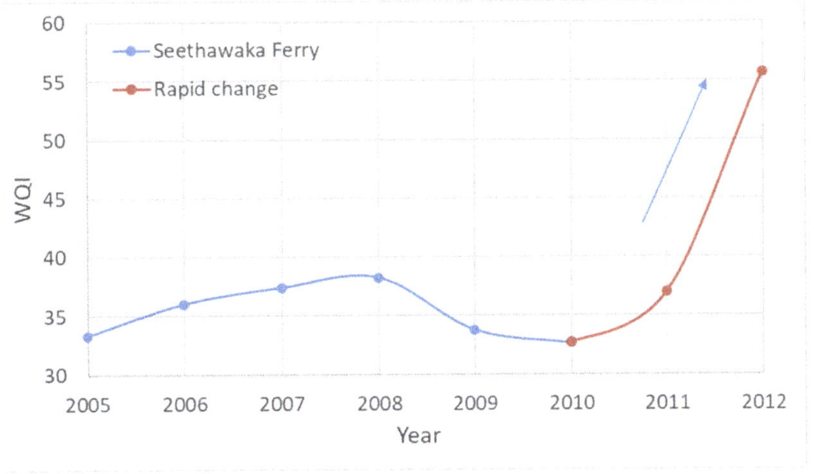

Figure 5. WQI variation at Seethawaka ferry sampling point.

3.3. Development of Simplified Equation for WQI

After analyzing the water quality index, a regression model was developed using the R Project for Statistical Computing. The water quality model was developed using data from 2005 to 2010, and the remaining data from 2011 to 2012 were utilized to validate the results. Table 4 presents the coefficients obtained for each parameter using the forward selection method. The interception point was found at −20.666.

Table 4. Coefficients for the water quality index parameters.

Parameter	Coefficient	Parameter	Coefficient
COD	0.8439	Phosphate	21.10
BOD	3.376	Chloride	0.00135
pH	6.619	Nitrate	0.03376
Temperature	0.135	EC	0.000938
DO	−1.758		

Therefore, the following equation was proposed as a regression model for the water quality index.

$$WQI = \begin{array}{l} -20.666 + 0.8439(COD) + 3.376(BOD) + 6.619(pH) + 21.10\left(PO_4^{2-}\right) + 0.135(T) + 0.00135(Cl^-) \\ -1.758(DO) + 0.000938(EC) + 0.03376\left(NO_3^-\right) \end{array} \quad (5)$$

It can be clearly seen that phosphate dominates the WQI for the Kelani River. Therefore, a deeper understanding of phosphate concentrations and their origin in the Kelani River should be required. However, a relatively insignificant contribution can be identified from chloride, electrical conductivity, and nitrate concentrations.

The validation of the simple equation for the WQI can be seen in Figure 6. The simple equation has a perfect match to the calculated WQI. Therefore, this can simply calculate the WQI for other years without bringing complex and lengthier calculations.

The obtained equation is limited to the selected Kelani basin. However, the results do not rule out employing the form of the equation for different basins. Regression coefficients can be derived based on their water quality parameters. Therefore, we encourage future studies to develop such relationships for other basins that can be critically assessed in a global context.

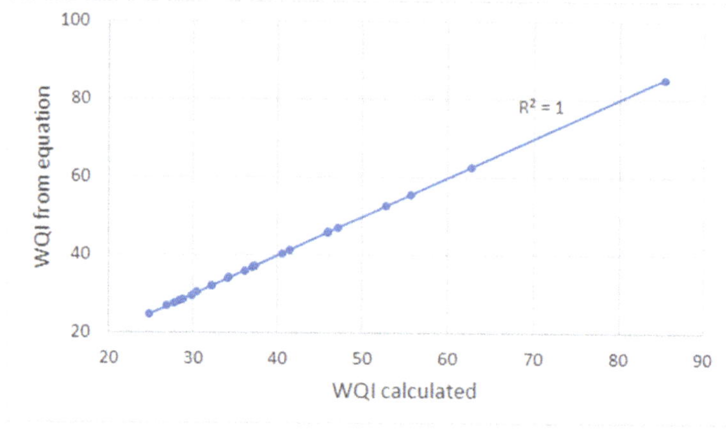

Figure 6. WQI validation from 2011 to 2012.

4. Conclusions

This study proposed a multiple linear regression model to predict the WQI for the Kelani River in Sri Lanka. The equation was validated using available water quality parameters of the Kelani River at distinct locations. The equation fits perfectly with the validation set with $R^2 = 1$. Furthermore, the equation signifies the critical water quality parameters and their contributions. On the other hand, the equation suggests the required water treatment strategies based on the contribution of each parameter. Thus, the required attention can be provided to protect the water quality of the river as it is the primary source of water supply to Colombo, the commercial capital of Sri Lanka.

The concentrations of DO, phosphate, COD, BOD, and nitrate were often above the standards at Raggahawatte Ela and Seethwaka ferry. The computed WQIs for the exact location validated these observations. Raggahawatte Ela was continuously reported to have the lowest water quality during the study period, while the Seethawaka ferry fell dramatically throughout the three years from 2010 to 2012. Due to the proximity of both places to an industrial zone, the surface water quality has degraded due to industrial effluent discharges. Thus, this study can be used to establish a water quality monitoring strategy, public awareness campaigns, and some policy decisions addressing the quality of water near industrial zones, such as imposing new legislation on industries that discharge effluents into natural streams.

Author Contributions: Conceptualization, R.M. and U.R.; methodology, R.M.; software, R.M.; formal analysis, R.M.; writing—original draft preparation, R.M.; writing—review and editing, R.M., D.P.P.M., H.M.A., M.P. and U.R.; supervision, U.R.; project administration, U.R.; funding acquisition, H.M.A., M.P. and U.R. All authors have read and agreed to the published version of the manuscript.

Funding: This research received no external funding.

Data Availability Statement: The water quality data can be freely downloaded as per the explanations given in the manuscript. In addition, all calculations can be requested from the corresponding author for research purposes.

Acknowledgments: The authors would like to acknowledge the support received from the Sri Lanka Institute of Information Technology, Sri Lanka to carry out this research work.

Conflicts of Interest: The authors declare no conflict of interest.

Appendix A

Table A1. Surface water pollution standards based on Sri Lankan water quality standards (SLS 614), WHO Standards, and BIS Standards.

Parameter	SLS 614/WHO/BIS Standard
pH	6.5–8.5
Electrical conductivity	300
Chloride	250
DO	5
COD	10
BOD	5
Phosphate	2
Nitrate	50
Temperature	25

All parameters are in milligrams per liter except pH, EC (μS/ cm), and Temperature (°C).

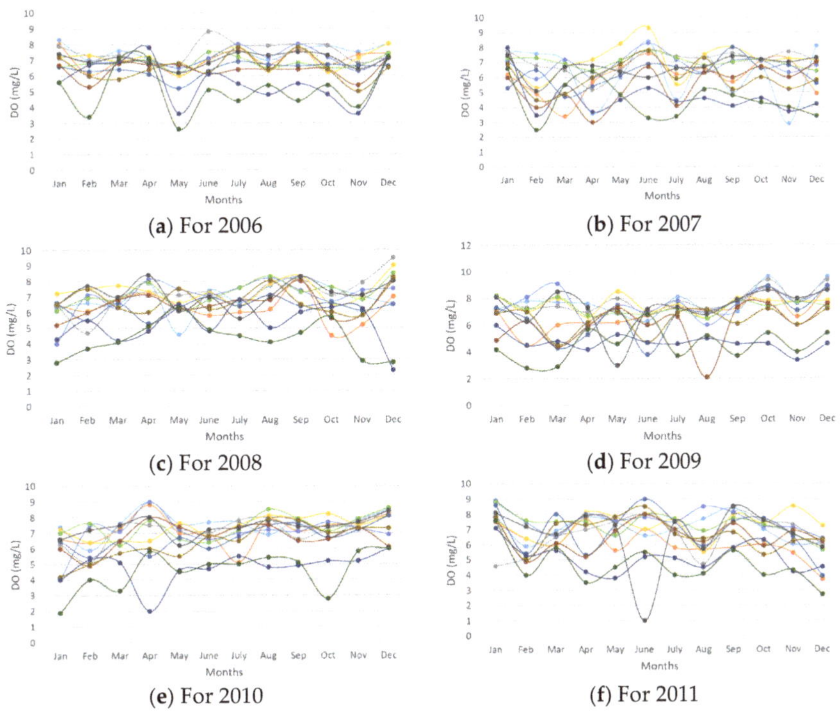

(**a**) For 2006 (**b**) For 2007 (**c**) For 2008 (**d**) For 2009 (**e**) For 2010 (**f**) For 2011

Figure A1. DO variation along the Kelani River basin: (**a**) For 2006; (**b**) For 2007; (**c**) For 2008, (**d**) For 2009; (**e**) For 2010; (**f**) For 2011.

References

1. Ni, X.; Wu, Y.; Wu, J.; Lu, J.; Wilson, P.C. Scenario analysis for sustainable development of Chongming Island: Water resources sustainability. *Sci. Total Environ.* **2012**, *439*, 129–135. [CrossRef] [PubMed]
2. Shu, J.; Ying, H. The economic analysis of sustainable utilization of agricultural water resources in Shandong province. *Energy Procedia* **2011**, *5*, 2120–2124. [CrossRef]
3. Abu-Zeid, M.A. Water and sustainable development: The vision for world water, life and the environment. *Water Policy* **1998**, *1*, 9–19. [CrossRef]
4. Biswas, A.K. Water for sustainable development in the 21st century. *Int. J. Water Resour. Dev.* **1991**, *7*, 219–224. [CrossRef]

5. Gartsiyanova, K.; Varbanov, M.; Kitev, A.; Genchev, S. Water quality analysis of the rivers Topolnitsa and Luda Yana, Bulgaria using different indices. In *Journal of Physics: Conference Series*; IOP Publishing: Bristol, UK, 2021; Volume 1960, p. 012018. [CrossRef]
6. Nizar, F.R.; Ghazi, R.M.; Awang, N.R.; Muhammad, M. Assessment of Kelantan River water quality using water quality index (WQI). In *IOP Conference Series: Earth and Environmental Science*; IOP Publishing: Bristol, UK, 2021; Volume 842, p. 012005. [CrossRef]
7. Mokarram, M.; Pourghasemi, H.R.; Huang, K.; Zhang, H. Investigation of water quality and its spatial distribution in the Kor River basin, Fars province, Iran. *Environ. Res.* **2022**, *204*, 112294. [CrossRef]
8. Satter, M.; Islam, M. Quality assessment of river water around Dhaka city. *Bangladesh J. Environ. Sci.* **2005**, *10*, 326–329.
9. John, V.; Jain, P.; Rahate, M.; Labhasetwar, P. Assessment of deterioration in water quality from source to household storage in semi-urban settings of developing countries. *Environ. Monit. Assess.* **2014**, *186*, 725–734. [CrossRef]
10. WHO/UNICEF Joint Monitoring Programme for Water Supply and Sanitation; World Health Organization; Water, Sanitation and Health Team. *Meeting the MDG Drinking Water and Sanitation Target: A Mid-Term Assessment of Progress*; World Health Organization: Geneva, Switzerland, 2004.
11. Astel, A.; Biziuk, M.; Przyjazny, A.; Namieśnik, J. Chemometrics in monitoring spatial and temporal variations in drinking water quality. *Water Res.* **2006**, *40*, 1706–1716. [CrossRef]
12. Xu, S.; Li, S.; Yue, F.; Udeshani, C.; Chandrajith, R. Natural and Anthropogenic Controls of Groundwater Quality in Sri Lanka: Implications for Chronic Kidney Disease of Unknown Etiology (CKDu). *Water* **2021**, *13*, 2724. [CrossRef]
13. Akhtar, N.; Ishak, M.S.; Bhawani, S.; Umar, K. Various Natural and Anthropogenic Factors Responsible for Water Quality Degradation: A Review. *Water* **2021**, *13*, 2660. [CrossRef]
14. Khatri, N.; Tyagi, S. Influences of natural and anthropogenic factors on surface and groundwater quality in rural and urban areas. *Front. Life Sci.* **2014**, *8*, 23–39. [CrossRef]
15. Liyanage, C.; Yamada, K. Impact of Population Growth on the Water Quality of Natural Water Bodies. *Sustainability* **2017**, *9*, 1405. [CrossRef]
16. Amarathunga, A.; Kazama, F. Impact of Land Use on Surface Water Quality: A Case Study in the Gin River Basin, Sri Lanka. *Asian J. Water Environ. Pollut.* **2016**, *13*, 1–13. [CrossRef]
17. Bandara, N. Water and wastewater related issues in Sri Lanka. *Water Sci. Technol.* **2003**, *47*, 305–312. [CrossRef]
18. Zubair, L. Challenges for environmental impact assessment in Sri Lanka. *Environ. Impact Assess. Rev.* **2001**, *21*, 469–478. [CrossRef]
19. Kumar, M.; Chaminda, T.; Honda, R.; Furumai, H. Vulnerability of urban waters to emerging contaminants in India and Sri Lanka: Resilience framework and strategy. *APN Sci. Bull.* **2019**, *9*, 57–66. [CrossRef]
20. Uddin, M.; Nash, S.; Olbert, A. A review of water quality index models and their use for assessing surface water quality. *Ecol. Indic.* **2021**, *122*, 107218. [CrossRef]
21. Son, C.; Giang, N.; Thao, T.; Nui, N.; Lam, N.; Cong, V. Assessment of Cau River water quality assessment using a combination of water quality and pollution indices. *J. Water Supply Res. Technol. Aqua* **2020**, *69*, 160–172. [CrossRef]
22. Bhat, S.; Pandit, A. Surface Water Quality Assessment of Wular Lake, A Ramsar Site in Kashmir Himalaya, Using Discriminant Analysis and WQI. *J. Ecosyst.* **2014**, *2014*, 724728. [CrossRef]
23. Kaurish, F.; Younos, T. Developing a Standardized Water Quality Index for Evaluating Surface Water Quality. *J. Am. Water Resour. Assoc.* **2007**, *43*, 533–545. [CrossRef]
24. Mahagamage, M.; Chinthaka, S.; Manage, P.M. Multivariate analysis of physico-chemical and microbial parameters of surface water in Kelani River basin. *Int. J. Multidiscip. Stud.* **2016**, *1*, 55–61. [CrossRef]
25. Abeysinghe, N.D.A.; Samarakoon, M. Analysis of variation of water quality in Kelani River, Sri Lanka. *Int. J. Environ. Agric. Biotechnol.* **2017**, *2*, 238965. [CrossRef]
26. The Ministry of Environment and Natural Resources and the Ministry of Water Supply & Drainage. 2022. Available online: http://www.waterboard.lk/web/images/contents/key_sections/policy_and_planning/downloads/policy_on_siting_of_high_polluting_industries.pdf (accessed on 19 April 2022).
27. Herath, H.; Weerahewa, J. An economic approach to manage industrial water pollution in river basins: Case of Kelani River Basin, Sri Lanka. In *XI World Water Congress—Water Resources Management in the 21st Century*; International Water Resources Association: Madrid, Spain, 2003; pp. 1–11.
28. Amiri, B.J.; Nakane, K. Modeling the linkage between river water quality and landscape metrics in the Chugoku district of Japan. *Water Resour. Manag.* **2009**, *23*, 931–956. [CrossRef]
29. Bora, M.; Goswami, D.C. Water quality assessment in terms of water quality index (WQI): Case study of the Kolong River, Assam, India. *Appl. Water Sci.* **2017**, *7*, 3125–3135. [CrossRef]
30. Huang, J.; Zhan, J.; Yan, H.; Wu, F.; Deng, X. Evaluation of the impacts of land use on water quality: A case study in the Chaohu Lake basin. *Sci. World J.* **2013**, *2013*, 329187. [CrossRef]
31. Shah, K.A.; Joshi, G.S. Evaluation of water quality index for River Sabarmati, Gujarat, India. *Appl. Water Sci.* **2017**, *7*, 1349–1358. [CrossRef]
32. Brown, R.M.; McClelland, N.I.; Deininger, R.A.; O'Connor, M.F. A water quality index—Crashing the physiological barrier. *Indic. Environ. Qual.* **1972**, *1*, 173–182.

33. Barakat, A.; El Baghdadi, M.; Rais, J.; Aghezzaf, B.; Slassi, M. Assessment of spatial and seasonal water quality variation of Oum Er Rbia River (Morocco) using multivariate statistical techniques. *Int. Soil Water Conserv. Res.* **2016**, *4*, 284–292. [CrossRef]
34. Kumari, M.; Tripathi, S.; Pathak, V.; Tripathi, B. Chemometric characterization of river water quality. *Environ. Monit. Assess.* **2013**, *185*, 3081–3092. [CrossRef]
35. Jo, E.-K.; Yang, C.-S.; Choi, C.H.; Harding, C.V. Intracellular signalling cascades regulating innate immune responses to Mycobacteria: Branching out from Toll-like receptors. *Cell. Microbiol.* **2007**, *9*, 1087–1098. [CrossRef]

fluids

MDPI

Article

Numerical Investigation of 3D Flow Properties around Finite Emergent Vegetation by Using the Two-Phase Volume of Fluid (VOF) Modeling Technique

Amina [1] and Norio Tanaka [2,*]

[1] Graduate School of Science and Engineering, Saitama University, 255 Shimo-Okubo, Sakura Ku, Saitama Shi 338-8570, Saitama, Japan; amina168civil@gmail.com
[2] International Institute for Resilient Society, Saitama University, 255 Shimo-Okubo, Sakura Ku, Saitama Shi 338-8570, Saitama, Japan
* Correspondence: tanaka01@mail.saitama-u.ac.jp

Abstract: This study predicts how the Free Surface Level (FSL) variations around finite length vegetation affect flow structure by using a numerical simulation. The volume of fluid (VOF) technique with the Reynolds stress model (RSM) was used for the simulation. Multizone Hexahedral meshing was adopted to accurately track the free surface level with minimum numerical diffusion at the water–air interface. After the validation, finite length emergent vegetation patches were selected based on the aspect ratio (AR = vegetation width-length ratio) under constant subcritical flow conditions for an inland tsunami flow. The results showed that the generation of large vortices was predominated in wider vegetation patches ($AR > 1$) due to the increase and decrease in the FSL at the front and back of the vegetation compared to longer vegetation patches ($AR \leq 1$), as this offered more resistance against the approaching flow. The wider vegetation patches ($AR > 1$) are favorable in terms of generating a large area of low velocity compared to the longer vegetation patch ($AR < 1$) directly downstream of the vegetation patch. On the other hand, it has a negative impact on the adjacent downstream gap region, where a 14.3–34.9% increase in velocity was observed. The longer vegetation patches ($AR < 1$) generate optimal conditions within the vegetation region due to great velocity reduction. Moreover, in all the AR vegetation cases, the water turbulent intensity was maximum in the vegetation region compared to the adjacent gap region and air turbulent intensity above the FSL, suggesting strong air entrainment over this region. The results of this study are important in constructing vegetation layouts based on the AR of the vegetation for tsunami mitigation.

Keywords: coastal vegetation; tsunami; aspect ratio (AR = vegetation width-length ratio); volume of fluid (VOF); Reynold's stress modeling (RSM); meshing

Citation: Amina; Tanaka, N. Numerical Investigation of 3D Flow Properties around Finite Emergent Vegetation by Using the Two-Phase Volume of Fluid (VOF) Modeling Technique. *Fluids* **2022**, *7*, 175. https://doi.org/10.3390/fluids7050175

Academic Editors: Prashanth Reddy Hanmaiahgari, Manish Pandey, Mohammad Amir Khan and Jaan H. Pu

Received: 20 April 2022
Accepted: 16 May 2022
Published: 18 May 2022

1. Introduction

Coastal vegetations play an essential role in disastrous floods such as tsunamis [1]. These vegetations are beneficial to minimize the velocity and energy of tsunamis [2]. The importance of coastal vegetation emerged after the 2004 Indian Ocean Tsunami (IOT) and 2011 Great East Tsunami Japan (GEJT), and various field investigations have been performed on the effectiveness of coastal vegetation in reducing tsunami flow energy [3,4]. To compare and measure the efficiency of coastal vegetation against field research, different experimental studies have also been carried out. The use of coastal vegetation as a tsunami countermeasure depends upon different factors, including density of vegetation [5,6], species [7,8], dimensions [9], alignments, and scale of vegetations [10]. Iimura and Tanaka [11] investigated the impact of tree vegetation density and found that the increment in vegetation density reduces the water level and velocity on the downstream side of the vegetation. Similarly, according to Pasha and Tanaka [12], the dense emergent infinite vegetation is more feasible for inland tsunami mitigation compared to the sparse emergent infinite coastal vegetation, but due to the land constraints in inland regions [13], building infinite dense coastal vegetation in the actual field is not realistic [14]. To replicate the real scenario, a finite length coastal vegetation patch for inland areas should be provided [15,16]. Therefore, to safeguard inland regions from catastrophic tsunami flow, a thorough investigation must be conducted to examine the flow changes surrounding finite length coastal vegetation. Although previous researchers have investigated the utilization of finite coastal vegetation with different gaps [17] and varying ground slopes through two-dimensional (2D) numerical simulation [18], some authors have studied in detail the behavior of flow analysis around discontinuous vegetation patches [19,20], while the authors in [21–25] studied internal flow characteristics and turbulent parameters without taking free surface movement into account. However, there has been less numerical research on the impact of three-dimensional (3D) flow behavior with free surface movement around a limited coastal vegetation patch by changing its aspect ratio (dimensions), i.e., length and width, for tsunami prevention. The results of this research can be used to better understand the role of inland coastal vegetation of a finite length in the event of major floods such as a tsunami by allowing researchers to compare flow characteristics between longer and wider vegetation patches. Hence, the present study is mainly focused on the numerical investigation of 3D flow behavior through longer (AR < 1) and wider (AR > 1) vegetation patches to determine the flow behavior of each kind with their flow affecting properties, based on the Aspect Ratio (AR = vegetation patch width/length) of the finite length vegetation. Furthermore, the research also includes a vast scope and broad applicability in clarifying the flow mechanism in the form of complex velocity patterns within and around the finite length vegetation.

The velocity patterns are nearly impossible to capture through an experimental investigation due to the limitations of instruments: laser sheets in a Particle Image Velocity Meter (PIV), for example, are not viable in the vegetation zone due to the presence of vegetation cylinders. Additionally, velocity sensors in the Electromagnetic Flow Meter (EFM) cannot be utilized as they can disturb the flow near the vegetation region and the bed region. This numerical study provides an advantage in bridging this gap. To accomplish the objectives through numerical investigation, we used a CFD tool called FLUENT, which runs a simulation using the Volume of Fluid model (VOF) with the Reynolds stress model (RSM). The flow properties are addressed in terms of three-dimensional free surface variations, velocity distribution, velocity contours, velocity vectors, and turbulence characteristics.

2. The Volume of Fluid Model (Air–Water) Two-Phase Flow

In 3D numerical simulations, a multi-phase model such as the volume of fluid (VOF) model is typically used to treat the free surface boundary and interaction between the two-phase flow (water and air). The VOF technique was initially suggested by Hirt and Nickhols [26]. VOF modeling was used by the majority of earlier studies to solve flow issues in stepped spillways and wide crested weirs [27–31]. However, few studies have been performed using VOF modeling and provide solutions mainly to basic rectangular and curved open channel flow problems [32,33]. Therefore, this study considers the vegetated open channel to check the performance of the VOF modeling.

By solving a single momentum equation, the VOF approach can model two or more immiscible fluids and control the fraction of each of the volume of the fluid in the entire domain. The fundamental principle of this kind of model is that two fluids do not interpenetrate. In this problem, the use of two fluids creates a new variable, i.e., the computational cell phase volume fraction. Consequently, regarding the field of air and water flow, a single set of air and water share a momentum equation and in each computational cell, the volume fraction of each fluid is followed across the whole domain. The volume fraction for all phases in each computational cell add up to (1) unity. The flow structure and flow properties for all variables are separated into phases and are indicated as volume-averaged values, such that the volume fraction of each phase is known at each location. Therefore, the variables and properties in each cell, depending upon the volume fraction value, are purely representative of a mixture of the phases. If the volume fraction of the q_{th} fluid in the cell is referred to as a_q, then the following three conditions are possible:

(a) $a_q = \mathbf{1}$ when the qth fluid fills the whole cell demonstrating that the cell is filled with water.
(b) $a_q = \mathbf{0}$ when the qth fluid does not occupy the whole cell means that the cell is filled with air.
(c) $\mathbf{0} < a_q < \mathbf{1}$ indicates that where the cell has an interface between the qth fluid, i.e., free surface between air and water layer.

If a_a = air volume fraction and a_w = water volume fraction, the free surface traceability of the water–air interface is performed by a continuity equation solution is provided by:

$$\frac{\partial \alpha_w}{\alpha_t} + \frac{\partial \alpha_w}{\partial x_t} = 0 \quad (1)$$

To evaluate the precise orientation of the free surface, i.e., the water–air interface, the VOF implicit scheme is used, and within each cell it has a linear slope, which is used to measure fluid advection via the corresponding cell face. The location of the linear interface to the center of each partially filled cell is therefore measured by the value of the volume fraction and its derivative within the cell.

3. Turbulence Modeling (RANS)

The Reynolds-averaged Navier–Stokes (RANS) equations, written in tensor notation, are used to define the flow behavior in vegetated channels.

$$\frac{\partial \overline{u}_i}{\partial x_i} = 0 \quad (2)$$

$$\frac{\partial \overline{u}_i}{\partial t} + \frac{\partial}{\partial x_j}\left(\overline{u_j u_i}\right) = -\frac{\partial \overline{p}}{\partial x_i} + v\frac{\partial^2 \overline{u}_i}{\partial x^2{}_j} - \frac{\partial}{\partial x_j}\overline{\left(u'{}_i u'{}_j\right)} \quad (3)$$

where overbar represents the Reynolds averaged value and $\overline{u_i}$ and $\overline{u_j}$ are the average and fluctuation velocities in the x_i direction, respectively. I = (1, 2, 3), indicating the longitudinal x, transverse y, and vertical z axis. Equations (2) and (3) may be calculated for the average value of velocity, pressure, and other variables when the turbulent correlations can be linked to average velocity or mean pressure, respectively. This is referred to as the turbulence modeling closure problem. Eddy viscosity, a concept developed by Boussinesq's, was utilized to simulate Reynold's stress, described by.

$$-\overline{u'{}_i u'} = v_e\left(\frac{\partial \overline{u}_i}{\partial x_j} + \frac{\partial \overline{u}_j}{\partial x_i}\right) - \frac{2}{3}k\delta_{ij} \quad (4)$$

In which, v_e represents the eddy/turbulent viscosity and $\boldsymbol{K} = \mathbf{0.5}\overline{u'{}_i u'{}_j}$ denotes the turbulent kinetic energy (TKE). The eddy viscosity is unknown, but through the dimensionless analysis it is given by.

$$v_e = C_\mu\left(\frac{K^2}{\in}\right) \quad (5)$$

Here, $C_\mu = \mathbf{0.09}$ is the coefficient of closure and $\in$ represents the dissipation rate of turbulent kinetic energy (TKE).

4. Material and Methods

4.1. Conditions for Numerical Simulation

4.1.1. Flow Conditions

The tsunami, which is a series of waves, continues to the shore, and runs up inland with a huge discharge [34]. On the other hand, if discontinuous vegetation patches are present in the path of flow, then the hydraulic conditions may change [35,36]. However, when the tsunami inundation enters in the inland regions, it does not contain any waves and propagates into the inland region with a long period. As a result, the tsunami inundation can be represented as a quasi-steady flow [37]. Many earlier researchers considered the flow in the inland area surrounding an inland vegetative forest in a steady and subcritical condition when modeling the tsunami flow. Thus, the flow parameters in this research were defined using Froude and initial water depth similarity, which assumes a subcritical and constant tsunami inland flow. During the 2004 Indian Ocean tsunami, the Froude conditions (F_r) were reported to be 0.64–1.04 in the inland region of Banda Aceh [38]. Furthermore, the tsunami flow around an inland forest in Miyagi Prefecture, Japan, was subcritical at numerous locations during the 2011 Great East Japan tsunami (GEJT), with varied Froude values in the range of 0.7–1 and an approximate tsunami depth of 7.3–8.3 m [39]. The vegetation is set to be placed inland in the present research. To simulate a tsunami flow on an actual scale, a model scale of 1/100 was used, and the Froude number in a channel was kept around 0.7 without any vegetation inserted in the rectangular channel. In this numerical research, the water depth (without a vegetation model) used to create the subcritical inland tsunami condition was 4.5 cm, against the initial Froude number of 0.70. Since the numerical domain used a two-phase model, i.e., (VOF), it is necessary to specify how much air is included above the water region. Salaheldin et al. [40] stated that there is no impact from the boundary at the top if air height to water height is kept above 0.5. In our current 3D simulation, the total air portion depth is maintained at 5.5 cm, resulting in a ratio of 1.22, which is sufficient to prevent any impact from the top of the domain boundary state.

4.1.2. Vegetation Conditions

The Japanese pine tree, with an average tree height of 15 m and a trunk diameter of 0.4 m, located in the Sendai Plain, was adopted as the tree species for the vegetation model. According to Tanaka et al., [41], a tree can be modelled as a rigid circular cylinder if the crown part of the tree is high above the tsunami height. The trees were thus modelled in 1/100 scale, utilizing rigid cylinders with a diameter of 0.4 cm placed in a staggered pattern, in accordance with the average diameter of the pine tree. The density of the vegetation is defined by S/d number [2]. There are three different vegetation densities based on this S/d number (where S represents the spacing between the rigid cylinders in a cross streamwise direction and d is the cylinder diameter). The sparse model is defined as having a S/d number of 2.13, while the intermediate and dense models have a S/d value of 1.03 and 0.25, respectively. In the current research, an intermediate density with a S/d value of 1.03 was used for the vegetation model. As stated in the introductory section, the construction of infinite vegetation in the inland areas would be impossible due to land constraints; therefore, to replicate the actual situation, a limited length vegetation was examined in this research to represent the inland vegetation. The main goal of the research was to examine the effect of change in the length and width of the vegetation model on the 3D flow characteristics, while accounting for free surface fluctuation based on the aspect ratio ($AR = W_y / W_x$, in which w_x and w_y are the vegetation length in the x-direction and width in the y-direction, respectively), which was considered as a changing parameter. Five different cases were considered with varying length and width of vegetation model, as shown in Figure 1. The vegetation models with five various configurations were placed separately in the channel to evaluate their performance in terms of flow resistance. The AR of the finite length vegetation in this study was in the range (0.2–5), as shown in Table 1. The AR of 0.2 and 0.3 indicate longer vegetation patches and the AR of 3 and 5 indicate wider vegetation patches configurations in the path of flow.

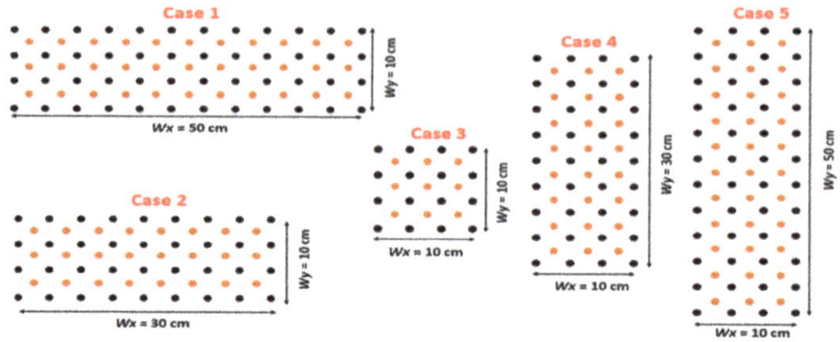

Figure 1. Layout of the domain for numerical simulation: Arrangement of the Vegetation patches.

Table 1. Hydraulic Conditions of Numerical Model: Note: W_x represents vegetation length; W_y depicts vegetation width; Fr shows Froude number ($Fr = U/\sqrt{gh}$, where U represents the initial flow velocity, g is the gravitational constant and h is the initial water depth); Dw stands for water depth; U is the initial velocity.

	Cases	AR	W_x (cm)	W_y (cm)	Fr	S/d	Dw (cm)	U (m/s)
Longer Vegetation Patch	1	0.2	50	10	0.7	1.03	4.5	0.465
	2	0.3	30	10	0.7	1.03	4.5	0.465
	3	1	10	10	0.7	1.03	4.5	0.465
Wider Vegetation Patch	4	3	10	30	0.7	1.03	4.5	0.465
	5	5	10	50	0.7	1.03	4.5	0.465

4.1.3. Measurement Locations

A rectangular channel 150 cm long and 70 cm wide was used to represent the current numerical domain, which included a vegetation model that covered the finite width of the domain. The top view of the channel domain is presented in Figure 2. Six specified points were selected to investigate the vertical distribution of stream wise velocities within the vicinity of the vegetation model ($x1$, $x2$, $x3$) and the adjacent gap region ($y1$, $y2$, $y3$). Furthermore, an important horizontal surface (at depth of z = 3.5 cm below the initial water depth) and two longitudinal sections (LS1 = 35 cm passed through the centre of channel domain or vegetation model and LS2 = 25 cm, passed through the adjacent gap region throughout the channel domain),as well as two cross streamwise sections (CS1, located within the vegetation model and CS2, located downstream of the vegetation model) were considered for the detailed investigation of the flow properties in the form of free surface level (FSL) distribution, velocity profiles, velocity vectors (flow movement), velocity and turbulent intensity contour plot distribution.

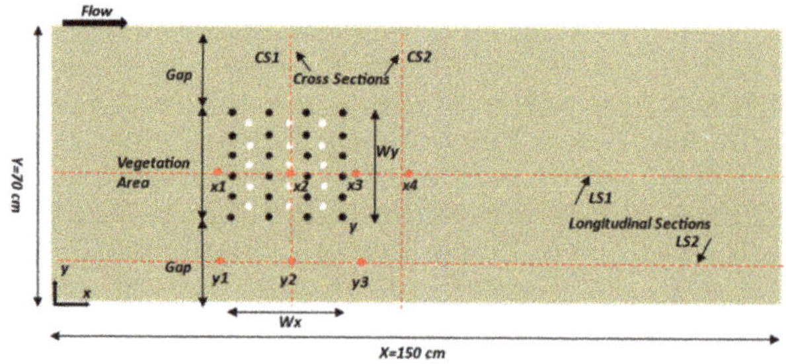

Figure 2. Top view of the vegetated channel setup and specified positions and sections. The red points depict the concerned locations, whereas the red dashed lines show the measurement sections.

4.2. Preprocessing and Postprocessing

4.2.1. Preprocessing

The simulation was carried out using the Computational Fluid Dynamics (CFD) tool, ANSYS. The numerical setup for this study consisted of two steps: preprocessing and postprocessing. In preprocessing, the geometry was modeled using a design modeler tool. A rectangular channel 150 cm long and 70 cm broad was created in a design modeler, together with a vegetation model (staggered arrangement) that occupied the domain's limited width. The arrangement of the vegetation model and the spacing in between the cylinders is shown in Figure 3a. The domain was then transferred to the mesh tool. Meshing is a critical stage in achieving high-quality results, as desired outcomes require proper meshing, which is described as the partition of the geometry into components, including cells and elements. Simulation precision, convergence and speed are all influenced by mesh size and density. More memory and time are required for fine meshing, whereas numerical diffusion in the results occurs when meshing is carried out incorrectly. Firstly, we tested various meshing methods (e.g., sweep meshing, tetrahedral primitive with numerous deformable components resulted in higher skewness values and in some cases negative volumes) on an open channel without vegetation to check the diffusion rate of water and air interface; however, we achieved convergence with reasonable computation time using multizone meshing with hexahedral components. Figure 3b,c depict the differences in results in terms of diffusion rate between the water and air volume fractions on a channel domain, when multizone and tetrahedral meshing are used separately. Tetrahedral meshing generated incorrect results and maximized numerical diffusion at the water–air interface, resulting in significant disparities. This was because tetrahedral meshing consisted of only tetra cells and did not well distribute the cells to define the boundary between air and water phase. It also created the poor-quality surface mesh around the vegetation cylinders. On the other hand, multizone meshing, which is a hybrid of hexahedral or brick and tetra elements, generally result in more accurate results at lower element counts than only tetra elements. Therefore, multizone meshing produces the best results with low diffusivity and can precisely track the water–air interface. Furthermore, the skewness and orthogonal quality criterion methods were used to evaluate the quality of the mesh. Skewness defines how close to the ideal (equiangular quad) a face or cell is. According to the definition, a value that corresponds to 0 indicates "Excellent", and 1 corresponds to the worst cells (degenerate) quality. For this present study, the equilateral volume-based skewness method was used to check the skewness. Under this method, the mesh is considered a good quality for 3D if its skewness value is less than 0.4. The present geometry mesh has a value of 0.16, which leads to excellent mesh quality. The mesh of the whole domain, especially around the cylinders, is shown in Figure 3d.

Finally, in order to acquire the best simulation results, a grid independency test was also conducted. The experimental results of vertical velocity distribution at position 1 (see Figure 3e were compared with the results of varying the mesh size of the numerical domain. Initially, the mesh grids of 0.4 million (coarse), 1.7 million (fine), and 2.8 million (finer) were investigated. A 7% change in vertical streamwise velocities between the coarse and fine grids was observed at position 1 and 2; however, the variance in findings by greater refinement was just 1%. Therefore, the fine mesh with 1.7 million grids was selected for the present study. It has been shown that the vertical stream wise velocity values for coarse grid are quite greater than those for medium and fine grid, in comparison with the experimental values, Figure 3e. This might be because a coarse mesh cannot properly recognize the boundaries of the vegetation cylinders, and so the resistance or drag produced by these structures could not be accurately calculated, resulting in significantly greater flow velocity magnitudes. On the other hand, a fine mesh (refinement method, particularly near the cylinders) solves this issue by estimating a realistic flow shape and ensuring computational precision.

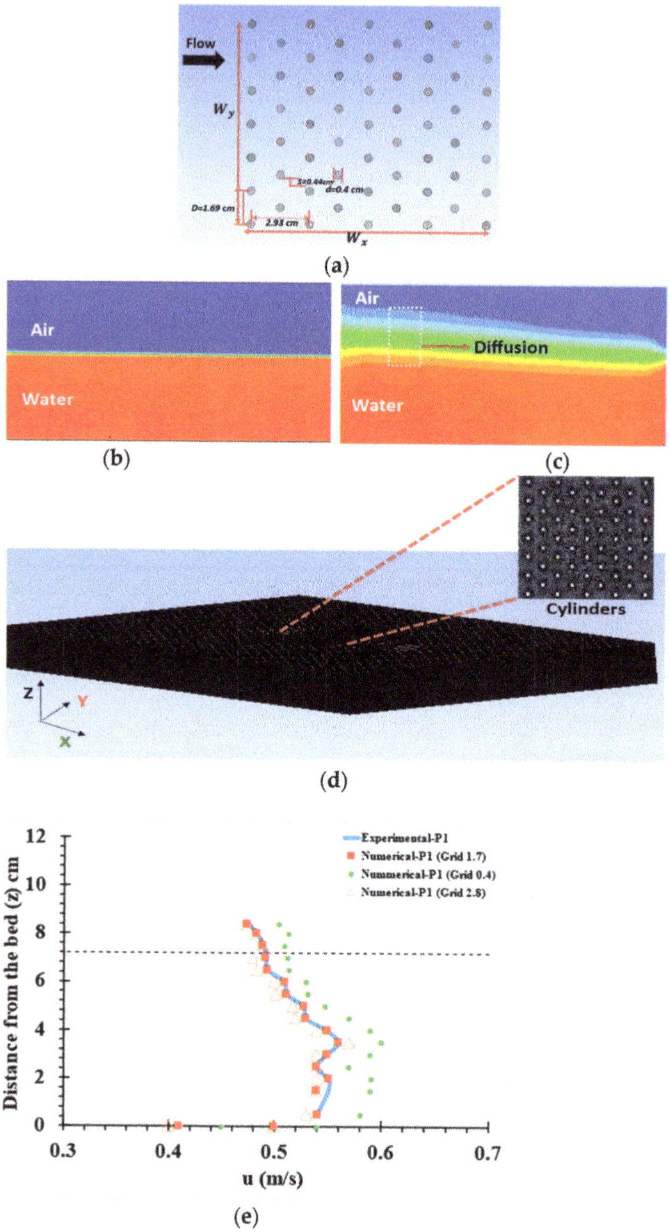

Figure 3. Preprocessing Setups. (**a**) Arrangement and spacing of cylinders: Comparison of Meshing techniques: (**b**) Multizone meshing (**c**) Tetrahedral meshing. (**d**) Mesh domain (**e**) Grid independency test.

4.2.2. Postprocessing

After meshing was complete, the next step was to set up the physics, which was carried out in post processing. In multiphase models, the volume of fluid (VOF) model was adopted, while the Reynolds stress model (RSM) was used for viscous modeling. The next step was to designate the phases; water liquid was selected as the primary phase, while air was defined as the secondary phase. The boundary conditions were than assigned to the faces of the geometry The velocity inlet boundary condition was given to the water inlet, and the pressure inlet was applied to the air inlet. The water and air density were set at 1000 (kg/m^3) and 1.225 (kg/m^3), respectively. The top boundary and outlet were both considered as pressure outlets, with the gauge pressure set to zero. However, just one outlet was defined so that the solution could use a flow level derived from the inner flow field. The boundary condition applied to the bed of domain, faces of cylinders are considered as a no-slip wall. The no-slip rule for viscous fluids in fluid dynamics means that the fluid will have zero velocity at the solid boundary, as many researchers have used this condition [42–44]. For side walls, a slip boundary condition was applied to avoid the effects of the side wall. The boundary conditions applied to the whole domain is shown in Figure 4.

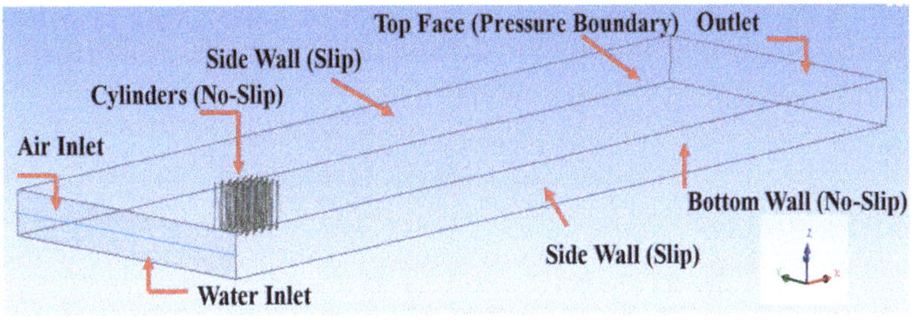

Figure 4. Postprocessing Setups: Boundary Conditions.

The solver for time was used as transient. The type for interface modeling for VOF was used as sharp/dispersed to avoid diffusion rate. Finally, for the pressure velocity-coupling method, a SIMPLE scheme was adopted and for spatial discretization, the third order upwind scheme was used. Under-relaxation factor values were considered as low for convergence criteria and residual smoothing. A fully implicit scheme was incorporated for the time incorporation. The normalized residual value was set at 1×10^{-6}; thus, when the solution's calculations were less than the specified number, it was considered converged. The standard initialization method was used for solution initialization. The user guide [45] contains descriptions of the algorithm, governing equations, and turbulence model.

5. Results and Discussion

5.1. Model Validation

A laboratory experiment was conducted to validate the results of the numerical model. The experimental setup was established in a laboratory channel of a rectangular cross-section with a length (L) of 500 cm, width (w) 70 cm, and a height of 50 cm at Saitama University, Japan. The vegetation was configured as an array of wooden cylinders in a staggered arrangement and precisely embedded on the channel bed at the center of the streamwise width, around 55 cm upstream of the channel. The dimensions of the vegetation model were chosen to be the same as those used in the current numerical model of Case No. 3 (Table 1), whereas the density of the vegetation model was considered sparse (S/d = 2.13). The boundary effect, also known as the side wall effect, has an influence on the channel flow structure when the blockage ratio, which measures how wide the vegetation patch front is in relation to how wide the channel itself is, exceeds a certain threshold value. According to previous studies, an utterly trivial impact of the side walls on the drag of a flat plate occurs when the obstruction ratio is equal to 5–6 percent [46]. To determine how the drag force decreases with increasing obstruction ratio (up to 40 percent), Okajima et al. [47] performed an experiment around a cylinder shaped like a rectangle, in which the flow rate was constant. They reported that the drag forces initially drop (obstruction ratio of 9 to 10 percent) and then rise as flow obstruction increases. The vegetation model (VM) studied in the present research has the capacity to enable water to flow through it, and this is mostly based on the density. As a result, in order to compute the blockage ratio, the front width (Wc) of the vegetation was estimated by multiplying the number of cylinders in each of the first two rows due to its staggered arrangement by the cylinder diameter. Then, the front width of the vegetation patch is divided by the channel's width. The considered vegetation model for experimental trial had a blockage ratio up to approximately 13%, which was under critical limit. Furthermore, it was also observed during the testing that no waves were reflected from the channel's side wall into the area under investigation during the experimental trials. As a result, it was assumed that the boundary effect had no influence on the results and the effect of the restriction was overlooked.

The initial water depth was 7.2 cm, which corresponded to an initial Froude value (Fr_o) of 0.73. The free surface level (FSL) was measured by point gauge at an interval of every 20 cm longitudinally along the centerline from the channel upstream to just behind the vegetation. Two positions were chosen (P1, located at front of the vegetation model and P2, located at back of the vegetation model) to measure the vertical distribution of streamwise velocity using an Electro Magnetic Flow Meter (EMFM). The velocity measurements were obtained by positioning the EMFM slightly above the flume bed and raising it vertically at an interval of 0.5 to 1 cm up to the FSL. The experimental setup of the vegetation model (Measurement Locations) and the arrangement of cylinders and resulting flow structure are presented in Figure 5a,b. To mimic the experimental setup for numerical modeling and avoid a large mesh grid pattern and lowering computing costs, a vegetation model in a channel with a length of only 90 cm and a width of 70 cm was built. The preprocessing and postprocessing procedures used for the validation domain were the same as those used for the present numerical model.

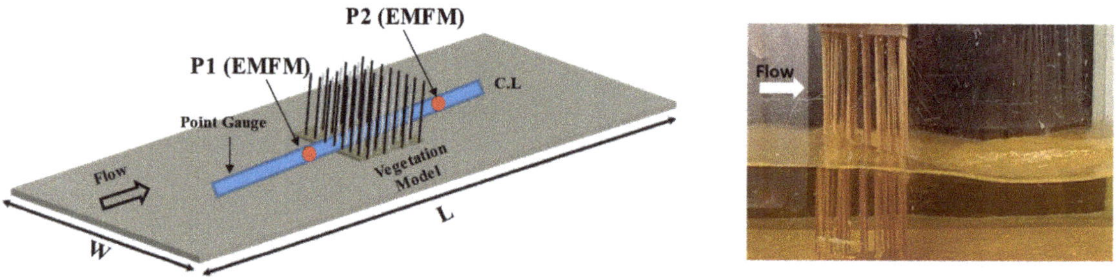

(a) **(b)**

Figure 5. Model Validation (**a**) Experimental Flume setup and specific locations for measurement of water level and vertical distribution of streamwise velocity; (**b**) Resulting flow structure.

The longitudinal distribution of a FSL calculated over the center portion of the domain for both the experimental and numerical results are presented in Figure 6a. The ordinate indicates the FSL in cm, while the abscissa represents the longitudinal distance along the channel. After installing the vegetation cylinders in the channel, for both the experimental and numerical results, the FSL was raised on the upstream side due to the vegetation model retardation and followed a decreasing trend in the downstream region of the vegetation side. This difference in elevation of the FSL between upstream and downstream of the vegetation developed the slope of the FSL inside the vegetation. Furthermore, Figure 6b depicts the results of the vertical distribution of streamwise velocity at defined locations from both the experimental and numerical findings. The streamwise velocity was considerably lower at the back (P1) of the vegetation patch compared to the front (P2) of the vegetation patch in both experimental and computational outcomes. The low velocity magnitudes of velocity at the back of the vegetation were caused by the retardation or blockage by the vegetation cylinders, which is beneficial for the mitigation of high tsunami magnitude. The numerical model outcomes are consistent with the findings in the experiment, demonstrating the model's validity. However, near the bed region, a difference in velocities was observed between the experimental and numerical findings. This discrepancy may be attributable to the fact that the EMFM's velocity sensor cannot be used in close proximity to the bed.

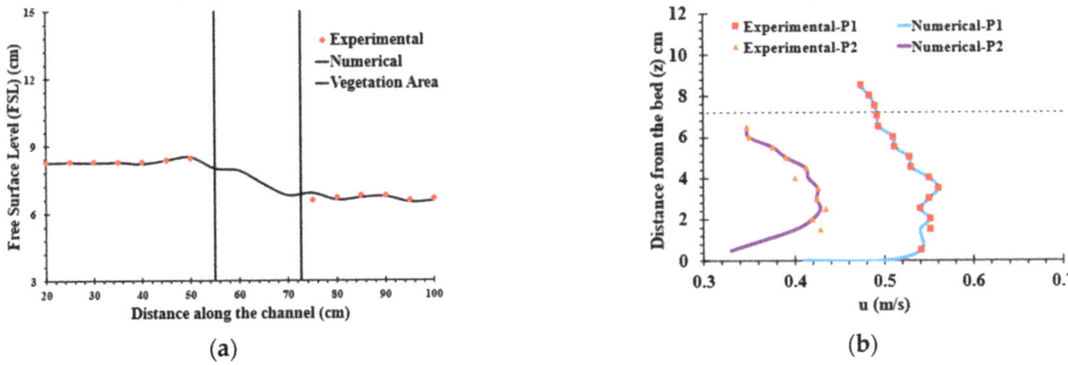

Figure 6. (**a**) Comparison of computed and experimental free surface level (FSL) profiles: (**b**) Comparison of computed and experimental vertical distribution of streamwise velocity.

5.2. Flow Characteristics

5.2.1. Free Surface Profiles

Vegetation in coastal areas is essential in reducing the velocity and energy of tsunami-generated floodwater. The flow properties around the coastal vegetation are highly dependent on external flow structure, i.e., free surface variations. Water reflection and vegetation resistance can reduce the flowing water energy, depth of inundation, region of inundation, and velocity behind the vegetation [12]. Therefore, this numerical study mainly focused on capturing the three-dimensional external flow structures, i.e., FSL variations around finite length emergent coastal vegetation of different configurations based on the AR and its impact on the flow behavior for tsunami mitigation. The two dimensional as well as three-dimensional vortices generation and free surface movement around the vegetation models of four cases (2, 3, 4, 5) are presented in Figure 7a–h. The free surface level (FSL) differences at the front, mid and behind the vegetation model are presented in 2D water–air volume fraction contours considering the longitudinal section (LS1), Figure 7a–d. The blue and red color represent the air and water, respectively. The 3D water volume fraction contours, i.e., Figure 7e–h are presented to understand the free surface vortices, overall shape, and the free surface wavy pattern in the vegetated open channel. The red color in the rectangular channel shows the water phase. By visualizing the contours, the formation of small and large-sized free surface vortices and their development process mainly depend on the vegetation elements configuration. The array of vegetation elements at the upstream side caused the flow to be irrotational. Retardation occurs due to the patch blockage impact and the drag interface discontinuity offered by the cylinders. The water gradually bleeds with minimum velocity through the vegetation area and travels rapidly at high velocity in the adjacent vegetation gap regions. The shear layer was formed by the significant difference in velocity between the vegetation patch and its gap area due to the FSL elevation difference. The shear layer then produces coherent structures or vortices due to Kelvin–Helmholtz instability. The vortices generated by the shear layer dominate the flux exchange (low and high velocity) between the vegetated and the gap region. The production of vortices and the exchange of mass flux at the interface between the vegetated and gap regions can be assumed to play a significant role in sediment transport into the vegetated plain [48]. These vortices have continuously grown from the upstream side towards the downstream side and gradually dissipate in all cases (2–5). In the longer vegetation patch (AR < 1), i.e., case (1) and case (2), it was found that the increment in elevation of FSL in front of vegetation was less and less steep slope was generated within the vegetation Figure 7a,b, resulting in the creation of vortices of small size vortices Figure 7e. Therefore, a longer vegetation patch would provide a smaller contribution to the overall water flow energy dissipation capabilities. Similarly, in case (3), i.e., (AR = 1), small sized vortices were also found when the length and width were same, as in Figure 7c,f. In addition, wider vegetation patches (AR > 1), i.e., case (4) Figure 7g and case (5) Figure 7d,h produced considerable resistance due to increased vegetation cylinders in the cross streamwise direction, and hence, maximum elevation of the FSL was observed. According to the findings of the Pasha and Tanaka [16] study, increasing the width of the vegetation patch results in a greater depth of water toward the front side of the vegetation, demonstrating that the results of the current study are in agreement. The maximum FSL results in a steep gradient slope inside the vegetation patch and minimum elevation of FSL was found on the vegetation patch downstream side. This higher difference in FSL upstream and downstream side of the vegetation patch resulted in the development of strong large vortices, Figure 7g,h. The maximum FSL was observed in the maximum AR, i.e., case (5), around 66% compared to the initial FSL, while this increment was observed to be 53% in case (4). Contrarily, fewer percentage differences were observed in lower AR cases, i.e., cases 1, 2, and 3, around 25%, 22%, and 20%, respectively. Thus, in this study, the wider vegetation patches significantly increased the FSL at the front side of the vegetation patch, which would help dissipate the flow energy behind the vegetation patch. These dramatically changed FSL heights in the upstream, inside, and behind the vegetation, and the development of vortices at the free surface were difficult to observe in those numerical studies where they assume water as a flat surface [21–23]. Therefore, VOF modeling (water–air interface) can well predict the FSL around the vegetation in an open channel.

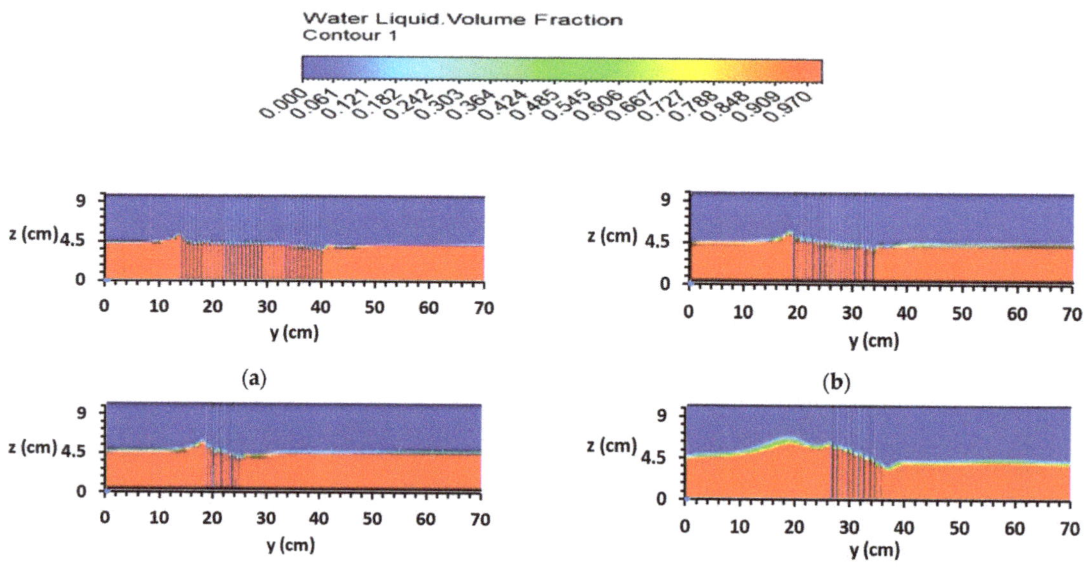

Figure 7. *Cont.*

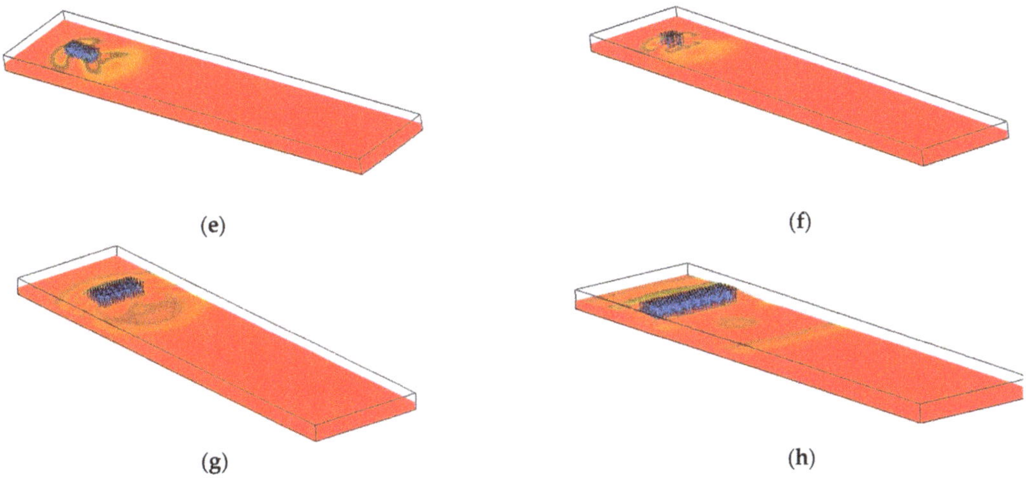

Figure 7. Two-dimensional (2D) Distribution of water–air volume fraction (**a**) Case 1, (**b**) Case 2, (**c**) Case 3, (**d**) Case 5: Three-dimensional (3D) distribution of water volume fraction (**e**) Case 2, (**f**) Case 3, (**g**) Case 4, (**h**) Case 5.

5.2.2. Velocity Distribution of Streamwise Velocity Profiles

The streamwise velocity distribution in the vertical direction around the finite length vegetation and their adjacent gap regions was investigated. The behavior of the streamwise velocity distribution is divided into three distinct regions that cover the whole array of vegetation cylinders in the vertical direction. (i) The region near the free surface is influenced by free surface oscillations. (ii) The middle region, found between the free surface and the bed region, where the vertical cylinders control the flow, and in that region, flow characteristics remain nearly constant in the vertical direction. (iii) The approaching bed region where the flow is extremely three-dimensional because of the bed interaction. The streamwise velocity at the centerline of the longitudinal section (L1) at the four specified points (X1–X4) for all the AR cases (1–5) is shown in Figure 8a–d. Point (X1) is located on the upstream side of the vegetation, and point (X2) is located at the center of the vegetation region. In addition, Point (X3) and (X4) are located at the vegetation region's downstream side. The x-axis represents streamwise velocity ($\boldsymbol{U_{veg}}$), while the y-axis indicates distance from the bed ($\boldsymbol{Z_w}$) to the FSL. Due to the presence of vertical stems, as well as the resistance given by the bed of the domain, the velocity is lower closer to the bed area of the domain. The maximum reduction in velocity was observed in wider vegetation patches (AR > 1), i.e., case (4) and case (5), at point (X1) due to the greater resistance provided by the cylinders in the crosswise direction and the increased height of the FSL, while less reduction in velocity was observed in the longer vegetation patch (AR < 1), i.e., case (1) and case (2). The results clearly demonstrate that wider vegetation patches (AR > 1) could effectively reduce the velocity in front of vegetation. In addition, near the free surface region in Figure 8a, the highest reduction in velocity was observed in comparison to the middle region, resulting from the difference in pressure caused by the rising FSL and the air region above the free surface [14]. At point (X2) in Figure 8b, which was located within the vegetation patch, the maximum reduction in velocity was observed in the longer vegetation patch (AR < 1), case (1) and case (2) due to continuous resistance provided by the vegetation cylinders in the longitudinal direction. In comparison, this reduction was low in wider vegetation patches (AR > 1), i.e., case (4) and case (5). As a result, a longer vegetation patch benefits the protection of the vegetation structure itself, which helps in reducing bed shear stress inside the vegetation structure. Additionally, the same trend of velocity reduction was observed at points (X3) and (X4) for all the cases in Figure 8c,d. These streamwise velocity profiles in the vertical direction within and around the vegetation were challenging to observe in the experimental study. This manifests the numerical simulation benefits.

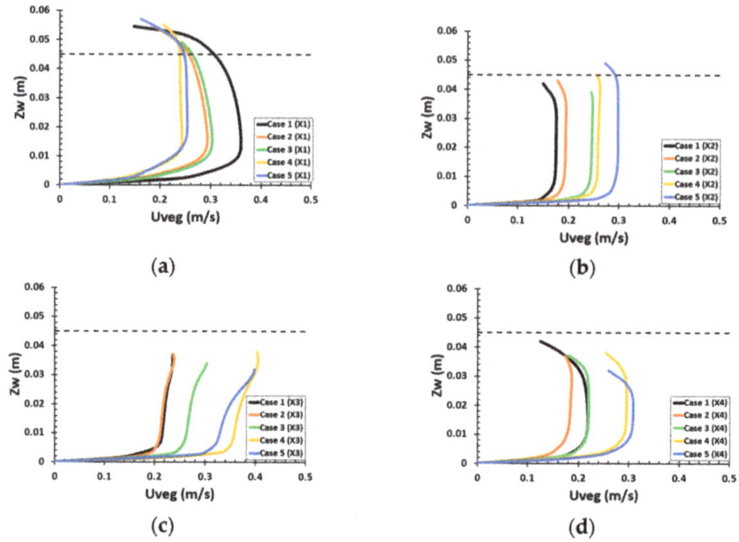

Figure 8. Vertical distribution of streamwise velocities in the vegetation region and downstream region at specified locations (**a**) X1; (**b**) X2; (**c**) X3; (**d**) X4. For the locations, see Figure 2.

For all the different AR cases (1–5), the streamwise velocity in the vegetation gap area is shown in Figure 9a–c. These points were considered on one side of the centerline in the gap region because the vegetation is located at the centerline, and conditions are symmetrical on both sides of the centerline. Point (Y1) is in the gap area on the upstream side of the vegetation, and point (Y2) is located at the center of the vegetation gap region, while point (Y3) is located downstream of the vegetation gap region. The x-axis depicted the streamwise velocity of the gap region (U_{Gap}), whereas the y-axis depicted the distance from the bed (Z_w) up to the FSL. It is observed in all the AR cases (1–5) that velocity profiles were lower in the upstream gap regions (Y1), marginally higher in the center of the gap region (Y2), increased more in the downstream of the gap region (Y3) and decreased to their minimum near the bed. Drop in velocities located within close vicinity to the bed were attributable to the bed's domain resistance. The streamwise flow velocity was observed to be low in all the aspect ratio cases (1–5) at point (Y1) Figure 9a on the upstream side of the vegetation gap region. This effect of low velocity was due to the sheltering effect of vegetation cylinders and water reflection, i.e., increased height of the FSL, which also affects the gap region. The maximum reduction in streamwise velocity was observed in a wider vegetation patch (AR > 1), i.e., case (5), at point (Y1), due to the increased width of vegetation in the crosswise direction, resulting in a more significant increase in the height of the FSL. The streamwise velocity of case (4) and case (5) is increased as it moves towards point (Y2), Figure 9b. The inflow of water from the vegetation zone into the gap leads to a large-scale eddy between the interface of the gap and the vegetation cylinders. The minimum length of vegetation along the longitudinal direction was also the main factor for increasing the velocity.

The streamwise velocity was further increased at point (Y3), Figure 9c, downstream of the vegetation gap region, and maximum velocity was observed in case (5) due to the fast diffraction of flow at the edges of the vegetation. From the above profiles, it can be shown that case (4) and case (5) provide the highest reduction in velocity upstream of the vegetation gap region, Figure 9a. Simultaneously, these configurations contribute to an increase in velocity downstream of the vegetation gap region Figure 9c, which may lead to erosion at the vegetation edges. However, in cases (1) and (2), where the vegetation length was greater than the vegetation width, the same pattern of velocities was observed at the upstream, middle, and downstream sides of the vegetation gap region, but the reduction in velocities was less due to the minimum FSL at the upstream side, and the increase in velocities was less due to the increased length of the vegetation patch.

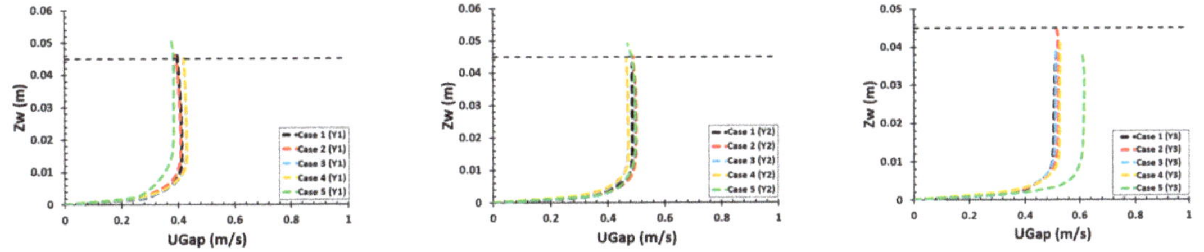

(a) (b) (c)

Figure 9. Vertical distribution of streamwise velocities in the gap regions at specifies locations (**a**) Y1; (**b**) Y2; (**c**) Y3. For the locations, see Figure 2. The black dashed lined shows the initial FSL.

5.2.3. Velocity Profiles along the Longitudinal Sections

The streamwise velocity distribution through the longitudinal sections (LS1 = y = 0.35 m) and (LS2 = y = 0.25 m), i.e., passed from the center of the channel and the gap region, for all the AR cases (1–5) is shown in Figure 10a,b. The dashed boxes depict the vegetation area along the longitudinal direction. In Figure 10a, all the cases followed an irregular pattern of velocities within the vegetation area. The longer the vegetation patch length, i.e., (AR < 1), the maximum irregularity is observed in the velocity profiles. A maximum reduction of 97% and 94.8% in velocities was observed within the vegetation patch in case (1) and case (2), respectively, as compared to the initial velocity without the vegetation. These configurations of vegetation patches could effectively reduce the bed shear stress inside the vegetation patch. When the vegetation patch of (AR $\geq$ 1) is considered, the velocity reduction in cases 3, 4, and 5 also led to approximately (40.02–53.50%).

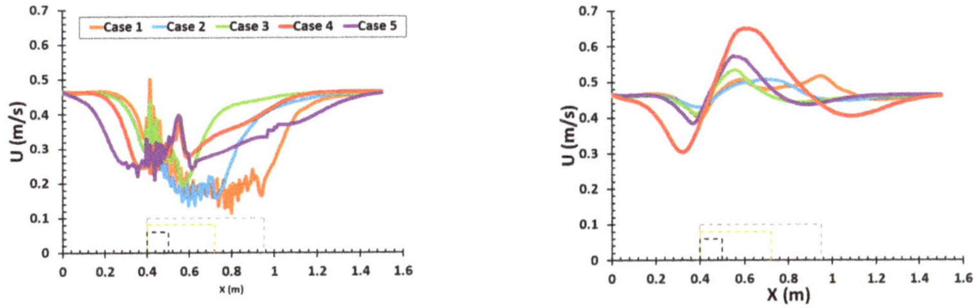

(a) (b)

Figure 10. Longitudinal distributions of streamwise velocities, (**a**) along the longitudinal section LS1 and (**b**) along the longitudinal section LS2. The dashed lines show the Vegetation length area (W_x).

The reduction in streamwise velocities upstream of the vegetation gap region was observed in all the AR cases due to increased FSL, Figure 10b. When the AR is increased, the greater the velocity reduction in the upstream of the gap region (around 11.83–32.7% reduction in velocities compared to initial velocity) was observed in cases 3, 4, and case 5. However, the smaller AR vegetation patch also leads to a reduction in velocity upstream of the vegetation gap area, which in case (1) and case (2) was approximately 7.1% and 14.1%, respectively. Contrarily, with the increase in the AR of vegetation patch, the greater the velocity increment downstream of the gap region (around 14.26–34.86% increment in velocities compared to initial velocity) was observed in case 3, 4, and case 5. While this increment was only 8.6% and 10.11% in the smaller AR case (1) and case (2), respectively, it suggests that the longer vegetation patch did not substantially contribute to the increase in velocity downstream of the vegetation gap area.

5.2.4. Simulated Contour Plot Distribution of Streamwise Velocity

The contour plot distribution of streamwise velocity on the horizontal *x*-*y* plane for all the cases is shown in Figure 11a–e. One horizontal surface was examined at depth z = 3.5 cm (below the free surface) to better understand the flow behavior. When the flow bleeds through the vegetation region, the reduction in stream wise velocities was observed at the front of the vegetation in all the cases (1–5). The lowest reduction was observed in longer vegetation patch configuration, i.e., case (1), case (2) and case (3), Figure 11a–c, whereas the highest reduction was seen in the wider patch configuration, i.e., case (4) and case (5), Figure 11d,e. This was due to the presence of the maximum number of vegetation cylinders in the path of flow. Regarding the inside of the vegetation region, the trend of reduction in velocities were different. Due to the continuous resistance provided by the vegetation cylinders, the highest reduction in velocity was observed in longer vegetation patches (AR < 1), considering the inside region of the vegetation patch. However, wider vegetation patches (AR > 1) also contribute to velocity reduction, but it produces a large, sheltered zone (area of low velocity), Figure 11d,e, downstream of the vegetation patch compared to the longer vegetation patches (AR < 1). Therefore, considering the tsunami mitigation capability point of view, the findings illustrate the significance of a wider vegetation patch, which could significantly increase the maximum velocity reduction area just behind the vegetation patch compared to the longer vegetation patch.

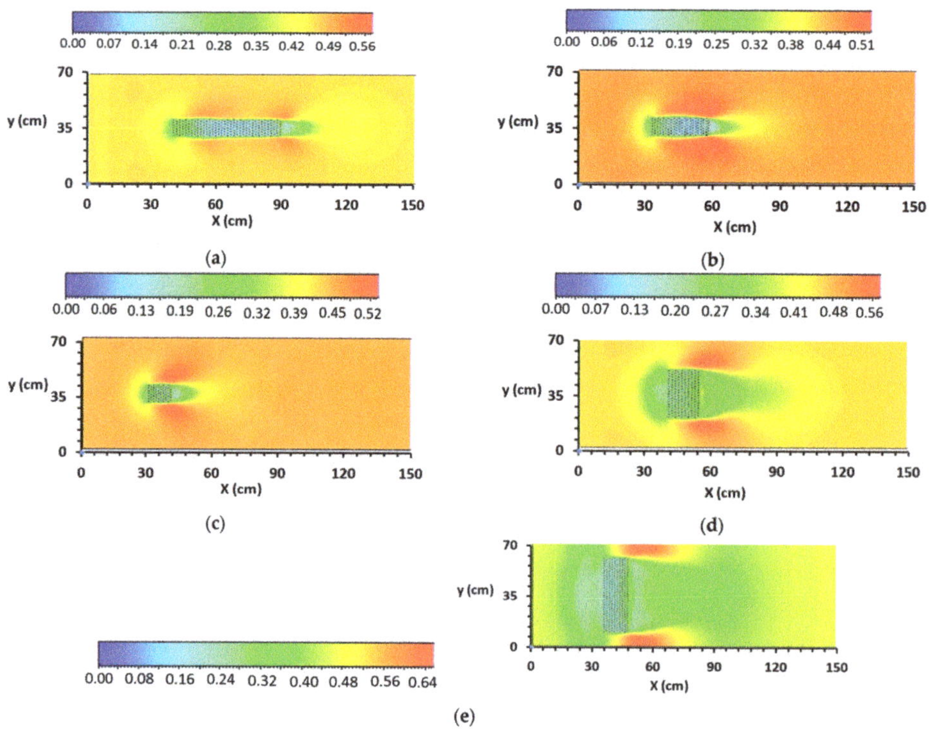

Figure 11. Contour plot distribution of streamwise velocity along the horizontal plane: (**a**) Case 1, (**b**) Case 2, (**c**) Case 3, (**d**) Case 4, (**e**) Case 5.

The sheltering effect of the longer vegetation patch, on the other hand, causes the velocity to decelerate in the nearby gap areas when compared to the wider vegetation patch. Therefore, the wider vegetation patches (AR > 1) contribute to generating a small zone of high-intensity velocity magnitude in the vegetation gap region, and the longer vegetation patches (AR < 1) lead to the formation of a large zone of low-intensity velocity magnitude. Therefore, an increased AR of the vegetation patch has a negative impact on the downstream gap region due to increased velocity magnitudes at the vegetation patch edges compared to the longer vegetation patch.

5.2.5. Flow Movement of Water and Air Phase

The schematic detail of the flow behavior between the air and water phase and generation of vortices at specified sections (CS1) and (CS2) for the only three AR cases (1, 3 and 5) are shown in Figure 12a–f. The *y*-axis represents the channel's crosswise direction, i.e., the channel's width and the *z*-axis represents the channel's total depth, including both the air and water region. Due to the slip wall presence, only the vegetated region and bed of the channel domain were accountable for the generation of vortices. The presence of vegetated and gap regions contributes to an inflection point along both the crosswise sections (CS1) and (CS2) in the streamwise velocity vectors. In section (CS1) near the vegetation region (water phase), i.e., the interface between the vegetated and gap region, flow divergence; that is, the movement of flow in a lateral direction, occurred in the regions of vegetation area towards the gap region, and, as a result, the downward movement of water flow occurred within the vegetation region, Figure 12a,c,e. Contrarily, flow convergence in section (CS2); that is, the movement of water flow, took place in the reverse direction just at the downstream side of the vegetation patch in a lateral direction, as the flow behind the vegetation again collided and, as a result, an upward movement of flow occurred in front of the downstream side of the vegetation patch Figure 12b,d,f.

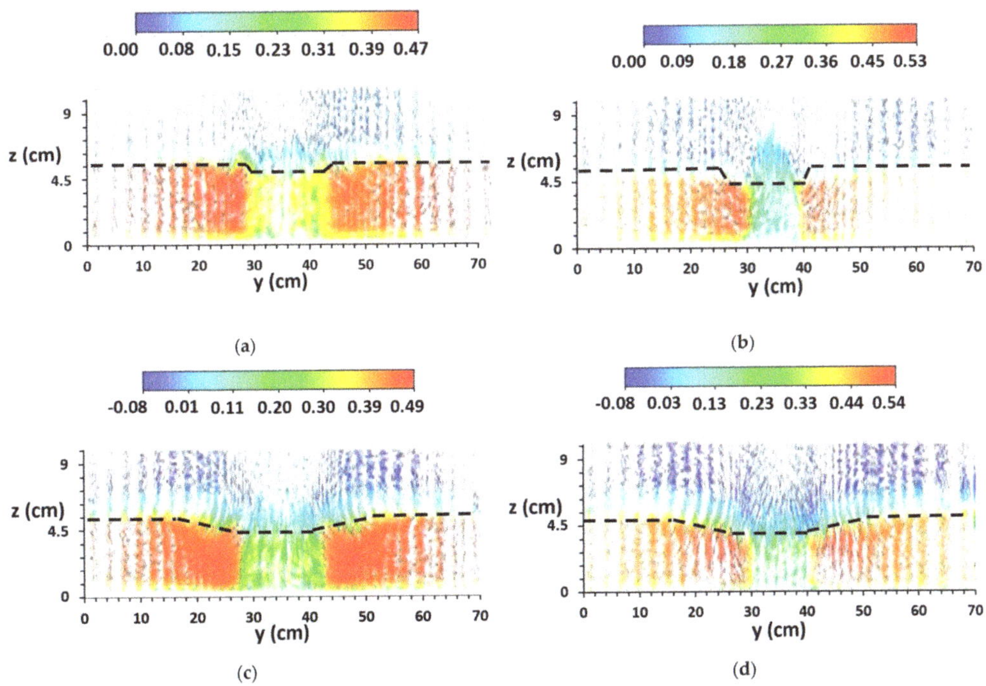

Figure 12. *Cont.*

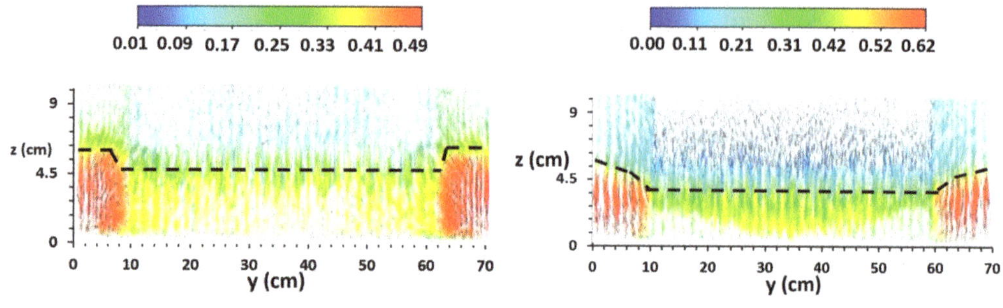

(e) (f)

Figure 12. Streamwise velocity vectors (m/s) along the specified cross sections CS1 and CS2 (**a**,**b**) Case 1; (**c**,**d**) Case 3; (**e**,**f**) Case 5. The black dashed lined shows the FSL (Free Surface Level).

Owing to the combination of the air and water layers, the inflection point in the streamwise velocities was seen near the free surface level in all three AR cases. The greater velocity difference of the two phases, i.e., air and water, contribute to a generation of strong vertical clockwise and counterclockwise vortices in the vertical direction Figure 13a,b. The combination of low-density air–liquid and high-density water–liquid was accountable for these vortices. However, these vortices are more prominent when the gap region is larger, i.e., case (1) and case (3). Above the FSL, a downward air movement occurs in the region above the vegetation area and the regions immediately behind the vegetation side. According to Rashedunnabi and Tanaka [14], the air movement that occurs in the downward direction in a vegetated open channel flow is the main reason for the air entrainment in the vegetation patch at the downstream side, which can be seen in the present numerical study.

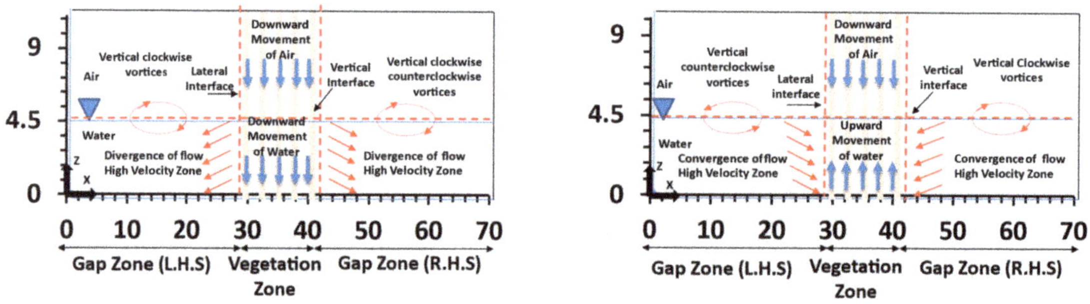

(a) (b)

Figure 13. Schematic diagram of general vortex behavior in an emergent vegetated channel of air and water behavior along with the cross-sections (**a**) CS1; (**b**) CS2 for all the cases.

5.2.6. Turbulent Intensity Distribution of Air–Water Phase

When the flow bleeds through the vegetation area, the flow becomes unstable in that region. The flow instability in the vegetation region leads to the turbulence level. In addition, the air region above the water region also affects the flow phenomenon in terms of air entrainment into the water region. Therefore, the estimation of the turbulence level of both the air and water regions is very important and can be easily understood by using VOF modeling. The turbulent intensity ($I = \sqrt{u'^2}/u_{avg}$) in the form of contour plots for the three AR cases (1, 3 and 5) only at the specified section (CS1) is shown in Figure 14a–c. On the *x*-axis is the width of the channel, and on the *z*-axis is the elevation, which includes both air and water regions. Figure 14a–c shows that the maximum water turbulence intensity below the free surface was observed both within the vegetation and near the bed of the channel domain. This was due to the vegetation cylinders resistance in the path of the water flow and the continuous resistance provided by the bed of the domain. Raupach et al. [49] and Ricardo et al. [50] found that the production of vortices within the vegetation leads (as seen in Figure 12a,c,e) to the dissipation of the turbulent energy. A considerable quantity of turbulence was detected near the vegetation edge (i.e., the boundary between the gap region and the vegetation region), which resulted from the rapid water moving towards the gap area (discussed in Section 5.2.5). The gap regions exhibited little resistance; thus, very little turbulence was recorded there. Therefore, the maximum turbulent intensity in the vegetation region showed that vegetation cylinders provide significant resistance in the path of the flow.

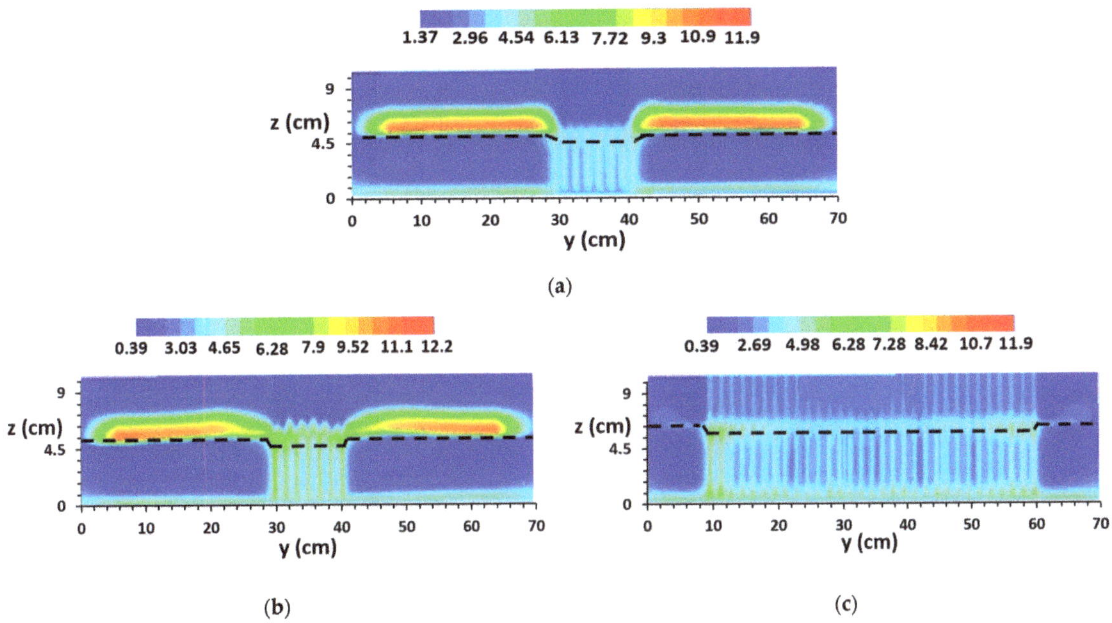

Figure 14. Turbulent intensities (%) along the cross-section CS1 (**a**) Case 1; (**b**) Case 3 and (**c**) Case 5.

Savelsberg and Van de water [51] found that turbulence of air at the free surface indicates air entrainment into the water region. Above the FSL, in the gap region, strong air turbulent intensity was observed in case (1), 11.9%, and case (3), 12.1%, indicating strong air entrainment over this region, Figure 14a,b. In the gap region of the maximum AR case (5), less than 3% of air turbulent intensity was gained over the free surface level due to constriction of the gap area. while a significant increase in air turbulent intensity above the vegetation region was found in case (5), i.e., wider vegetation patch, Figure 14c. A study by Cain [52] revealed that air entrainment occurred in a turbulent flow due to irregular vortices' formation. In the present study, these vortices were formed at the free surface level, as discussed in the above section. In addition, Chanson [53] showed that if the turbulent properties are high enough to transcend both gravity and surface tension impact, the air bubbles can be entrained. Therefore, a significant amount of air bubbles forming in a maximum aspect ratio can result from the higher turbulent intensity of air above the vegetation cylinders. Previous studies have also shown that higher generation of turbulent structures can promote the flow of air bubbles into the vegetation region [54,55], which can play an important role in energy dissipation [14,56]. Therefore, the maximum air intensity above the wider vegetation patch region (AR > 1) can result in greater energy reduction relative to that of longer vegetation patch configuration (AR < 1).

6. Conclusions

In this research, a three-dimensional numerical model was established using ANSYS to investigate inland tsunami flow in a vegetated open channel. The two-phase volume of fluid (VOF) model was used to track the free surface combined with the Reynolds stress model (RSM). The accuracy of the free surface was checked using a variety of meshing methods; however, multizone meshing was the most effective in correctly tracking the free surface with the smallest amount of diffusion, compared to the tetrahedral meshing. The present numerical model has successfully captured reasonable outcomes and has better correspondence with the experimental results in terms of free surface and velocity profiles.

The numerical model was then utilized to investigate the 3D flow behavior through the longer (AR < 1) and wider (AR > 1) vegetation patches to compare the flow behavior of each type under subcritical flow conditions depending on the Aspect Ratio (AR = vegetation patch width/length). The results showed that the maximum free surface level (FSL) at the front and minimum FSL behind the vegetation increases with the increase in AR > 1 (in the case of the wider vegetation patch), which consequently results in the generation of large size vortices. On the other hand, with the decrease in AR $\leq$ 1 (in the case of the longer vegetation patch), less FSL is observed at the front of the vegetation, which consequently results in the generation of small size vortices. Considering the tsunami mitigation point of view, the wider vegetation patch configurations (AR > 1) give favorable conditions in terms of the large area of low velocity directly downstream side of the vegetation patch, whereas it has a negative impact on the adjacent downstream gap region due to increased high velocity (around 14.3–34.9% increment in velocities compared to initial velocity), which would increase bed shear stress at the edges of the vegetation. The longer vegetation patch configurations (with AR < 1) gives favorable conditions due to maximum velocity reduction within the vegetation patch (around 94.8–97.14% decrement in velocities), which would decrease bed shear stress and protect the vegetation structures itself. However, it has a disadvantage in the downstream vegetation region due to the generation of a small, sheltered area of low velocity. In addition, the present numerical model successfully captured the flow movement of water and air region around the vegetation patches. The turbulent intensity distribution of air and water region showed that the turbulence of air at the free surface suggests strong air entrainment into the water region, which can result in greater energy reduction. The maximum air intensity above the wider vegetation patch region (AR > 1) was observed compared to the longer vegetation patch configuration (AR < 1).

These results are beneficial and provide basic information for considering the suitable design of finite length emergent vegetation based on the Aspect Ratio (AR). Therefore, in the future, more computational research should be conducted to analyze the flow properties against the finite length emergent vegetation, with further varying configurations and with some angled gaps to overcome the high-velocity zone at the edges of the vegetation patch (observed in the wider patch configuration) and other ground conditions.

Author Contributions: Conceptualization, A. and N.T.; methodology, A.; software, A.; validation, A.; writing—original draft preparation, A.; review, N.T.; supervision, N.T. All authors have read and agreed to the published version of the manuscript.

Funding: This research received no external funding.

Institutional Review Board Statement: Not applicable.

Informed Consent Statement: Not applicable.

Data Availability Statement: Not applicable.

Acknowledgments: The authors acknowledge the support of MEXT scholarship (A from the Japanese Ministry of Education, Culture, Sports, Science, and Technology (Monbukagakusho).

Conflicts of Interest: The authors declare no conflict of interest.

References

1. Shuto, N. The effectiveness and limit of tsunami control forests. *Coast. Eng. Jpn.* **1987**, *30*, 143–153. [CrossRef]
2. Fadly, U.; Murakami, K. Study on reducing tsunami inundation energy by the modification of topography based on local wisdom. *J. Jpn. Soc. Civ. Eng. Ser. B3* **2013**, *68*, 66–71. [CrossRef]
3. Kathiresan, K.; Rajendran, N. Coastal mangrove forests mitigated tsunami. *Estuar. Coast. Shelf Sci.* **2005**, *65*, 601–606. [CrossRef]
4. Yanagisawa, H.; Koshimura, S.; Goto, K.; Miyagi, T.; Imamura, F.; Ruangrassamee, A.; Tanavud, C. The reduction effects of mangrove forest on a tsunami based on field surveys at Pakarang Cape, Thailand and numerical analysis. *Estuar. Coast Shelf Sci.* **2009**, *81*, 27–37. [CrossRef]
5. Harada, K.; Kawata, Y. Study on tsunami reduction effect of coastal forest due to forest growth. *Ann. Disaster Prev. Res. Inst.* **2005**, *48*, 161–165.
6. Irtem, E.; Gedik, N.; Kabdasli, M.S.; Yasa, N.E. Coastal Forest effects on tsunami run-up heights. *Ocean Eng.* **2009**, *36*, 313–320. [CrossRef]
7. Tanaka, N. Vegetation bio shields for tsunami mitigation: Review of effectiveness, limitations, construction, and sustainable management. *Landsc. Ecol. Eng.* **2009**, *5*, 71–79. [CrossRef]
8. Tanaka, N.; Jinadasa, K.; Mowjood, M.I.M.; Fasly, M.S.M. Coastal vegetation planting projects for tsunami disaster mitigation: Effectiveness evaluation of new establishments. *Landsc. Ecol. Eng.* **2011**, *7*, 127–135. [CrossRef]
9. Ismail, H.; Abd Wahab, A.K.; Alias, N. Determination of man- grove forest performance in reducing tsunami run-up using physical models. *Nat. Hazards* **2012**, *63*, 939–963. [CrossRef]
10. Thuy, N.B.; Tanimoto, K.; Tanaka, N. Flow and potential force due to runup tsunami around a coastal forest with a gap, Experiments and numerical simulations. *Sci. Tsunami Hazards* **2010**, *29*, 43–69.
11. Iimura, K.; Tanaka, N. Numerical simulation estimating effects of tree density distribution in coastal forest on tsunami mitigation. *Ocean Eng.* **2012**, *54*, 223–232. [CrossRef]
12. Pasha, G.A.; Tanaka, N. Undular hydraulic jump formation and energy loss in a flow through emergent vegetation of varying thickness and density. *Ocean Eng.* **2017**, *141*, 308–325. [CrossRef]
13. Tanaka, N. Effectiveness and limitations of coastal forest in large tsunami: Conditions of Japanese pine trees on coastal sand dunes in tsunami caused by Great East Japan Earthquake. *J. Jpn. Soc. Civil. Eng. Ser. B1* **2012**, *68*, II_7–II_15. [CrossRef]
14. Rashedunnabi, A.H.M.; Tanaka, N. Effectiveness of double-layer rigid vegetation in reducing the velocity and fluid force of a tsunami inundation behind the vegetation. *Ocean Eng.* **2020**, *201*, 107142. [CrossRef]
15. Iimura, K.; Tanaka, N. Dangerous zone formation behind finite-length coastal forest for tsunami mitigation. *J. Earthq. Tsunami* **2013**, *7*, 1350034. [CrossRef]
16. Pasha, G.A.; Tanaka, N. Effectiveness of Finite Length Inland Forest in Trapping Tsunami-Borne Wood Debris. *J. Earthq. Tsunami* **2016**, *10*, 1650008. [CrossRef]
17. Thuy, N.B.; Tanimoto, K.; Tanaka, N.; Harada, K.; Iimura, K. Effect of open gap in coastal forest on tsunami run-up-investigations by experiment and numerical simulation. *Ocean Eng.* **2009**, *36*, 1258–1269. [CrossRef]
18. Nandasena, N.A.K.; Tanaka, N.; Tanimoto, K. Perspective of coastal vegetation patches with topography variations for tsunami protection in 2D numerical modeling. *Annu. J. Hydraul. Eng. JSCE* **2008**, *52*, 138. [CrossRef]
19. Vandenbruwaene, W.; Temmerman, S.; Bouma, T.J.; Klaassen, P.C.; De Vries, M.B.; Callaghan, D.P.; Van Steeg, P.; Dekker, F.; Van Duren, L.A.; Martini, E.; et al. Flow interaction with dynamic vegetation patches: Implications for bio geomorphic evolution of a tidal landscape. *J. Geophys. Res. Earth Surf.* **2011**, *116*, F01008. [CrossRef]
20. Irish, J.L.; Weiss, R.; Yang, Y.; Song, Y.K.; Zainali, A.; Marivela Colmenarejo, R. Laboratory experiments of tsunami run-up and withdrawal in patchy coastal forest on a steep beach. *Nat. Hazards* **2014**, *74*, 1933–1949. [CrossRef]
21. Ghani, U.; Anjum, N.; Pasha, G.A.; Ahmad, M. Numerical investigation of the flow characteristics through discontinuous and layered vegetation patches of finite width in an open channel. *Environ. Fluid Mech* **2019**, *19*, 1469–1495. [CrossRef]
22. Hasnain, T.; Usman, G.; Naveed, A.; Ghufran, A.P. 3D Numerical modeling of flow characteristics in an open channel having in-line circular vegetation patches with varying density under submerged and emergent flow conditions. *J. Hydrol. Hydromech.* **2022**, *70*, 128–144.
23. Muhammad, A.; Usman, G.; Naveed, A.; Ghufran, A.P.; Muhammad, K.U.; Afzal, A. Investigating the flow hydrodynamics in a compound channel with layered vegetated floodplains. *Civ. Eng. J.* **2020**, *6*, 860–876.
24. Maji, S.; Pal, D.; Hanmaiahgari, P.R.; Gupta, U.P. Hydrodynamics and turbulence in emergent and sparsely vegetated open channel flow. *Environ. Fluid Mech.* **2017**, *17*, 853–877. [CrossRef]
25. Stoesser, T.; Kim, S.J.; Diplas, P. Turbulent Flow through Idealized Emergent Vegetation. *J. Hydraul. Eng.* **2010**, *136*, 1003–1017. [CrossRef]
26. Hirt, C.W.; Nichols, B.D. Volume of fluid (VOF) method for the dynamics of free boundaries. *J. Comput. Phys. Citeseer* **1981**, *39*, 201–225. [CrossRef]
27. Chen, Q.; Dai, G.; Liu, H. Volume of Fluid Model for Turbulence Numerical Simulation of Stepped Spillway Overflow. *J. Hydraul. Eng* **2002**, *128*, 683–688. [CrossRef]

28. Xiangju, C.; Chen, Y.; Luo, L. Numerical simulation of air-water two-phase flow over stepped spillways. *Sci. China Ser. E Technol. Sci.* **2006**, *49*, 674–684.
29. Zhenwei, M.; Zhang, Z.; Zhao, T. Numerical simulation of 3-D flow field of spillway based on VOF method. *Procedia Eng.* **2012**, *28*, 808–812. [CrossRef]
30. Eghbalzadeh, A.; Javan, M. Comparison of Mixture and VOF models for numerical simulation of Air-Entrainment in skimming flow over stepped spillways. *Procedia Eng.* **2012**, *28*, 657–660. [CrossRef]
31. Zhan, J.; Zhang, J.; Gong, Y. Numerical investigation of air-entrainment in skimming flow over stepped spillways. *Theory Appl. Mech. Lett.* **2016**, *6*, 139–142. [CrossRef]
32. Ramamurthy, A.S.; Qu, J.; Vo, D. Volume of fluid model for an open channel flow problem. *Can. J. Civ. Eng* **2005**, *32*, 996–1001.
33. Patel, T.; Gill, L. Volume of fluid model applied to curved open channel flows. *WIT Trans. Eng. Sci* **2006**, *52*, 361–369.
34. Carrier, G.; Wu, T.T.; Yeh, H. Tsunami run-up and draw-down on a plane beach. *J. Fluid Mech.* **2003**, *475*, 79–99. [CrossRef]
35. Yang, Y.; Irish, J.L.; Weiss, R. Impact of patchy vegetation on tsunami dynamics. *J. Waterw. Port C-ASCE* **2017**, *143*, 04017005. [CrossRef]
36. Zainali, A.; Marivela, R.; Weiss, R.; Yang, Y.; Irish, J.L. Numerical simulation of nonlinear long waves in the presence of discontinuous coastal vegetation. *Mar. Geol.* **2018**, *396*, 142–149. [CrossRef]
37. Ali, H.M.; El-Gendy, M.M.; Mirdan, A.M.H.; Ali, A.A.M.; Abdelhaleem, F.S.F. Minimizing downstream scour due to submerged hydraulic jump using corrugated aprons. *Ain. Shams Eng. J.* **2014**, *5*, 1059–1069. [CrossRef]
38. Fritz, H.M.; Borrero, J.C.; Synolakis, C.E.; Yoo, J. 2004 Indian Ocean tsunami flow velocity measurements from survivor videos. *Geophys. Res. Lett.* **2006**, 33, L24605. [CrossRef]
39. Tanaka, N.; Yagisawa, J.; Yasuda, S. Breaking pattern and critical breaking condition of Japanese pine trees on coastal sand dunes in huge tsunami caused by Great East Japan Earthquake. *Nat. Hazards* **2013**, *65*, 423–442. [CrossRef]
40. Salaheldin, T.M.; Imran, J.; Chaudhry, M.H. Numerical Modeling of Three-Dimensional Flow Field Around Circular Piers. *J. Hydraul. Eng* **2004**, *130*, 91–100. [CrossRef]
41. Tanaka, N.; Yasuda, S.; Iimura, K.; Yagisawa, J. Combined effects of coastal forest and sea embankment on reducing the washout region of houses in the Great East Japan tsunami. *J. Hydro–Environ. Res.* **2014**, *8*, 270–280. [CrossRef]
42. Lee, J.; Suh, J.; Sung, H.J.; Pettersen, B. Structures of turbulent open-channel flow in the presence of an air-water interface. *J. Turbul.* **2014**, *13*, N18. [CrossRef]
43. Bayon, A.; Toro, J.P.; Bombardelli, F.A.; Matos, J.; Lopez-Jimenez, P.A. Influence of VOF technique, turbulence model and discretization scheme on the numerical simulation of the non-aerated, skimming flow in stepped spillways. *J. Hydro-Environ. Res.* **2017**, *19*, 137–149. [CrossRef]
44. Naik, B.; Khatua, K.K.; Wright, N.; Sleigh, A.; Singh, P. Numerical modeling of converging compound channel flow. *ISH J. Hydraul. Eng.* **2018**, *24*, 285–297. [CrossRef]
45. *Fluent Inc User's Guide*; Release 19.2; ANSYS, Inc.: Canonsburg, PA, USA, 2020.
46. Okamoto, S.; Okamoto, T. Theoretical study of blockage effect of wind-tunnel on wake of two-dimensional flat plate normal to plane wall. *Trans. Jpn. Soc. Aero. Space Sci.* **1984**, *27*, 134–144.
47. Okajima, A.; Yi, D.; Kimura, S.; Kiwata, T. The blockage effects for an oscillating rectangular cylinder at moderate Reynolds number. *J. Wind Eng. Ind. Aerodyn.* **1997**, *69*, 997–1011. [CrossRef]
48. White, B.L.; Nepf, H.M. A vortex-based model of velocity and shear stress in a partially vegetated shallow channel. *Water Resour. Res.* **2007**, *44*, 1–15. [CrossRef]
49. Raupach, M.R.; Finnigan, J.J.; Brunet, Y. Coherent eddies and turbulence in vegetation canopies: The mixing-layer analogy. *Bound.-Layer Meteorol* **1996**, *78*, 351–382. [CrossRef]
50. Ricardo, A.M.; Koll, K.; Franca, M.J.; Schleiss, A.J.; Ferreira, R.M.L. The terms of turbulent kinetic energy budget within random arrays of emergent cylinders. *Water Resour. Res* **2014**, *50*, 4131–4148. [CrossRef]
51. Savelsberg, R.; Van De Water, W. Turbulence of a free surface. *Phys. Rev. Lett* **2008**, *100*, 034501. [CrossRef]
52. Cain, P. Measurements Within Self-Aereated Flow on Large Spilway. Ph.D. Thesis, University of Canterbury, Christchurch, New Zealand, 1978.
53. Chanson, H. Air-water flows in water engineering and hydraulic structures. Basic processes and metrology. *Dam. Eng.* **2004**, *1*, 3–16.
54. Baldy, S. A generation-dispersion modal of ambient and transient bubbles in the close vicinity of breaking waves. *Geophys. Res.* **1993**, *98*, 18. [CrossRef]
55. Ma, G.F. Multiscale Numerical Study of Turbulent Flow and Bubble Entrainment in the Surf Zone. Ph.D. Thesis, University of Delaware, Newark, DE, USA, 2012.
56. Thompson, E.; Piecuch, P.R.; Merrifield, C.G.; McCreary, M.A.; Firing, J.P. Turbulent flow field and air entrainment in laboratory plunging breaking waves. *J. Geophys. Res. Ocean* **2016**, *121*, 6762–6778. [CrossRef]

fluids

MDPI

Technical Note

Sediment Transport Patterns of Channels on Tidal Lowland

Achmad Syarifudin [1], Alfrendo Satyanaga [2,*], Martin Wijaya [3], Sung-Woo Moon [2] and Jong Kim [2]

1 Department of Civil and Environmental Engineering, Universitas Bina Darma, Palembang 30264, Indonesia
2 Department of Civil and Environmental Engineering, Nazarbayev University, Nur-Sultan 010000, Kazakhstan
3 Department of Civil Engineering, Parahyangan Catholic University, Jl. Ciumbuleuit No. 94, Bandung 40141, Indonesia
* Correspondence: alfrendo.satyanaga@nu.edu.kz

Abstract: Many reclaimed areas in Indonesia have abandoned swampland or idle land which is attributed to various factors. One of the main factors is the unsuitability of the exiting flow system in this area since the condition of the canals and water structures in this area has not been rehabilitated for a long time. No study has been carried out to investigate the suitable model for simulating the appropriate criteria for assessment of erosion within the channel on Tidal lowland in Indonesia. This study focuses on the investigation of erosion occurring within the Rural Channel and Main Drainage Channel on Tidal lowland in Palembang, Indonesia which becomes the originality of this manuscript. The erosion was attributed to the accumulation of sediment transport within the channel of the reclaimed tidal delta region Telang I. The results of the research on the P8-13S scheme show that equilibrium on the accumulation of sediment transport in the channel was observed in the Rural Channel and Main Drainage Channel on average ranging from 3,301,859 m^3 to 3,349,103 m^3 while the average sedimentation ranged from 809,232–898,467 m^3. This study is very important in minimizing the possible erosion near riverbank.

Keywords: rural channel; main drainage channel; sediment transport pattern; tidal lowland and erosion

Citation: Syarifudin, A.; Satyanaga, A.; Wijaya, M.; Moon, S.-W.; Kim, J. Sediment Transport Patterns of Channels on Tidal Lowland. *Fluids* **2022**, 7, 277. https://doi.org/10.3390/fluids7080277

Academic Editors: Mohammad Amir Khan, Jaan H. Pu, Prashanth Reddy Hanmaiahgari and Mehrdad Massoudi

Received: 27 May 2022
Accepted: 4 August 2022
Published: 15 August 2022

1. Introduction

Sediment transport is critical for all aquatic habitats, but particularly lagoons, where the volume and direction of suspended matter has a direct effect on the turbidity of the water [1–3]. It has the potential to alter primary production and other ecological processes within a system [4]. Sediments play a critical role in sculpting the terrain where rivers meet the sea. Inadequate sedimentation to the sea may result in coastal erosion, leading in the loss of land and tidal wetlands, necessitating coast protection and saltmarsh or beach nourishment methods [5]. Furthermore, human action occasionally modifies the coastline, frequently unintentionally altering previously identified sediment transport channels and coastal morpho dynamics [6]. Juez et al. [7] proposed a model for simplification of coupling technique for modelling transient shallow flow and movable bed to reduce the computational cost and time. Zordan et al. [8] concluded that bed shear stress and Reynolds stresses are correlated with sediment entrainment for longer periods, such that it can be inferred that they are associated to the distal transport of sediment. The main responsible for sediment dislodging is found to be the vertical component of the fluid movement, which presents a high peak of vertical mean velocity at the front of the currents.

Potential land for farming in Indonesia includes around 150 million ha. Some parts are located within swamp area, which cover an area of around 33 million ha. This swamp land is situated within tidal swamp with area of 21 million ha and normal swamp with area of 12 million ha. These swamps land are distributed across the islands of Kalimantan, Sulawesi, Sumatera and Papua, covering an area of about 10 million hectares each [9].

The Indonesian government has reclaimed some of the swamp areas. The reclamation by private sector is around 1.9 million ha while the reclamation by the community is around

2.1 million ha with a total reclamation of 4 million ha [10]. However, only few areas of these reclaimed lands are utilized to generate crops. The productivity of the land is around 3 t/ha which is utilized by the government. A lack of attention to Operations and Maintenance has resulted in a micro-scale repair effort in tertiary channels without a sluicegate structural framework, which means that farmers are doing their own micro-scale maintenance on their own initiative [11].

Based on an inventory study of swamp area data in the western and eastern regions, it is concluded that 0.8 million ha of 1.8 million ha reclaimed area is abandoned swampland or idle land, which is attributed to various factors [12]. One of the main factors is the unsuitability of the exiting flow system in this area since the existing water system network is not optimal. The other important factor is related to the condition of the canals and water structures in this area which has not been rehabilitated for a long time. In addition, the maintenance of the channels is not optimal [13].

The originality of this study is related to the investigation of the suitable model for simulating the appropriate criteria for assessment of erosion within the channel on Tidal lowland in Indonesia. The criteria to determine the stability within the channel is related to no erosion and sedimentation observed in the channel with the equilibrium condition of the channel. Although erosion and sedimentation exist, but it should be merely temporary material movements in the channel which does not affect the general condition of the channel. No study has been carried out to investigate the suitable model for simulating the appropriate criteria for assessment of erosion within channel on Tidal lowland in Indonesia. This study focuses on the investigation of erosion occurring within the Rural Channel and Main Drainage Channel on Tidal lowland in Palembang, Indonesia.

2. Site Overview and Methodology

According to this research, the area under investigation is located in Delta Region Telang I in the province of South Sumatra, which is characterized by swampy terrain and little rainfall. With the second generation of the double-grid design layout (Rib System), this region was also reclaimed, along with Telang II, Delta Saleh, and Sugihan, as the second generation [14]. A future open channel system has been designed for the Delta region Telang I [15]. In addition to the main line (which is also used for navigation), the system includes secondary channels and tertiary channels. The site location in this study is depicted in Figure 1 [16].

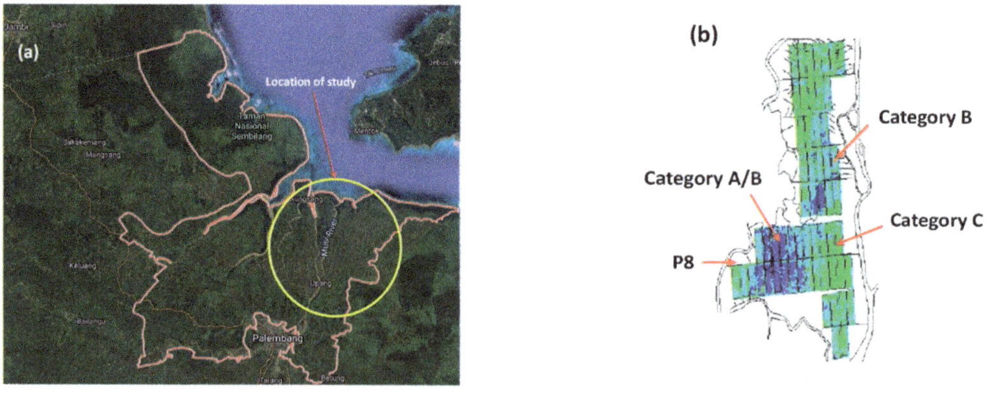

Figure 1. (**a**) The investigated location in this study; (**b**) Layout of the investigated primary channel 8 (P8).

Telang I lies between 020 29 and 020 48′ latitude and 1040 30 and 1040 52′ east longitude, according to GPS. As a result, the country's northern and southern borders are bounded by the Strait of Bangka and Telang I respectively. In contrast, the eastern region, which includes the river Musi, and the western region, which includes the river Telang I [17].

Telang I is a hydrologically significant location since it is bordered by tidal rivers. The Musi river runs through the eastern section of the territory, while the Telang river runs through the western part of the region. According to the Land and Water Management Tidal Lowlands (LWMTL) [11], the Bangka Strait borders the southern portion of the province, whilst the river borders the northern part of the territory. As a result of these factors and others, hydrology can be affected on a local and global scale by the surrounding channels, the water levels in each channel, the sluicegate's operation, tides, and other environmental factors.

The numerical studies in this work were carried out using Mike-11 for simulating each prototype of flap-gate based on two-dimensional (2-D) finite element calculations. MIKE-11 mathematical modelling technique was developed in the Danish Hydraulic Institute. The MIKE11 software (Version 2009, Danish Hydraulic Institute, Kopenhagen, Denmark) is based on an implicit finite difference scheme solution of the Saint Venant equations [18]. Equations (1) and (2) are used for hydrodynamic model simulated by the software.

$$b|\frac{dh}{dt} + \frac{dQ}{dx} = q \tag{1}$$

$$\frac{dQ}{dt} + \frac{d}{dx}\left(\beta\frac{Q^2}{A}\right) + gA\frac{dh}{dx} + \frac{gn^2|Q^2|}{AR^{\frac{3}{4}}} = 0 \tag{2}$$

where A the flow area, b is the width of channel, h is the stage, Q is the discharge, R is the hydraulic radius, n is the roughness coefficient, β is the momentum distribution coefficient, q is the lateral inflow rate per unit length.

3. Numerical Analyses Results and Discussions

Research on channel stability has been carried out under various scenarios to determine a suitable pattern for channels in tidal swamp areas. The stable criterion is that there is no erosion or sedimentation in the channel with equilibrium channel conditions. Although erosion and sedimentation occur, it is only a momentary material transfer to the channel without affecting the condition of the channel in general.

Tidal swamps in the delta Telang I where the study was conducted consists of a 26,680 ha of reclamation area. This area was opened in 1976 and it was developed as transmigration area for people from Java which began in the early 1980s [19]. In the development stage, this area was built with a water system and supporting infrastructure for the development of tidal swamp farming business [20].

The length of the primary channel P8-13S is 19,071 m. The secondary channel is divided into two: Rural Channels (RC) and the Drainage Channel or Main Drainage Channel (MDC) which is located on the boundary of the business area II. The length of the RC secondary channel is 3267 m and MDC is 3807 m. the area of one secondary block is 256 hectares.

The water system in this area works based on the concept of a one-way flow system where tidal water flows into the RC and then it is released through the MDC. At present, the RC and MDC both have functions as water intake and drainage channels. This one-way water management system will provide good prospects if it is equipped with water gates (flap-gate, stop-log or block) as a water regulator that can be controlled (Noor, 2004).

On the secondary channel P8-13S, a survey was carried out on the longitudinal and transverse directions based on topographic and water level measurements. Most of the tertiary canals in P8 are still unmodified, only certain part of this area has been modified including the Southern part of the area with 13 secondary block (P8-13S). In general one secondary block has the same area of ±256 ha with a total of 17 tertiary canals channel bar (Figure 2).

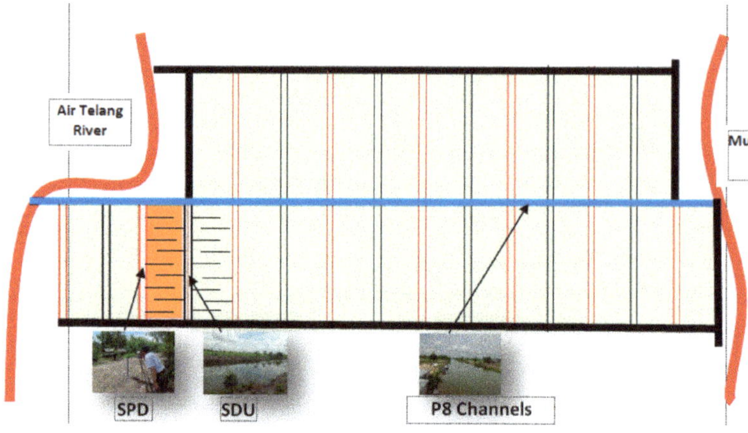

Figure 2. Schematic diagram of water system at P8 Telang I.

Many tertiary channels in P8 are not connected into the RC. There are only a small number of channels that penetrate to the RC. Almost all of channels penetrate to the secondary Main Drainage Channel (MDC). All secondary channels, both RC and MDC in line P8 initially were not supported by water control structures. Only until 2010, four sluice-gates were built in SDU 13-S secondary channels and 5 are in RC13-South.

3.1. P8-13S Channel Hydraulic Model Analysis and Simulation

Simulations of morphological changes in both Primary 8 (P8) channels and SPD and SDU secondary channels, were performed using MIKE-11 2D software [18]. The initial condition in the numerical simulation was determined based on the elevation of the water level in the RC and MDC channels in P8-13S. The elevation of the water level in the Rural Channel was recorded for 24 h starting from 8:00 am in the morning until 8:00 am in the next day [21]. The elevation of the water level in the channel was −0.20 m recorded at 08:00 am and the elevation of the peak water level was 1.60 m recorded at 13.00 pm. Then the elevation of the water level decreased to the minimum value of −0.61 m at 03:00 am in the following morning. (Figure 3).

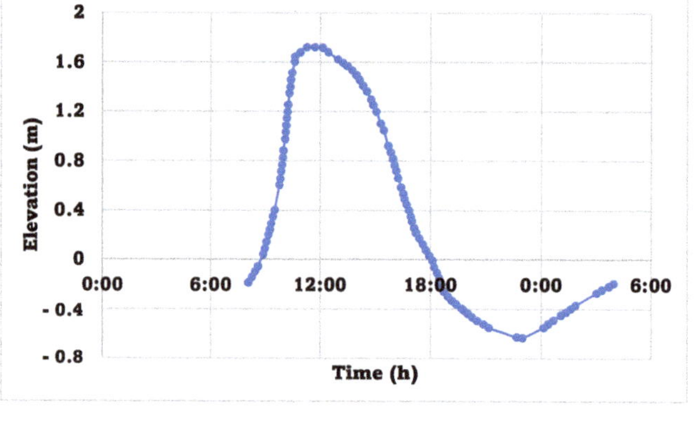

Figure 3. Water level elevation in the Rural Channel.

Likewise, the elevation of the water level in the SDU channel was recorded for 24 h starting from 08:00 am in the morning until 08:00 am in the next day. The level of water level in the channel was −0.28 m at 08:00 am and the elevation of the peak water level was 1.63 m at 13.00 pm. Then the water level decreased to the minimum value of −0.50 m at 3:00 am in the next morning. A boundary condition in the numerical simulation was determined based on the water level in the mouth of the P-8 channel. The water level in the mouth of the P8 channel was recorded for 24 h starting at 08:00 am in the morning until 08:00 am in the next day. The elevation of the water level in the channel was 2.80 m at 08.00 am and the peak water elevation was 4.60 m at 13:00 pm. Then the water level decreased to the minimum value of 2.00 m at 3:00 am in the next morning. (Figure 4). Calibration for sediment movement was conducted based on P8 channel parameters. The width of P8 channel is 50 m and the height or depth of this channel is 4.80 m. (Figure 5). All boundary conditions are incorporated in the numerical model. Table 1 presents RC and MDC schemes without sluicegate structure, RC and MDC schemes with sluicegate structure, and RC, MDC schemes and tertiary channels with sluicegate structure [22].

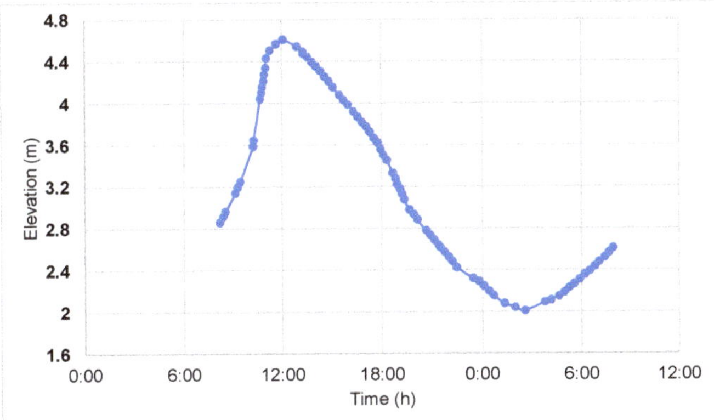

Figure 4. The water level at the mouth of the P-8 channel.

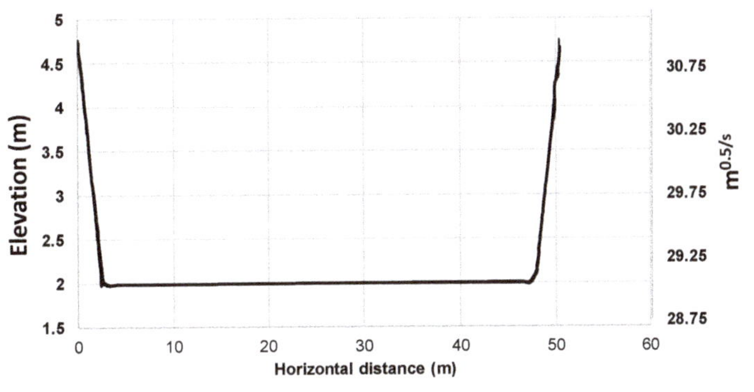

Figure 5. Sediment transport model simulation results.

Table 1. Parametric studies carried out in this study.

Scenario	O&M (%)	Works
I	25	RC and MDC without sluicegate
II	50	RC and MDC with sluicegate
III	75	RC, MDC and tertiary canals all with sluicegate

3.2. *Accumulation of Sediment Transport in the Channel*

The state of the study region prior to 2012 is presented in Scenario 1. The state of the study area after 2012 is presented in Scenario 2 [23]. This scenario models P8, RC, or

MDC without the use of a sluicegate, with the assumption that channel maintenance is not correctly executed. The Manning roughness coefficient (n) is chosen to be 0.025 $s/m^{1/3}$ in this calculation. The channel in this situation still has grass in it, which indicates that the flow is obstructed, which means that the flow velocity in the channel is decreased [15].

This scenario consists of eight primary channels, one RC channel, one MDC channel, and two tertiary channels. In both the tertiary and secondary channels, sluicegate structures are not available. The function of the channel as a supply or as a sewer cannot be recognized without the use of a sluicegate structure. The water supply routes are used during high tides, and the drainage channels are used during low tides. Figure 6 depicts a network system in the scenario described above [24].

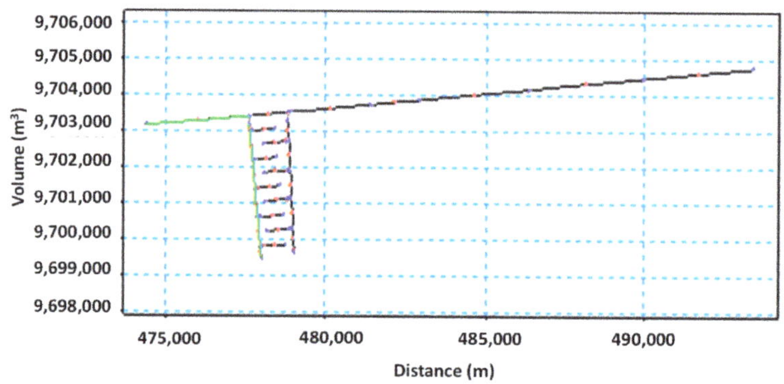

Figure 6. Schematic diagram of channel models for sediment movement in RC without sluicegate of the scenario I (OM 25%).

A distance of 3200 m from the beginning of the P8 channel is shown in Figure 7 as sediment movement in the RC and the P8 channel. A total of 4,224,089.34 m^3 was eroded before the Rural Channel (RC) sedimentation began at 871. As a result, no sediment movement occurred up to a distance of 3650 m from the channel's end.

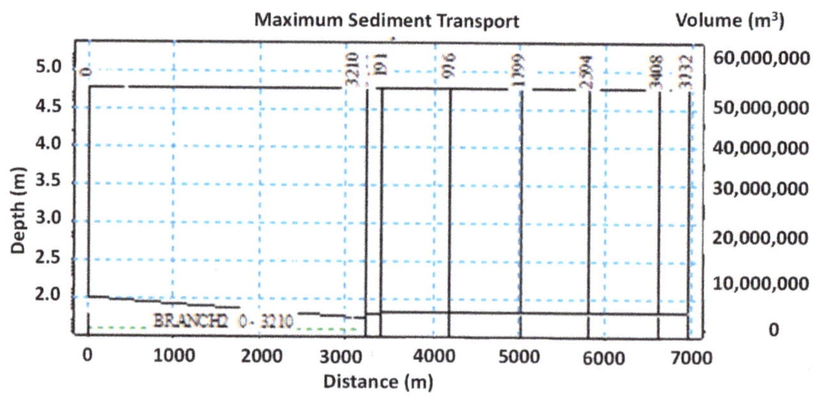

Figure 7. Sediment movement in RC without flap-gate.

Primary 8 channel and MDC have sluicegate structures that are not present in secondary and tertiary networks. The channel's ability to serve as either a supply or a sewage cannot be determined without the presence of a sluicegate structure. At high tide, all waterways are water supply routes, while at low tide, all channels are drains. For example, Figure 8 depicts the accumulation of sediment movement that occurs along the P8 channel and MDC without a sluicegate 3200 m downstream of the Telang river erosion, which totaled 4,224,089.33 m^3, and 4500 m downstream of the MDC, where no sedimentation had occurred prior to the start of the MDC. Up to a distance of 3650 m, the MDC channel did not have any sediment flow.

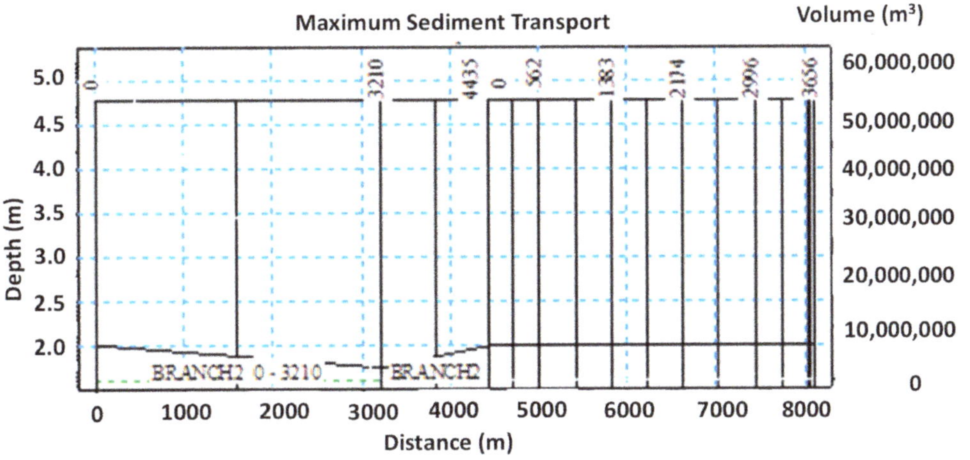

Figure 8. Sediment movement in MDC without sluicegate.

Due to maintenance, the waterway in Scenario II is in a better condition than in the prior scenario. To put it another way, it is anticipated that there will be no grass along the channel's path because of its shape. Some of the secondary channels' cross sections have the same dimensions as those in the tertiary channels. It was decided that in this case the manning coefficient n would be bigger than in scenario I. Manning n is assumed to have a coefficient of 0.033 s/m$^{1/3}$ in this case. Primary 8 and RC networks are shown schematically in Figure 9. To put it another way, the sluice gate is located in the Rural Channel (RC). At high tide, all waterways are water supply routes, while at low tide, all channels are drains.

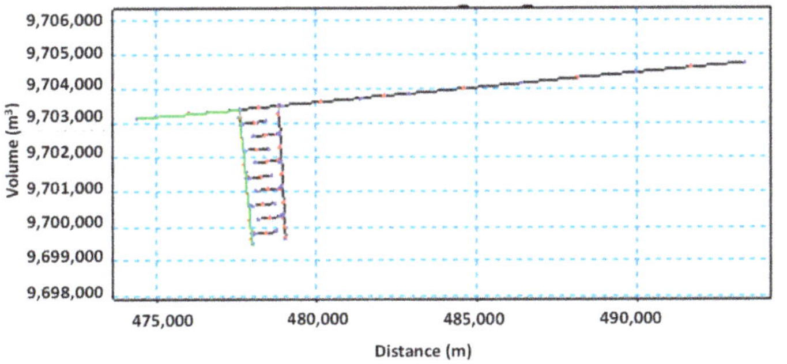

Figure 9. Schematic diagram of channel models for sediment movement in RC with sluicegate on scenario II (O&M 50%).

A total of 4,360,350.28 m^3 of erosion was caused by sediment transport in the P8 channel and Rural Channels (RC) located 3200 m from the P8 channel's starting. A total of 1,090,087.57 m^3 of sediment was deposited at the start of the Rural Channel (RC). The Rural Channel had no sediment movement up to a distance of 3650 m from the channel's end. A total of 4,496,611.23 m^3 of sediment was transported by P8 channel and MDC over a 3200 m distance from the P8 channel's beginning, according to Figure 10. At a distance of 4500 m, sedimentation happened once more in the P8 channel. Sediment is not moving until the end of the channel, which is 3650 m away from the MDC meeting point in the canal.

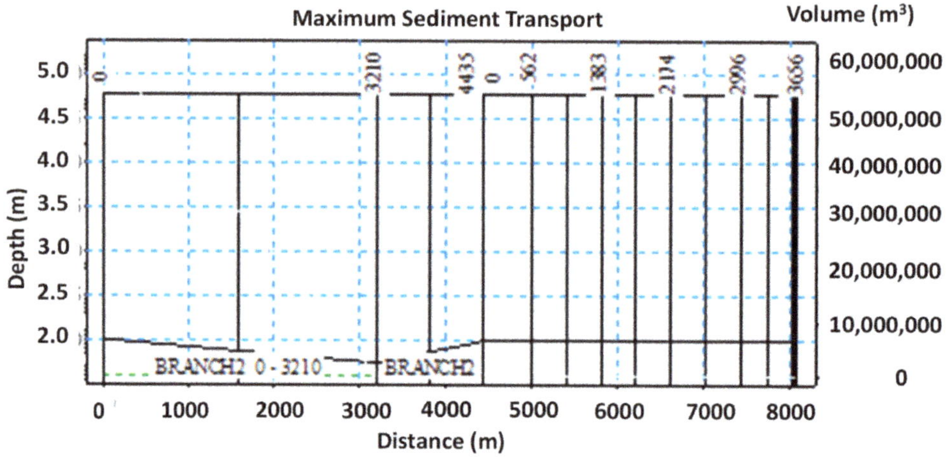

Figure 10. Sediment movement in MDC is sluicegate.

Compared to scenarios I and II, Scenario III assumes that the channel is well-maintained. As a result, the channel's shape is based on the original plan, and no new grass is expected to sprout along its length. Both the secondary and tertiary channels have the same cross sectional area. *n*, the manning coefficient, is higher in this situation than in either scenario I or II [15]. The of coefficient of manning (*n*) for this scenario is 0.035 $s/m^{1/3}$.

Figure 11 depicts sediment transit buildup in channels P8, RC, and tertiary channel 1. (tc-1). 3200 m in length from the P8 channel's inception to its end, 4,360,350.28 cubic meters (m^3) of erosion was recorded. Sediment deposition totaled a total of 953,826.62 cubic meters from the beginning of channel RC to the beginning of channel TC-1's third-generation (a distance of 200 m). The terminus of the tc-1 channel saw sedimentation of 3,470,461.48 m^3 from the beginning to a distance of 750 m.

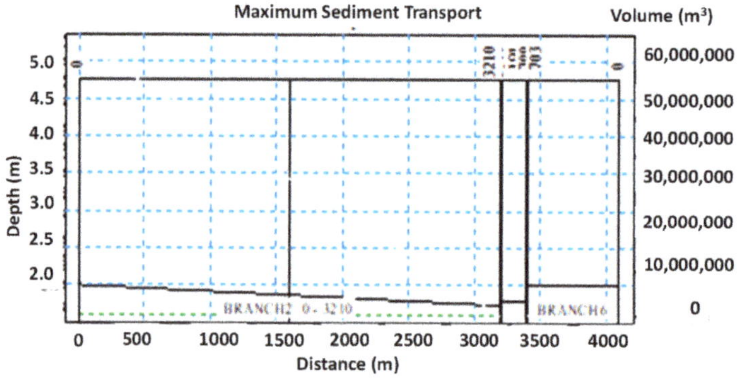

Figure 11. Movement of sediment in the tc-1 tertiary channel has a sluicegate.

In the RC secondary channel and the tc-3 tertiary channel, sluicegate structures have a roughness coefficient of 0.35. A total of 4,496,611.27 m^3 of cumulative erosion occurred over a distance of 3200 m from the P8 channel to the RC channel. A total of 953,826.62 m^3 was deposited in the RC channel from its inception to the beginning of the first tc-3 tertiary channel. The terminus of the tc-3 channel experienced sedimentation of 3,222,571.38 m^3

from the beginning to a distance of 750 m. Figure 12 is a network system scheme of Primary channel 8 (P8), RC and tertiary channel 5 (tc-5). The RC secondary channel and tc-5 tertiary channel contain the valve door construction, with a roughness coefficient of 0.035 in the channel.

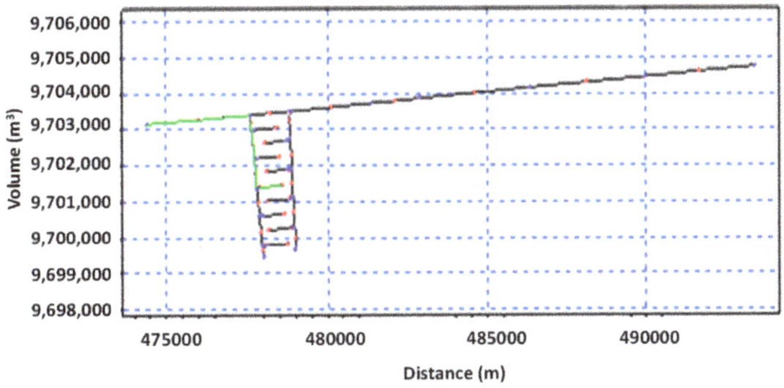

Figure 12. Cross section of channel models for sediment movement in channels tertiary tc-5 there is a sluicegate.

The P8 channel, the Reverse Channel (RC), and the Tertiary Channel (5, or tc-5) all followed a similar pattern of sediment transport, with a total of 4,496,611.27 m^3 of channel erosion occurring over a distance of 3200 m from the beginning of the P8 channel in the Telang river to the beginning of the RC, and then 1800 m from the beginning of the RC to the beginning of the tc-5 tertiary channel. For this construction, the roughness coefficient in the channel is taken to be 0.035. It is placed in the secondary channel and the seventh tertiary channel.

Figure 13 shows that sediment transport occurred in the P8 channel, the RC channel, and the tertiary channel 7 (tc-7), with channel erosion totaling 4,709,912.01 m^3 at a distance of 3200 m between the beginning of the P8 channel and the beginning of the RC channel, and sedimentation totaling 867,615.37 m^3 between the beginning of the RC channel and the beginning of the tc-7. Figure 13 shows that sediment transport occurred in Between the beginning of the tc-7 channel and the end of the channel, sedimentation totaled 3,222,571.38 m^3 and the end of the channel, sedimentation totaled 3,222,571.38 m^3. The beginning of the tc-7 channel up to a distance of 750 m and the end of the channel, sedimentation totaled 3,222,571.38 m^3. For the sluicegate structure, it is positioned both in the RC secondary channel and in the tc-9 secondary channel, with a roughness coefficient in the channel taken at 0.0355 for the RC secondary channel and 0.0355 for the tc-9 secondary channel.

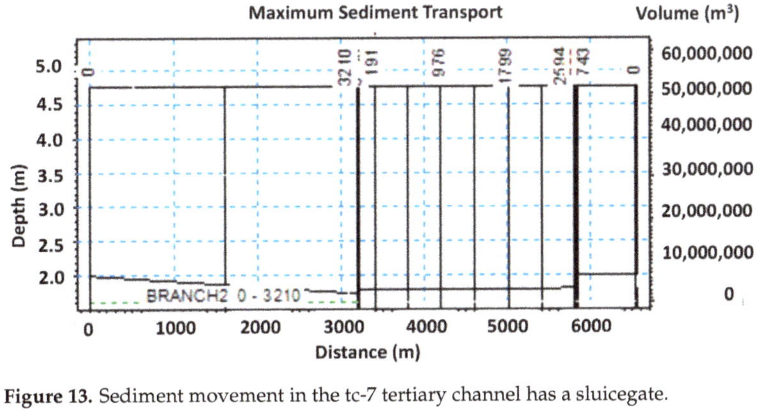

Figure 13. Sediment movement in the tc-7 tertiary channel has a sluicegate.

4. Conclusions

Based on the three scenarios carried out in the simulation using MIKE-11 2D program, it is observed that Scenario III with a 75% O&M model (RC, MDC and tertiary channels have sluicegate) is considered as the most suitable condition for the P8-13S scheme prototype model. P8 channel erosion occurred with an average volume of 3,599,979.28 m^3 while sedimentation occurred in the Rural Channel (RC) with an average volume of 963,836.56 m^3. An average tertiary channel erosion amounted to 3,251,972.30 m^3. The P8 channel eroded with an average volume of 3,419,184.75 m^3 and sedimentation occurred at Main Drainage Channels (MDC) and the tertiary channel was the same value as the magnitude of the erosion at P8 which was an average of 3,419,184.75 m^3. The channel conditions in the P8-13S scheme of the Telang I tidal area can be said to have reached an equilibrium.

Author Contributions: A.S. (Alfrendo Satyanaga), A.S. (Achmad Syarifudin), S.-W.M. and J.K. conceptualized the study; A.S. (Alfrendo Satyanaga), A.S. (Achmad Syarifudin) and M.W. implemented data processing under the supervision of S.-W.M. and J.K.; the original draft of the manuscript was written by A.S. (Alfrendo Satyanaga), A.S. (Achmad Syarifudin), S.-W.M. with editorial contributions from M.W. and J.K.; the funding acquisition was made by A.S. (Alfrendo Satyanaga). All authors have read and agreed to the published version of the manuscript.

Funding: This research was funded by the Nazarbayev University Research Fund under Grants 11022021CRP1512 and Social Policy Grant.

Institutional Review Board Statement: Not applicable.

Informed Consent Statement: Not applicable.

Data Availability Statement: All data are available upon request.

Acknowledgments: This research was supported by the Nazarbayev University Research Fund under Grants 11022021CRP1512 and Social Policy Grant. The authors are grateful for this support. Any opinions, findings, and conclusions or recommendations expressed in this material are those of the author(s) and do not necessarily reflect the views of the Nazarbayev University.

Conflicts of Interest: The authors declare that the research was conducted in the absence of any commercial or financial relationships that could be construed as a potential conflict of interest.

References

1. Nearing, M.A.; Xie, Y.; Liu, B.; Ye, Y. Natural and anthropogenic rates of soil erosion. *Int. Soil Water Conserv. Res.* **2017**, *5*, 77–84. [CrossRef]
2. Alewell, C.; Borrelli, P.; Meusburger, K.; Panagos, P. Using the USLE: Chances, challenges and limitations of soil erosion modelling. *Int. Soil Water Conserv. Res.* **2019**, *7*, 203–225. [CrossRef]
3. Wang, B.; Zheng, F.; Guan, Y. Improved USLE-K factor prediction: A case study on water erosion areas in China. *Int. Soil Water Conserv. Res.* **2016**, *4*, 168–176. [CrossRef]
4. Satyanaga, A.; Wijaya, M.; Zhai, Q.; Moon, S.-W.; Pu, J.; Kim, J.R. Stability and Consolidation of Sediment Tailings Incorporating Unsaturated Soil Mechanics. *Fluids* **2021**, *6*, 423. [CrossRef]
5. Pu, J.H.; Wallwork, J.T.; Khan, M.A.; Pandey, M.; Pourshahbaz, H.; Satyanaga, A.; Hanmaiahgari, P.R.; Gough, T. Flood Suspended Sediment Transport: Combined Modelling from Dilute to Hyper-concentrated Flow. *Water* **2021**, *13*, 379. [CrossRef]
6. Wallwork, J.T.; Pu, J.H.; Kundu, S.; Hanmaiahgari, P.R.; Pandey, M.; Satyanaga, A.; Khan, M.A. Review of Suspended Sediment Transport Mathematical Modelling and Experimental Studies. *Fluid* **2022**, *7*, 23. [CrossRef]
7. Juez, C.; Murillo, J.; García-Navarro, P. A 2D weakly-coupled and efficient numerical model for transient shallow flow and movable bed. *Adv. Water Resour.* **2014**, *71*, 93–109. [CrossRef]
8. Zordan, J.; Juez, C.; Schleiss, A.J.; Franca, M.J. Entrainment, transport and deposition of sediment by saline gravity currents. *Adv. Water Resour.* **2018**, *115*, 17–32. [CrossRef]
9. Syarifudin, A. The influence of Musi River Sedimentation to The Aquatic Environment. *MATEC Web Conf.* **2017**, *104*, 10104026. [CrossRef]
10. Tallar, R.Y.; Suen, J.P. Aquaculture Water Quality Index: A low-cost index to accelerate aquaculture development in Indonesia. *Aquacult. Int.* **2016**, *24*, 295–312. [CrossRef]
11. Land and Water Management Tidal Lowlands (LWMTL); South Sumatera Province. *Operasi dan Pemeliharaan Jaringan dengan Perkumpulan Petani Pemakai Air (P3A)*; ARCADIS-Euroconsult in Cooperation with Kimpraswil, Department Pertanian, Sriwijaya

University and Local Government South Sumatera; Indonesia, Rikjkswaterstaat: Utrecht, The Netherlands; UNESCO-IHE: Paris, France, 2004.
12. Tallar, R.Y.; Satyanaga, A. Incorporating rainwater-harvesting and retention basins design into urban development paradigms in Greater Bandung, Indonesia. In *Water and Urban Development Paradigms*; Feyen, B., Shannon, K., Neville, M., Eds.; CRC Press: London, UK, 2008; p. 712.
13. Syarifudin, A. The Effect of Channel Sedimentation in the Tidal Swamp Area on the Typology of Land A/B Delta Telang I Banyuasin Regency. In Proceedings of the 5th AvoER Proceeding Seminar, Palembang, Indonesia, 28 November 2013.
14. Houterman, J.; Djoeachir, M.; Robiyanto, H.S.; van Steenbergen, F. *Water Resources Management During Transition and Reform in Indonesia Toward an Integrated Perspective on Agricultural Drainage, Agriculture and Rural Development*; The International Bank for Reconstruction and Development Agriculture & Rural Development Department: Washington, DC, USA, 2004.
15. Syarifudin, A.; Momon, S.I.; Arie, S.M.; Yazid, M.; Suryadi, F.X. Erosion on the Secondary Canal of Reclaimed Agriculture Tidal Lowlands Telang I Banyuasin Regency. *J. Clean Energy Technol.* **2014**, *2*, 1. [CrossRef]
16. Euroconsult; Arcadis; Indec & Associates Ltd.; PT. Trans Intra Asia; PT. Necon Ciptajasa; PT. Binatama Wirawredha. *Integrated Swamps Development Project IBRD Loan 3755–IND, Final Report O&M Strengthening in ISDP*; Republic of Indonesia, Ministry of Settlement and Regional Infrastructure, Directorate General of Rural Development: Jakarta, Indonesia, 2000.
17. Euroconsult; PT. BIEC International; PT. Trans Intra Asia. *Telang and Saleh Agricultural Development Project, Drainage Development Component, O&M Manual*; Republic of Indonesia, Ministry of Public Works, Directorate General of Water Resources Development: Jakarta, Indonesia, 1996.
18. Harsono, E. *Prospect of the Development of Swamp Areas in Indonesia*; 60 Years of the Department of Public Works: Jakarta, Indonesia, 2005.
19. Hartoyo, S.; Sumarjo, G.I.; Robiyanto, H.S.; Schult, B.; Suryadi, F.X. Potential and constrains of water management measures for tidal lowlands in South Sumatra. Case study in a pilot area Telang I. In Proceedings of the 9th Inter-Regional Conference on Water Environment, Envirowater, Concept for Water Management and Multifunctional Land Uses in Lowlands, Delft, The Netherlands, 17–19 May 2006.
20. Danish Hydraulic Institute (DHI). *MIKE-11. A Modeling System for Rivers and Channels, User Guide*; DHI Software: Kopenhagen, Denmark, 2007; pp. 15–17.
21. Robiyanto, H.S. Water management technologies on tidal wetlands in Indonesia in a multidimensional perspective. In Proceedings of the National Seminar the Role and Prospects of Development of Wetlands in National Development, Jakarta, Indonesia, 2006.
22. Syarifudin, A.; Momon, S.I.; Arie, S.M.; Yazid, M.; Suryadi, F.X. Technical Approach of Erosion and Sedimentation on Canal (Case study in Delta Telang I, Banyuasin, South Sumatra Province). In Proceedings of the International Workshop on Sustainable Management of Lowland for Rice Production 2012, Banjarmasin, Indonesia, 27–28 September 2012.
23. Suprianto, H.; Ravaie, E.; Irianto, S.G.; Susanto, R.H.; Schultz, B.; Suryadi, F.X. Land and water management of tidal lowlands: Experiences in Telang and Saleh, South Sumatra. *Irrig. Drain.* **2010**, *59*, 317–335.
24. Syarifudin, A.; Dewi, S. A Scouring Patterns Around Pillars of Sekanak River Bridge. *J. Phys. IOP Conf. Ser.* **2019**, *1167*, 012019. [CrossRef]

fluids

MDPI

Systematic Review

Hydraulic Flushing of Sediment in Reservoirs: Best Practices of Numerical Modeling

Yong G. Lai [1,*], Jianchun Huang [1] and Blair P. Greimann [2]

[1] Technical Service Center, U.S. Bureau of Reclamation, Denver, CO 80225, USA; vhuang@usbr.gov
[2] Stantec Inc., 410 17th Street Suite 1400, Denver, CO 80202, USA; blair.greimann@stantec.com
* Correspondence: ylai@usbr.gov; Tel.: +1-303-445-2560

Abstract: This article provides a comprehensive review and best practices for numerically simulating hydraulic flushing for reservoir sediment management. Three sediment flushing types are discussed: drawdown flushing, pressure flushing, and turbidity current venting. The need for reservoir sediment management and the current practices are reviewed. Different hydraulic drawdown types are described in terms of the basic physical processes involved as well as the empirical/analytical assessment tools that may be used. The primary focus has been on the numerical modeling of various hydraulic flushing options. Three model categories are reviewed: one-dimensional (1D), two-dimensional (2D) depth-averaged or layer-averaged, and three-dimensional (3D) computational fluid dynamics (CFD) models. General guidelines are provided on how to select a proper model given the characteristics of the reservoir and the flushing method, as well as specific guidelines for modeling. Case studies are also presented to illustrate the guidelines.

Keywords: reservoir sedimentation; sediment flushing; sediment sluicing; hydraulic flushing; numerical model; reservoir sustainability

Citation: Lai, Y.G.; Huang, J.; Greimann, B.P. Hydraulic Flushing of Sediment in Reservoirs: Best Practices of Numerical Modeling. *Fluids* **2024**, *9*, 38. https://doi.org/10.3390/fluids9020038

Academic Editor: Jaan H. Pu

Received: 30 December 2023
Revised: 25 January 2024
Accepted: 26 January 2024
Published: 1 February 2024

1. Introduction

Reservoirs have been built on natural rivers to provide various benefits such as hydropower, flood protection, water supply, recreation, and navigation. They increase water surface level upstream and reduce flow velocities, leading to undesirable consequences upstream, downstream, and within the reservoir [1]. One consequence is reservoir sedimentation—a large amount of inflowing sediment is blocked by the dam, causing deposition within the reservoir and the starvation of sediment supply downstream. Reservoir sedimentation in general reduces the reservoir capacity [2], increases the risk of water intake plugging [3], and alters ecology negatively [4]. It was estimated that reservoir sedimentation has led to an annual worldwide storage loss of 0.5 to 1% relative to the initial design capacity [5–9]. Reservoir sustainability, therefore, has been becoming a prominent topic in the last decade—it calls for actions to either remove sediment out of a reservoir or to pass it through or around it to maintain reservoir capacity at a constant in the long term. Failure to manage reservoir sediment today can be consequential in the future: dams may eventually lose their benefits and need to be decommissioned—a substantial cost to future generations.

As an example, most U.S. dams are approaching the age of 100 years—the typical design lifespan allocated for reservoir sedimentation. The 100-year outlet level is being reached and even exceeded at some reservoirs, impacting water delivery or leading to the loss of reservoir functions. Some of the examples at the U.S. Bureau of Reclamation (USBR) include the Paonia Reservoir in Colorado, Buffalo Bill Dam in Wyoming, Black Canyon Dam in Idaho, Elephant Butte Reservoir and Summer Reservoir in New Mexico, Arrowrock Reservoir in Idaho, and Horseshoe Dam in Arizona [10,11]. Seven additional examples were discussed in the U.S. Army Corp of Engineers study [12].

Reservoir sustainability may be partially or fully achieved by adopting proper sediment management measures. Among them are upstream watershed management, sediment bypass, hydraulic flushing, and mechanical dredging [13]. Upstream watershed management involves land-use practices to reduce sediment delivery to the reservoir; sediment bypass diverts sediment through upstream tunnels to the downstream before it reaches the dam; and hydraulic flushing is to flush (or sluice) sediment within and near the dam to the downstream through increased water flows [13,14]. Various types of dredging may also be used, such as the mechanical, hydraulic, and hydro-suction means.

This article focuses primarily on hydraulic flushing as it is the most commonly adopted method, economically attractive, and technically effective in managing sedimentation [13,15]. See [16,17] for reviews on the topic; additionally, hydraulic flushing examples were discussed in [13,17–20], among others. In this article, the term "flushing" is defined as the removal of deposited sediment, while "routing" is the process of keeping sediment moving through the reservoir. The term hydraulic flushing refers broadly to both flushing and routing using the flowing water. This article is further limited to the numerical modeling aspect of hydraulic flushing. There exists a large body of literature related to the laboratory and field studies of hydraulic flushing [5,16,21–23]. Numerical modeling has also been carried out. However, there are no general guidelines available for the modeling of sediment flushing. A recent review [24] concluded that "existing codes can successfully simulate sediment management, but because each code has limitations, they require seasoned judgment in their choice, application, and interpretation".

In the following, a comprehensive review is presented, and the best-practice guidelines are then provided on the topic of numerically modeling hydraulic flushing. Case studies are also used to illustrate various guidelines, as well as their performance in practical usage.

2. Literature Review: Empirical/Analytical Analyses

2.1. *Hydro Flushing Types*

Three types of hydraulic flushing are considered: drawdown flushing, pressure flushing, and turbidity current venting. Drawdown flushing is carried out by lowering the reservoir pool elevation, partially or fully, through releasing waters out of the reservoir. Pressure flushing refers to the process by which sediment is moved by opening the low-level outlets while keeping the reservoir water at a constant level well above the outlets. Finally, turbidity current venting refers to sediment routing that vents out the suspended fine sediments while they are moving towards the dam during a large storm event. The three types are illustrated in Figures 1–3.

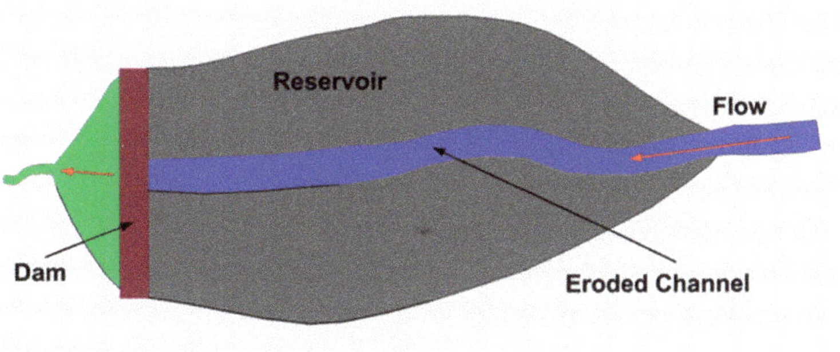

Figure 1. Sketch of an eroded sediment channel in typical drawdown flushing (created by authors [5]).

An eroded flushing channel is usually generated for a typical drawdown flushing operation; it constitutes a major portion of the deposited sediment removed (Figure 1). For pressure flushing, however, the eroded area is usually limited to a scour cone near the outlet (Figure 2) and the sediment volume flushed is much smaller than the drawdown flushing. A comparison of the two was discussed in [13,25]. For the turbidity current as in Figure 3, the dense sediment flows into a less turbid reservoir and plunges when certain conditions are met. Once plunged, the turbidity current usually attaches to the bed and moves as an undercurrent propagating down the reservoir slope.

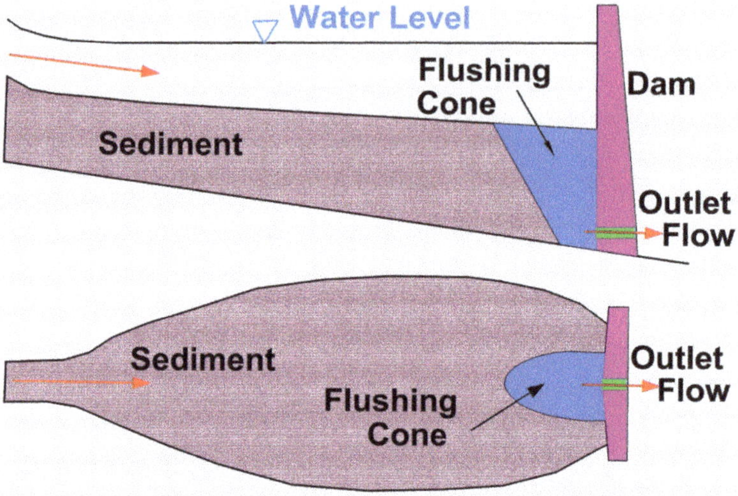

Figure 2. Sketch of a typical scour cone developed in pressure flushing (created by authors [26]).

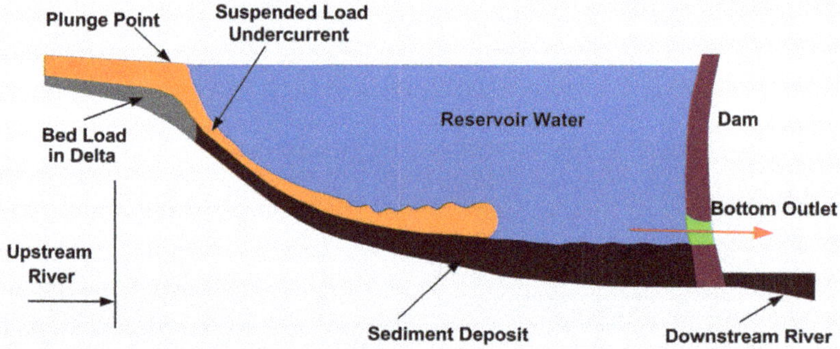

Figure 3. Sketch of a turbid undercurrent movement for a typical turbidity current venting event (created by authors [27]).

2.2. Drawdown Flushing

Drawdown flushing is operated by lowering the reservoir pool elevation to create an increased flow velocity, resulting in the erosion of sediment deposit from the reservoir. This technique has been widely used and is most effective among the three hydraulic types (particularly so for narrow reservoirs). However, drawdown flushing may require a large volume of water released out of the dam; sometimes, it may require the entire reservoir to be emptied [16] which may not be feasible for some reservoirs. When poorly managed, sediment from the upstream delta may move towards deeper portion of the reservoir and then be deposited, rather than be flushed out of the dam [23]. Therefore, drawdown operation needs to be studied carefully.

Previous studies have established the following general favorable conditions for carrying out drawdown flushing [5,21]:

- Steep longitudinal slope;
- Narrow valleys with steep banks;
- High flow velocity to mobilize and transport sediment;
- Low-level gates large enough to pass flows;
- Strongly seasonal flow patterns.

Specific guidelines were also proposed to achieve flushing success at a particular reservoir. For example, [23] presented three factors to consider: total capacity of the reservoir (CAP), mean annual runoff to the reservoir (MAR), and mean annual sediment inflow to the reservoir (MAS). Low ratios of CAP/MAR (<0.1) and CAP/MAS (<30) were recommended to ensure the success of drawdown flushing. Other criteria were also recommended. For example, [28] suggested that the CAP/MAR ratio be less than 1/50, [29] recommended the CAP/MAR ratio be less than 1/25, while [21] concluded that the CAP/MAR ratio should not exceed 4%.

The efficiency of drawdown flushing depends on many factors that are listed below:

- Reservoir geometry: width, depth, and area-capacity table.
- Reservoir sediment: size and gradation.
- Incoming flow: annual flow and flow hydrograph.
- Incoming sediment: rate (or concentration), size, and gradation.
- Outlets: location, invert elevation, and opening size.
- Reservoir operation: rules, gate-opening-duration constraints, and downstream gravel-bed channel constraints.

It is noted that only a narrow width similar to the natural channel width will be eroded, not the entire reservoir width, when drawdown flushing is applied to a reservoir much wider than the original river width. Such an example is shown in Figure 4 at the Paonia Reservoir, Colorado.

Figure 4. A photo of Paonia Reservoir in Colorado after the reservoir was drawn down: a channel incised through a portion of the reservoir sediments.

The concept of sustainable reservoir capacity—which is the storage volume that may be sustained by hydraulic flushing in a long term—is useful and has been discussed since [16]. If a reservoir is narrower than the width of a self-formed channel produced by the drawdown flushing, the reservoir is sustainable as all incoming sediment may be hydraulically flushed. If the flushing channel width is much narrower than the reservoir width, the sustainable reservoir capacity will be much smaller than the design storage.

An estimate of the sustainable reservoir capacity is possible by using empirical/analytical analyses by first estimating the long-term bed profile and bank-side slope. The long-term channel profile may be computed from the initial bed profile and the difference between the dam outlet and initial bed elevation at the dam. The channel bank-side slope may be computed from the dry sediment density. The flushing channel width may be estimated using empirical relations such as the one recommended by Atkinson [16]: $B = 12.8Q_f^{0.5}$ (B is the width in meter and Q_f is the flushing discharge in m^3/s). This relation is applicable when the reservoir sediment has reached near the dam. It is cautioned that the flushing channel width may have high variability. Randle et al. [14], for example, found that the channel width varied significantly both spatially and temporally. The channel might be narrower where it comes into contact with cohesive sediment and wider where it contacts coarser non-cohesive sediment. The channel in the upstream half of the reservoir can become highly braided and change course daily in the process of reworking the non-cohesive delta sediment.

Four sediment transport stages are generally involved in a drawdown flush:

1. When the low-level outlets are first opened, a high velocity flow is generated and fine deposits are entrained close to the outlets, resulting in a short period of high sediment concentration outflow. This stage is similar to pressure flushing.
2. After local deposits are removed, the velocity is not sufficiently high to move the remaining sediment. This stage is similar to the final stage of pressure flushing.
3. As the reservoir level is lowered further, the sediment deposit at the reservoir upstream is entrained. At this stage, the entrained upstream sediments move towards the downstream and eventually out of the dam; in the process, coarse ones may redeposit in the reservoir.
4. In the final stage, when the water level is at its lowest level, previous reservoir deposits may be resuspended and transported out of the dam.

2.3. Pressure Flushing

Pressure flushing has been widely used for reservoirs where water storage is important and the inflowing sediment rate is relatively small. A key benefit of the pressure flushing is that much less water is released. Sediment removal, however, is usually limited to the vicinity of the outlets [30,31], as illustrated in Figure 2. Pressure flushing is often used to clean sediment and debris near water intake outlets. In general, the pressure flushing schedule adopted—the timing, duration, and release rate—may impact significantly the flushing process.

Our current practice of pressure flushing is mostly empirical. The flushing design is based mostly on empirical relations, as reviewed in [32]. Upstream of a gate in an unbounded reservoir, potential flow analytical solutions were derived by [33]. They proposed that the maximum velocity (u_{max}) upstream of the gap where the flow depth was much greater than the gap diameter (D) might be computed by:

$$\frac{u_{max}}{U_0} = 1 - \left[1 + 0.25(D/x)^2\right]^{-0.5}$$

In the near gap zone of $x/D < 2$, where x is the distance along the centerline from the gap, D is the gap diameter, and U_0 is the average velocity within the gap. The maximum velocity is proportional to x^{-2} for x/D greater than 2. The analysis demonstrated that the flow velocity decreased rapidly upstream of the gap and, therefore, the effect of the gap on erosion was limited to a relatively small area upstream of the gap.

Powell and Khan [34,35] studied the flow field upstream of a bounded gap with both fixed and mobile beds. They analyzed a circular gap and the flow field was found to be similar to the unbounded gap. A similar generalized equation was derived as:

$$\frac{u_{max}}{U_0} = 1 - \left[1 + a(D/x)^b\right]^{-c}$$

The *a*, *b*, and *c* values were found to be 0.332, 1.679, and 0.515 for the fixed bed, and 0.145, 1.493, and 0.913 for the mobile bed.

A conceptual sketch of the scour cone upstream of a gap is shown in Figure 5. The scoured area is assumed to form at depth D_s below the gap invert. There is a flat area projecting from the wall and then the scoured cone projects upstream at a constant angle θ.

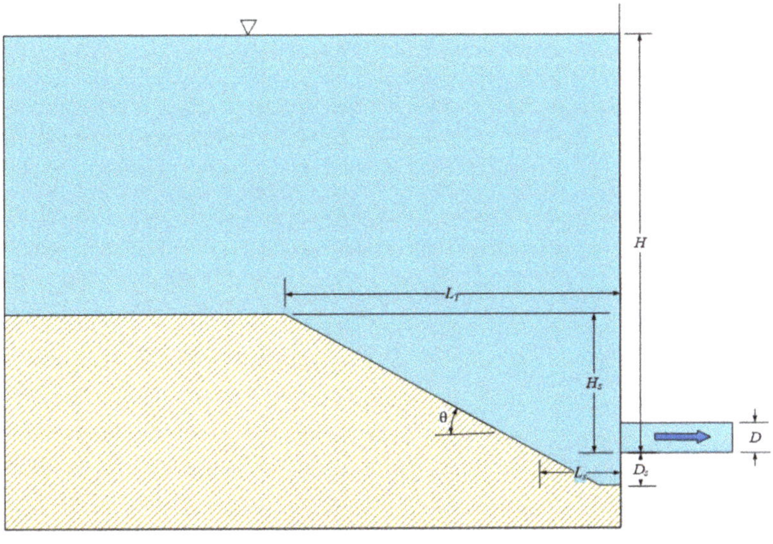

Figure 5. A conceptual sketch of the profile view for an equilibrium scour upstream of a gap.

Fathi-Moghadam et al. [36] performed a series of physical modelling studies on non-cohesive sediment scouring under pressure flushing upstream of circular gaps. The equations developed were based upon laboratory data and did not incorporate field data. It is difficult to scale both the geotechnical processes and the fluid processes simultaneously, and laboratory relationships are not generally applicable to field situations.

More empirical equations have been developed by, e.g., [34,35,37–39]. Refer also to [32] for a more detailed review. It is sufficient to comment that the empirical equations are useful in designing and evaluating expected scour in pressure flushing, but the following limitations need to be taken into consideration:

- The empirical equations may not apply outside of the range of parameters used in their development.
- They do not describe all the characteristics of the scour hole.
- They apply only for simple geometric conditions. Other structures added may alter the scour significantly.
- The sediment size and cohesive properties were not scaled from the field to the laboratory. The non-dimensional parameters related to the sediment size may be quite different in the laboratory and in the field.

2.4. Turbidity Current Venting

Turbidity current may be formed if the incoming sediment is fine but the concentration is high. This is the result of sediment-laden flow plunging beneath clear water in the reservoir. Once plunged, the turbidity current usually moves along the reservoir bed towards the dam as an undercurrent (Figure 3). Such a bottom-moving turbidity current may be vented out of reservoir if there are low-level outlets opened at the right time. While the turbidity current is being vented, the reservoir deposits on the bed may also be entrained and removed out of the reservoir. Venting can be an effective sediment management technique that has been studied extensively [11].

Empirical and analytical analyses have been carried out and much information is available for understanding the turbidity current characteristics such as the plunging criteria, mixing rate, and entrainment from the upper clear water layer. The need to analyze and predict the turbidity current has also prompted the development of empirical/analytical tools. Representative works were documented by [40,41]. A large amount of information contained in [40] was used to develop an empirical tool DCURL at USBR [42].

The plunging point is an important parameter that might be determined empirically. In a narrow reservoir, the plunging flow will form a line across the width. When a sediment-laden flow discharges into a wide reach, the turbid water may extend into the reservoir as an irregular tongue-like current which may shift from one side of the impoundment to the other. The location of the plunge point may be determined by the balance between the inflow momentum, the pressure gradient across the interface separating the river and reservoir waters, and the resisting force. The location is also influenced by morphological factors (bed slope, bed friction, and cross-sectional area). It can move several kilometers in a few hours in response to the dynamic flows due to a storm or the hydropower operation. For example, the water depth at the plunge point may be estimated using the densimetric Froude number at the plunging point as [5]:

$$F_p = \frac{V_i}{\sqrt{g' h_p}};\ g' = \frac{\rho_i - \rho_a}{\rho_a} g$$

where V_i is the depth-averaged velocity of the incoming flow, h_p is the depth at the plunging point, and ρ_a and ρ_i are the densities of the ambient and the incoming waters, respectively. Both laboratory and field data showed that F_p had a value of about 0.78, although other values were also reported [5]. Different values of F_p were mostly associated with the different assumed cross-section geometries. See [43] for a summary of the various F_p values.

Entrainment and mixing of the ambient clear water into the incoming turbid water occurred both in the plunging zone and after the turbidity undercurrent formed. The entrainment processes into the undercurrent are well suited to laboratory studies and have been carried out for simple cases such as [43,44]. Entrainment and mixing within the plunging zone have received less attention, although up to 80% of the total entrainment might occur in this zone [40]. Initial mixing at the plunging point is complex and empirical equations have been derived and discussed (e.g., [40,43,45]). Later, Fleenor [46] developed a new method to compute the initial entrainment, which varied from 30 to 300 percentage of the initial inflow. The study described the process of initial entrainment in the plunging zone, parameterized mixing, and developed and validated a computer tool that fully accounted for the initial mixing over a range of flows and conditions expected in natural water bodies.

Entrainment after the plunging zone has been analyzed extensively. One-dimensional density underflows are governed by equations similar to the backwater equations of an open channel flow. Detailed mathematical derivations and treatments are available in the literature such as [47]. Analytical solutions to the simplified equations are often used by the empirical/analytical tools. In such models, the entrainment rate for the underflow appeared in the volume conservation law, as follows [40]:

$$\frac{\partial A}{\partial t} + \frac{\partial V_{b0} A}{\partial x} = E V_{b0} B$$

where A is the cross-sectional area, B is the channel width, V_{b0} is the revised velocity downstream of the plunging point, and E is the entrainment rate. A number of entrainment relations were developed and used. For example, Ashida and Egashira [48] used the following expression:

$$E = 0.0015 R_i^{-1};\ R_i = \frac{\rho_i - \rho_a}{\rho_a} \frac{g h_{b0}}{V_{b0}^2}$$

where h_{b0} is the water depth at the plunging point. See [49] for other alternatives.

$$E = 0.5\eta^3 C_k C_b^{1.5} R_i^{-1}$$

3. Literature Review: Numerical Models

Numerical models which solve the partial differential equations (PDE) of the basic conservation laws have been widely used in hydraulic flushing studies, as the empirical/analytical methods have limitations. The model complexity varies widely, ranging across 1D, 2D, and 3D. A literature review is provided below concerning these models, as a comprehensive review has not been carried out, although limited reviews are available [24,50].

3.1. One-Dimensional Numerical Models

One-dimensional numerical models require the computing resources and are thus widely used for long-term simulations. However, care needs to be taken as 1D models have their own limitations; for example, they are primarily for run-of-the-river or narrow reservoirs where flow is highly channelized and transverse mixing is well accomplished [50]. Previous 1D modeling studies are reviewed below.

Morris and Hu [51] used the 1D HEC-6 to simulate sediment flushing in the Loíza Reservoir in Puerto Rico. The numerical study indicated that the conversion of the reservoir from the continuous high-pool operation to the low-pool operation during flood periods had the potential to reduce the sediment trap efficiency by 65%.

Chang et al. [52] used a 1D FLUVIAL-12 model to evaluate the feasibility and effectiveness of drawdown flushing during a flood. A series of reservoirs were examined on the North Fork Feather River in Northern California. The numerical modeling demonstrated that that sediment could be flushed and the reservoir capacity could be maintained by adopting an extended drawdown flushing operation. It was further shown that the reservoir drawdown operation and reservoir operation could be controlled such that no sand would be deposited on the gravel bed downstream of the reservoirs for fish habitat benefits.

Liu et al. [53] reported a 1D numerical modeling to simulate the 2001 sediment flushing operation at two reservoirs in a series—the Dashidaira and Unazuki reservoirs on the Kurobe River in Japan. The model computed the bed evolution, suspended sediment concentration, and the sediment volume flushed from or deposited in the two reservoirs.

Ahn et al. [54] applied the 1D GSTAR4 to simulate sediment flushing in the Xiaolangdi Reservoir on the Yellow River, China, as well as in the Lewis and Clark Lake on the Missouri River, USA. Useful results were reported.

Guertault et al. [55] applied the 1D Mage-AdisTS to simulate the 2012 sediment flushing at the Genissiat Reservoir on the French Upper Rhone River. During the flushing, water and sediment were released at different levels to maintain the average downstream concentration below 5 g/L throughout the operation. Flushing outlets were located at three different levels and varied gate opening schedules were used to achieve the desired result. The model reproduced the sediment concentration at the three vertical outlets well.

Boyd and Gibson [56] used the 1D HEC-RAS model to simulate the 2014 flushing of the Spencer Dam reservoir, about 40 miles upstream of the confluence of the Niobrara River and the Missouri River. The reservoir has been flushed twice annually, spring and fall, for the last 60 years for sediment management. The numerical model was calibrated using the measured data. It was reported that the model underpredicted the delta scour by about 50% and over-predicted the peak downstream sediment concentration. The authors attributed the mismatch to the channel widening process that was not captured by the model. The same 1D model was also applied to the hydraulic flushing studies in other reservoirs, such as those reported in [57].

Brignoli [58] applied the SRH-1D model to simulate the controlled sediment flush at Isolato and Madesimo Reservoirs in Italy. The Madesimo Dam is on the Scalcoggia River, a tributary of Liro River, and about 1.4 km upstream of the confluence of Scalcoggia and Liro rivers. Isolato Dam is about 1.4 km upstream of the same confluence in the Liro River. The numerical model was used to predict the sediment impact downstream of the two dams and at the confluence. A satisfactory agreement between the computed and observed depositional patterns was obtained.

Huang et al. [11] updated the SRH-1D model to evaluate the sediment flushing operation at the Paonia Reservoir on the Muddy Creek, Colorado. A flushing plan was designed to lower the reservoir pool in early spring and flush the sediment during the spring runoff. Model modifications included the input of the user-defined reservoir operation rules. The model was first calibrated using 3 years of field data under the existing condition, and was then applied to predict the sediment management effects under different reservoir operation plans. The goal was to maximize the sediment flushing while maintaining the reservoir fill of water from the spring snowmelt. The timing of when sluicing was stopped and the storage of water was begun was based on the forecasted spring snow melt volume. The modeling showed that the reservoir trap efficiency depended strongly on the exceedance percentage of the inflow volume forecasted. For 2016, for example, the trap efficiency was 20% with the 90% exceedance, but 3% (net erosion) when the exceedance was 10%.

3.2. *Two-Dimensional Numerical Models*

Two-dimensional depth-averaged numerical models provide a more detailed representation of the reservoir hydraulic characteristics which might be missed in cross-sectionally averaged 1D models. Two-dimensional models are more general than the 1D ones in that they are applicable to both narrow and wide reservoirs and whether sediment deposition and resuspension vary or not across the channel.

Olsen [59] reported a 2D numerical model for sediment flush studies. The numerical model SSIIM solved the 2D depth-averaged equations for the flow and the 3D convection–diffusion equation for the sediment concentration. The mesh was made adaptive in the vertical direction and changed according to the water and bed levels. The numerical model was evaluated by comparing with the physical model data obtained for the Kali Gandaki Hydropower Reservoir in Nepal. Later, the 3D version of SSIIM was also reported [60].

Dewals et al. [61] used the WOLF 2D model to simulate sediment flush in an unknown reservoir in India. The 2D modeling was based on the multi-block rectangular mesh. The model predicted that only a narrow channel was generated by the flushing that was unable to be extended to the broad part of the reservoir width. The results demonstrated the ability of the model to simulate the sediment entrainment well during flushing; no measured data, however, were available to validate the numerical results.

Boeriu et al. [62] reported case studies using the 2D version of Delft3D for reservoir drawdown studies. Erosion was calculated differently for cohesive and non-cohesive beds, and the bed level was computed using the Exner equation. Case studies were reported in an unnamed reservoir in Sri Lanka where multiple days, 5 to 10, of drawdown flushing were simulated. Modeling was also performed at the Koga reservoir in Ethiopia with a 35-day drawdown. No measured field data, however, were available to validate the model results.

Chen and Tsai [63] developed a 2D model to simulate the sediment flushing efficiency of the A-Gong-Dian Reservoir in southern Taiwan. The reservoir was wide and the dam length was 2.4 km. The reservoir received waters from both Joushui River and Wanglai River. The flushing efficiency simulated by emptying the reservoir was found to match that based on the laboratory physical modeling study. The numerical modeling predicted the erosion upstream of the outlet on the Joushui river side and the deposition on the Wanglai river side. Based on the study, relocation of the outlet was proposed towards the Wanglai river side to improve the flushing efficiency.

Iqbal et al. [64] used the 2D BASEMENT model to simulate the sediment flushing processes. The model was based on the finite volume technique to solve the 2D shallow water equations on an unstructured triangular mesh. The transition from subcritical to supercritical during the rapid sediment flushing process was handled by solving the

Riemann problem at cell interfaces using the Godunov scheme. The model was used to simulate two sediment flushing cases: a 1:40 physical model of the reservoir at the Gulpur Hydropower Plant on the Poonch river in the Pakistan-administrated Kashmir and the laboratory flushing experiment of [65]. The model was found to reproduce the bed longitudinal and lateral erosions as well as the flushed sediment volume.

Chaudhary et al. [66] used MIKE21C to simulate reservoir flushing at a proposed reservoir on the Dibang River in East Asia. The reservoir will collect water from the Dri and Tangon Rivers. The 1D MIKE 11 was used to calculate the long-term sediment distribution in the reservoir, while MIKE21C was used to simulate sediment flush under a proposed drawdown plan. Reservoir flushing was carried out through the low-level spillway and the downstream water surface was from the flow rates over the spillway. The numerical model estimated the amount of sediment that would be flushed out under various flushing schemes.

Stillwater Sciences [67] reported the use of the SRH-2D model to simulate the sediment processes under the dam removal scenarios at the Matilija Dam within the Ventura River watershed in southern California. One-dimensional sediment transport modeling was conducted first to determine the downstream project impact within the Matilija Creek and Ventura River. Later, it was determined that 2D modeling was needed to provide more detailed data in areas having potential flood risks.

SRH-2D has also been adopted to study several reservoir drawdown scenarios in the A-Gong-Dian Reservoir, Taiwan [68]. The study examined and then determined the main factors that would influence the flushing efficiency. Recommendations were developed on how to increase the sediment flushing efficiency. Modeling studies confirmed the benefits by lowering the initial water level, creating narrower gorge-like geometry by partitioning and modifying the operation rules related to the flushing duration and the release rate.

3.3. Two-Dimensional Turbidity Current Model

Turbidity current modeling is complex and empirical models have been widely adopted. Most existing numerical models cannot be used to simulate turbidity current transport in reservoirs once the sediment has plunged to the bottom. A few advanced models, however, have been reported to overcome the limitations of the empirical/analytical tools and the existing 2D models; they range across 2D laterally averaged models, 2D layer-averaged models, and 3D models.

The 2D laterally averaged model has been widely used for stratified flows in reservoirs, particularly in the field of water quality modeling. For reservoirs, when the variation in key variables over the depth is more important than the lateral changes, 2D equations may be derived by integrating the 3D equations laterally across the reservoir to obtain the so-called laterally averaged equations. Wells and Gordon [69] have shown that such models may be adequate for some reservoirs.

The limitations of the laterally averaged 2D models, however, are similar to the 1D models: they are suitable primarily for relatively narrow reservoirs where the water surface level does not vary appreciably and there are no significant lateral inflows or outflows. A key advantage of the 2D models over the 1D ones is that the vertical stratification may be simulated directly. Laterally averaged models may be classified as with or without hydrostatic assumption. A widely used 2D, laterally averaged, hydrostatic, hydrodynamic, and water quality model is the so-called CE-QUAL-W2 [70]. This model assumes that vertical velocities are sufficiently small so that the vertical momentum equation may be reduced to the hydrostatic condition. The model has been widely adopted to simulate narrow reservoirs such as [71,72].

For wide or general reservoirs, layer-averaged 2D models are more appropriate for the turbidity current modeling. A complete set of layer-averaged governing equations was reported in [47]. Only a handful of 2D layer-averaged models have been reported that can deal with unsteady and non-conservative turbidity currents. One of these was the work of Bradford and Katopodes [73], who studied turbidity undercurrents in the deep-sea environment. A high-resolution, total-variation-diminishing, finite-volume numerical model was developed to capture the current front using the predictor–corrector time-stepping scheme. Mesh was used to represent the geometry suitable to deep sea applications. The model was verified by comparison with the experimental data of turbidity currents driven by uniform and non-uniform sediment. Groenenberg et al. [74] reported a 2D model that used a combination of the explicit fractional-step MacCormack scheme and a high-resolution shock-capturing technique. The model was based on a rectangular mesh and used a second-order finite-difference approximation. The model was verified using multiple laboratory cases; reasonable agreements with the measured data were reported.

A general 2D layer-averaged turbidity current model was developed by Lai et al. [75], and applications were reported in [76]. An extensive number of verification cases was documented which highlighted the applicability range of the model. Later, a case study will be presented and discussed using the model. We believe that the 2D layer-averaged model is a good compromise between the empirical/analytical models that are over-simplified and the 3D models that are still at the research stage and yet to become practical.

3.4. *Three-Dimensional Numerical Models*

Three-dimensional numerical models have also been reported in simulating reservoir drawdown flushing. Some models are limited to hydrostatic assumption, for which an extensive review was presented in [77]. In the following, only the non-hydrostatic 3D models are reviewed. We will use the term "3D CFD (computational fluid dynamics) models" to refer to this model category. Note that the full Navier–Stokes (NS) equations are solved with such models along with an appropriate turbulence model. Our experience suggested that only a 3D CFD model would be beneficial in obtaining more accurate results than the 2D layer-averaged models, at least for hydraulic flushing modeling studies.

Ghoreishi and Tabatabai [78] reported the use of a 3D CFD model to simulate the reservoir sediment flushing experiment carried out by [65]. The experiment was conducted in a rectangular flume which had dimensions of 50.0 m long, 2.44 m wide, and 1.52 m high, and the sediment was paved in a 9 m reach upstream of the dam. The numerical model predicted the channel erosion near the dam well; however, it did not reproduce the observed channel erosion further upstream longitudinally. The authors attributed the mismatch to the assumption of the rigid lid for the water surface. Temporally, the 3D model reproduced the erosion patten in the initial stage of the flushing well and showed differences in the later stages. Results indicated that the model performed well for the pressure flushing process, but less good for the later drawdown processes.

Haun and Olsen [79] applied the 3D SSIIM to predict reservoir sediment flushing processes. The model used the structured mesh in the horizontal plane and adaptive grid in the vertical direction. Only one vertical cell was adopted in the shallow area, while up to eleven cells were used in the deep area. The model was applied to simulate a physical model case conducted at the Kali Gandaki hydropower reservoir in Nepal. Later, Olsen and Haun [80] updated SSIIM to include the bank failure module to improve the prediction of channel widening processes during reservoir flushing. The model domain was divided into water, soil, and slide cells. The soil domain used a 2D horizontal mesh, while the water and slide domains were based on a 3D mesh. The bank failure algorithm was tested for the 2014 reservoir flushing event at the Bodendorf reservoir, Austria. The model was able to reproduce the number and magnitude of the slides well, although the locations were not always correct. The updated model worked well for thick sediment layers, but instabilities occurred for thin layers. SSIIM was also used in [81] to simulate the flushing efficiency of the Schwarzenbach Reservoir in the Black Forest, Germany, under both the partial and full drawdowns.

Esmaeili et al. [82] employed the SSIIM model to simulate the 2012 sediment flushing operation at the Dashidaira Reservoir in Japan. The sediment flushing operation was performed at the site through the bottom outlets annually in the early rainy season from 1991. The model was first calibrated by reducing the difference between the computed and measured total flushed-out sediment volume. The simulated reservoir bathymetry after flushing was compared with the measured data. The results showed that the 3D model simulated reasonably well the flushing channel evolutionary pattern. Modeling found that finer sediment was entrained and flushed out earlier than the coarser sediment. They later presented additional modeling studies using the same model by adding additional artificial discharges during the free-flow state [83], increasing the drawdown speed, and adding an auxiliary longitudinal channel.

Our review found that numerical modeling of the pressure flushing processes is rarely carried out. It is possibly because only 3D CFD modeling is adequate for the pressure flushing simulation. An attempt was reported by Ermilov et al. [84], who used the TELEMAC-MASCARET package to simulate the pressure flushing scenarios. Model results were compared with the physical model data. It was shown that 3D modeling was capable of simulating sudden sediment removal in a schematized reservoir under the pressure flushing operation. The model reproduced the typical scour cone shape upstream of the flushing gate; locally varying flow features were also captured. The simulated bed scouring changes were in good agreement with the results of the physical model study. It was also found that the 3D model results were sensitive to the numerical model parameters adopted.

Three-dimensional CFD modeling was also carried out by Lai and Greimann [32] to simulate the pressure flushing process at the Cherry Creek Reservoir, Denver, Colorado. The numerical model results were compared with the field data and good results were obtained. Further discussion of the case will be presented later in this article.

3.5. Nested Approach

It has been commented that the so-called *Nested Approach* has been adopted for the numerical modeling of hydraulic flushing and can be beneficial. Often, a comprehensive reservoir management/sustainability study may adopt the nested approach in which 1D, 2D, and 3D models are applied conjunctively. In this approach, 3D modeling is applied in a subset zone of the 2D model, and 2D modeling is a subset of the 1D model domain. Often, the coarser models provide the boundary conditions to the refined ones. The nested approach has the best potential to address an extensive list of study questions and may be adopted for large projects.

Castillo et al. [85] provided an example of the nested approach to understand the changes expected in the Paute River, Ecuador, after the Paute-Cardenillo Dam was constructed. Then, 1D, 2D, and 3D models were used together to answer various study questions related to the sediment transport and flushing consequences. The 1D model was used to estimate the long-term reservoir sedimentation; the 2D model was used to simulate the 72 h hydraulic flushing operation; and the 3D model was applied to investigate the sediment transport details when the bottom outlets were operated. The study demonstrated that different model simulations were needed to achieve appropriate resolutions in the prediction of the sedimentation and flushing operation. The reported computing times of all models by Castillo et al. [85] may shed light on the model performance. The 2D modeling of a 72-h period took about 24 h in computing time for the entire reservoir, while 3D modeling of the same period required above 1600 h. The run time for the 1D model was not reported; we estimate that a 1D model would require only an hour to run a 100-year simulation.

4. Guidelines

In this section, general and specific guidelines are presented based on years of experience in research, development, and application at USBR.

4.1. General Guidelines

General recommendations are first presented on how to select a 1D, 2D depth-averaged, or a 3D CFD model.

For 1D models, a key advantage is that a minimal amount of computing time is required so that a long-term modeling study may be conveniently carried out. A list of general guidelines is provided below for using the 1D model:

- Recommended for projects whose study questions require long-term simulations (e.g., >10 years); a 100-year modeling study has been routinely carried out.
- More appropriate for evaluating alternative operational options or design strategies when many simulations are needed.
- Study questions that may be answered by 1D models include: reservoir sedimentation and storage loss, long-term flushing efficiency for different flushing alternatives, reservoir sustainability impacts of reservoir operation, long-term sediment impact downstream of the reservoir, and quantification of the uncertainty of the model results, among others.
- One-dimensional modeling can often start from the pre-impoundment geometry and use the historical inflows and reservoir operations as the inputs for the model calibration. Measured longitudinal profiles and/or the reservoir sedimentation volumes at different times may often be used for the model calibration.
- Limitations of 1D modeling should be kept in mind, such as:
    - High uncertainty exists when the reservoir is wide and geometry is not a single channel. However, 1D model accuracy improves significantly with a narrowing reservoir channel. A reservoir is considered narrow when the width ratio of the largest reservoir cross section to the narrowest or drawdown section is less than 4 to 5.
    - The reliability of the 1D results increases with the pool level lowering depth. The highest accuracy is achieved when the flow during the flushing is of the run-of-the-river type.

For 2D depth-averaged models, the computing time may be much longer than 1D models. Therefore, the spatial extent and simulation time duration may be limited. Specific guidelines/comments are as follows:

- Spatial and time limitations: the longitudinal length along the river is not higher than, e.g., 20 km, and the simulation time is often limited to a single drawdown event or no more than a few years.
- Two-dimensional modeling is generally applicable to most reservoirs, and particularly recommended if the reservoir width is large (more than 4 to 5 times of the drawdown width) or if the assumption of a constant water level across the cross sections is not valid. Tow-dimensional modeling is highly recommended if the delta evolution will be simulated where the delta moves into lateral tributaries and margins.
- Two-dimensional modeling is needed where multiple gates are used for drawdown across the dam and when gates are operated differently.
- Potential limitations:
    - The flushing-induced channel erosion may be underpredicted significantly if the bed consists of cohesive materials, unless the cohesive properties are properly taken into consideration and the model is properly calibrated.
    - Channel erosion during drawdown may be under-predicted significantly if the erosion is mainly contributed by bank erosion or knickpoint process.

Three-dimensional CFD models refer to those which solve the Reynolds-Averaged Navier–Stokes (RANS) equations without the use of the hydrostatic assumption. General guidelines are provided below:

- Three-dimensional models are the most general and applicable to all types of hydraulic flushing.
    - In contrast, 1D and 2D models have restrictions due to the various model assumptions adopted. For example, 1D and 2D models are generally not applicable to pressure flushing, as velocity is assumed to be uniform throughout

the pool depth with such models. For pressure flushing, 3D modeling is probably the only viable option.

- Few adjustable model parameters are needed for 3D CFD models, so model calibration is not critical, at least for the flow field processes. Please refer to [86] for a recent review of the 3D CFD models on sediment modeling. Three-dimensional CFD models, therefore, are highly recommended if the flow field, particularly in the vertical direction, is important to answer the study questions.
    - Model input parameters that may impact the model accuracy include mesh size, time step, and turbulence model.
- It is cautioned that most existing 3D CFD models adopted similar sediment theory and equations to the 2D depth-averaged models [75].
    - Therefore, any model errors associated with the sediment theory/equations would not be improved by using the 3D models.
- An important 3D model limitation: the runtime can be very high and days or even weeks have been reported in case studies. Therefore, both the spatial extent and time duration may need to be much reduced from the 2D models.
- Other general guidelines:
    - Model results may be sensitive to the mesh resolution; in general, a mesh sensitivity study should be carried out.
    - The selection of a turbulence model may not be critical in applications (though it has been reported to be significant in the literature, but mostly for theoretical studies), and may be treated as a secondary issue.
    - The accuracy of the scour and sediment predictions is additionally dictated by the empirical sediment equations adopted by the 3D model; therefore, sediment input parameters need to be carefully selected.
    - The bed sediment in front of the bottom gates is often cohesive for reservoir management applications; care should be taken to ensure that the model has cohesive bed modeling capability and that the cohesive erosion properties are adequately measured with known uncertainty ranges. Proper model parameter sensitivity studies may be important in obtaining statistically meaningful results.

4.2. *Specific Guidelines*

4.2.1. Drawdown Flushing Modeling

- One-dimensional or two-dimensional models are recommended for drawdown flushing simulation unless the pool water level is too high and the sediment processes are similar to pressure flushing.
- The reservoir narrowness should be used as a guide for the choice of 1D or 2D models (discussed previously). In theory, there is no width restriction with the 2D models.
- The reliability of 1D and 2D model results near the dam is influenced significantly by the amount of pool level lowering—the lower the pool level, higher the accuracy.
- Three-dimensional modeling may still be needed for the early stage of the drawdown flushing operation.

Two channel erosion processes should be recognized during drawdown: progressive and retrogressive. Progressive erosion is characterized by the emergence of an eroded channel at the upstream of the reservoir, and erosion moves downstream towards the dam and finally reaches the dam while the reservoir is being emptied. Channel formation is via fluvial processes of increased sediment-carrying capacity. Progressive erosion is often initiated when reservoir drawdown is achieved through low-level outlets and the drawdown rate is not rapid. This type is relatively easier to simulate than the retrogressive one. Retrogressive erosion is characterized by a zone of high slope and fast erosion that is moving upstream. The point of slope change has the highest erosion rate (commonly called the knickpoint). Retrogressive erosion may initiate in instances where sediment deposits are deep and near the dam and drawdown rate is rapid or an initial steep slope is created. This erosion type has been observed both in the laboratory and in the field [5]. Retrogressive erosion is much more difficult to simulate as knickpoint erosion may not be incorporated in most numerical models.

4.2.2. Turbidity Current Venting Modeling

- Empirical/analytical models may be used first at the beginning of a project to gain an overview of the turbidity current processes and obtain an estimate of general parameters such as the plunging point location and mixing characteristics. Not all turbid water would plunge to the reservoir bottom and form an undercurrent. Field observations and data may also be needed to gain an understanding of the turbidity current characteristics.
- Two-dimensional layer-averaged models are recommended for most projects as recent two-dimensional models are more general than the existing one-dimensional models and the computing time is becoming reasonable with the availability of fast modern-day PCs. In particular, turbidity current modeling is often event-based, so the simulation time is finite. If 1D models are used, ensure that the river resembles the run-of-the-river type.
- Three-dimensional CFD models may also be used and they are accurate. However, the runtime of most 3D models may be prohibitively long; 3D models are yet to become practical.

4.2.3. Pressure Flushing Modeling

- Only 3D CFD models may be appropriate for pressure flushing simulation. One-dimensional and two-dimensional models are not recommended unless the specific case study questions warrant their use.
- Only a portion of the reservoir pool near the pressure flushing bottom outlets needs to be simulated. Use of the entire reservoir is generally unnecessary and serves only to increase the computing runtime. The reason is that the scour cone is usually small and limited to the front part of the outlets. Away from the outlets, flow velocity is low and there is little sediment movement.

5. Case Studies

Case studies are provided in this section to shed light on the use and performance of the numerical models, as well as the use of the guidelines.

5.1. One-Dimensional Modeling at Paonia Reservoir, Colorado

SRH-1D model has been applied to several drawdown sediment flushing studies at USBR. Herein, the modeling at the Paonia reservoir, Colorado, was chosen to illustrate the selection, use, and performance of 1D modeling.

The Paonia Reservoir on the Muddy Creek is located about 27 km northeast of Paonia, Colorado. The reservoir has a surface area of 1.35 km^2, a total design capacity of 25.8 million-m^3, and is considered narrow (4.8 km long and 0.3 km wide). Hydrology is characterized by spring snowmelt with summer thunderstorms. The estimated average annual sedimentation rate is about 0.125 million m^3 per year; nearly 25% of the reservoir capacity has been lost. In 2010, the outlets were blocked by sediment and debris, leading to the start of a comprehensive study to design a sediment flushing plan to solve the sedimentation problem. Field drawdown flushing was exercised and able to pass sediment through the reservoir. However, it was insufficient to sustain the reservoir storage. In 2014, the dead pool of the reservoir was completely filled (Figure 6).

Figure 6. Paonia Dam water diversion outlet structure partially blocked by sediment (USBR photo taken on 11 November 2014).

A numerical modeling study was initiated, using the SRH-1D model, to evaluate both the short-term and long-term plans [10,11]. Modeling provided answers about the reservoir sedimentation rate, trap efficiency, sediment release concentration, and other variables. One-dimensional modeling was adequate for the case, as the reservoir is relatively narrow and it was chosen as the study questions were related to the long-term effect of the flushing alternatives. The set of user-definable reservoir operation rules include (a) the minimum and maximum reservoir releases at different reservoir water levels, (b) timing to start reservoir filling according to the forecast incoming flows, (c) constraints on the releases imposed by the gate and spillway capacity, and (d) constrains on the ramping rate of the releases.

The model inputs included cross-sectional geometry, incoming flow and sediment rates, downstream water surface elevation or discharge, bed sediment size and gradation, flow roughness coefficient, sediment transport capacity equation, and other model-specific inputs. The model was first calibrated using the data collected in the time period of 11 June 2013 to 30 June 2015. The measured sediment load at the reservoir exit was used for the model calibration. It was found that the reservoir channel bed profile and the outflowing suspended concentration were predicted reasonably by the model. The calibrated model was then used for both the short-term and long-term simulations for model application predictions. The impact of various reservoir operational rules was evaluated.

5.2. Two-Dimensional Depth-Averaged Modeling of Drawdown Flushing on the Klamath River, Oregon

Two-dimensional modeling has been routinely carried out at USBR for numerous sediment transport projects. Herein, the study of the drawdown processes at the Copco-1 Reservoir on the Klamath River, Oregon is presented—a 2D modeling study of hydraulic drawdown flushing. The study was carried out as part of a much larger effort at USBR to support the Secretarial Determination on Klamath Dam Removal and Basin Restoration [87].

Four dams, JC Boyle, Copco-1, Copco-2, and Iron Gate on the Klamath River, were studied for possible decommissioning. Numerical modeling was carried out at the Copco-1 Reservoir to address two study questions: (a) channel erosion during drawdown, and (b) the estimate of suspended sediment concentration released to the downstream. The drawdown process was event-based (a relatively short time period) and the reservoir was relatively wide (erosion was expected to have lateral variation)—these two facts led to the selection of 2D modeling for answering the study questions. The modeling study was further used to aid in determining the strategies for revegetating the reservoir area and recovering a functional riparian corridor after dam decommissioning. The downstream sediment release would help to determine the proper timing, duration, and release rate of the drawdown design. Note that extensive 1D modeling was also carried out to answer other long-term questions, as documented in [87].

Copco-1 Dam, constructed in 1918, is 38.4 m high; the reservoir is 7.5 km long, and has 10.4 m average depth and 41.6 million-m^3 storage capacity. The 2D modeling began with the selection of a model domain (entire reservoir selected) and the generation of a 2D mesh (10,504 mixed quadrilaterals and triangles)—they are shown Figure 7a. Measured bathymetric and terrain data were used as the initial reservoir terrain before the drawdown (Figure 7b). Main model inputs included (a) the Manning's roughness coefficient of 0.03; (b) bed and subsurface layer sediment data (the top layer consisted of mostly silt and clay and the bottom layers of coarse gravel); and (c) the historical upstream flow rate and zero sediment rate (as the majority was wash load that simply passed through).

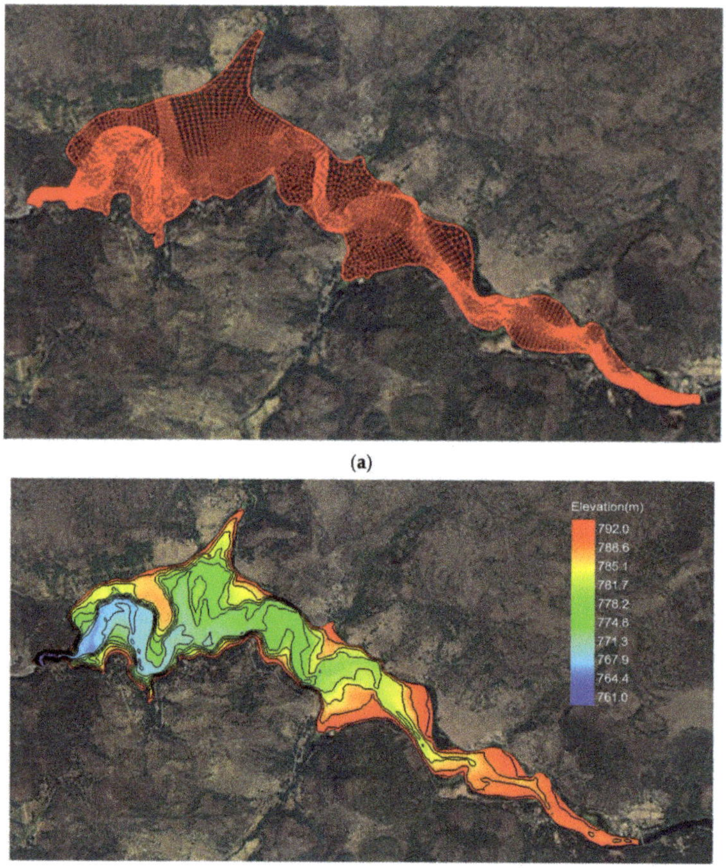

Figure 7. (**a**) 2D model domain and mesh; (**b**) reservoir initial bathymetry/terrain for the numerical modeling.

Initially, the reservoir was filled with water to an elevation of 793.4 m (inversion of the spillway). Drawdown was then initiated through water release at the low-level gates. The release rate was 0.914 m/day but varied according to the gate constraint. The modeling used seven sediment size classes, with one size representing the cohesive material smaller than 0.0625 mm in diameter. Three simulations were carried out, corresponding to three hydrological scenarios (Dry-Year, Average-Year, and Wet-Year). Each run started on 15 November and ended on 15 May of the following year, a duration of six months.

In the following, the model results corresponding to the Average-Year, named the baseline, are presented. The predicted reservoir water surface elevation and discharges into and out of the reservoir are displayed in Figure 8. It can be seen that the reservoir elevation was lowered to below 762 m within one month. However, only under the relatively dry year could the reservoir water level be maintained at such a low level. The reservoir would be filled with water quickly with the Wet-Year hydrology. The predicted sediment concentration released to the downstream from Copco-1 is shown in Figure 9 for the three hydrological scenarios. The predicted concentration did not differ substantially between the Dry-Year and Average-Year, as both flows were sufficient to mobilize the majority of the reservoir deposits. There was, however, a noticeable difference between the Wet-Year and the other two. This might be explained by the fact that the reservoir water level remained low for the Dry- and Medium-Year but not so for the Wet-Year. Low water elevation led to high flow velocity and sediment carrying capacity. With the Dry- and Medium-Year simulations, the predicted sediment concentration pulse had an average peak of about 6000 ppm (occasionally exceeding 7000 ppm) and a duration of about 1.5 months. With the Wet-Year, the average pulse peak was lowered to 4000 ppm (occasionally exceeding 6000 ppm). After 45 days of the drawdown, sediment concentration fell to a low level (about a few hundred ppm).

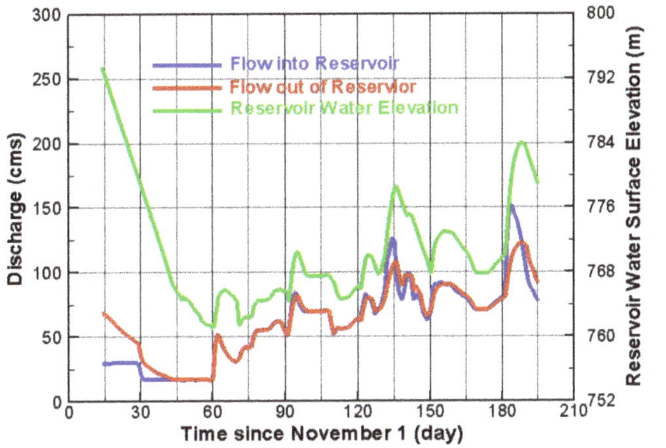

Figure 8. Simulated reservoir water surface elevation and discharges into and out of the reservoir under the baseline scenario.

One of the study questions involved understanding erosion channel formation due to the drawdown. As discussed earlier, two channel erosion forms may occur: retrogressive and progressive. The numerical modeling of Copco-1 showed that the channel process belonged to the progressive type, as shown in Figure 10. Progressive erosion was expected to occur, given the assumed drawdown conditions according to the analyses in [88,89]. The reservoir pool level was near its lowest on 29 December, while 14 May was the end of the simulation. The pre-dam geomorphology of the reservoir area was delineated in [87] and is plotted in Figure 11. The predicted bed elevation and the net eroded depth are compared with the initial top bed layer thickness and bed elevation in Figure 12. It is seen that the model-predicted channel erosion pattern and thalweg agreed well with the pre-dam channel. The majority of the reservoir deposits within the pre-dam channel had eroded after 45 days of the drawdown, particularly for the upstream half of the reservoir (note that the results are inaccurate near the dam). Incision into the bottom bed layer was also predicted for the upstream half of the reservoir six months after the drawdown. In the upstream zone 1 and 2 areas, channel incision decreased with increasing flow into the reservoir (wet year). The trend, however, was reversed in zones 4 and 5, where incision increased with an increasing flow.

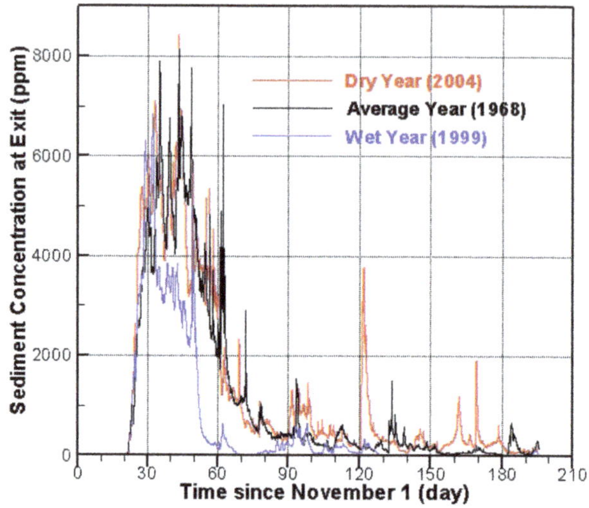

Figure 9. Predicted sediment concentration from the drawdown gate of Copco-1 under the three hydrological scenarios.

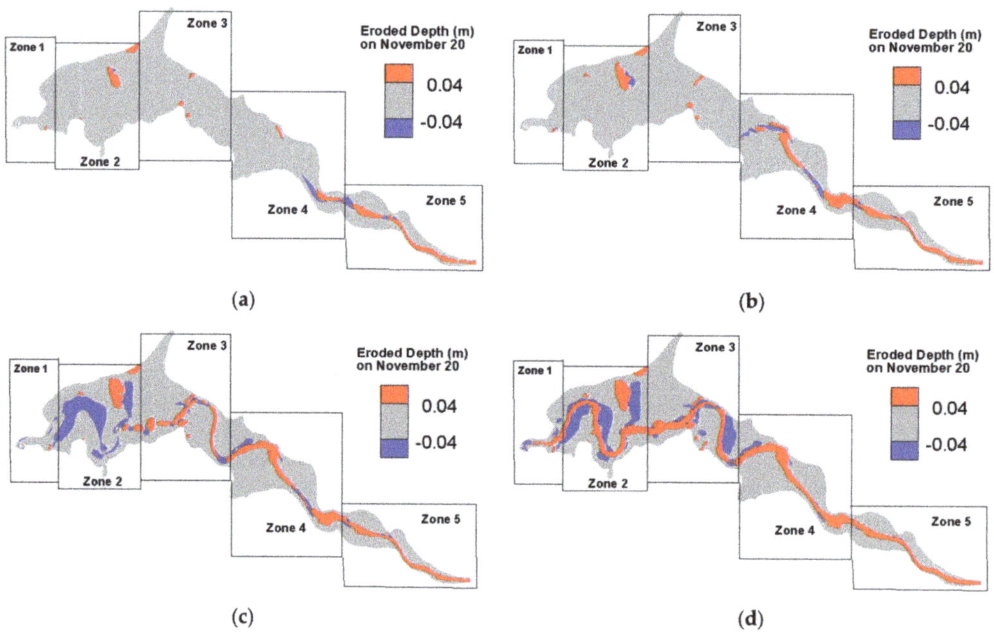

Figure 10. (**a**) From 25 November 2008; (**b**) 30 November 2008; (**c**) 10 December 2008; (**d**) 14 May 2009. Predicted erosion/deposition pattern during the drawdown of Copco 1 reservoir under the Average-Year hydrology (2008) and Medium-Erosion bed sediment.

Deposition was predicted in the pre-dam floodplain area in the downstream half of the reservoir. It was particularly visible in the open area near the narrow canyon. These model results provided the data needed to make the decision on how revegetation and habitat restoration would be planned after dam decommissioning.

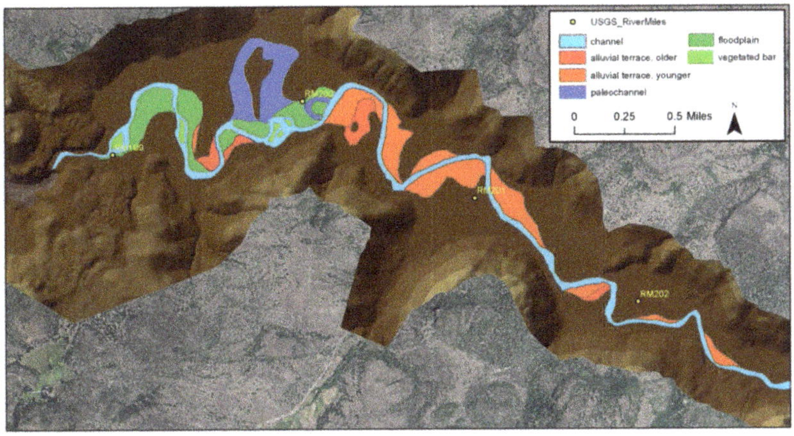

Figure 11. Geomorphic map of the river corridor prior to the construction of the Copco I Dam.

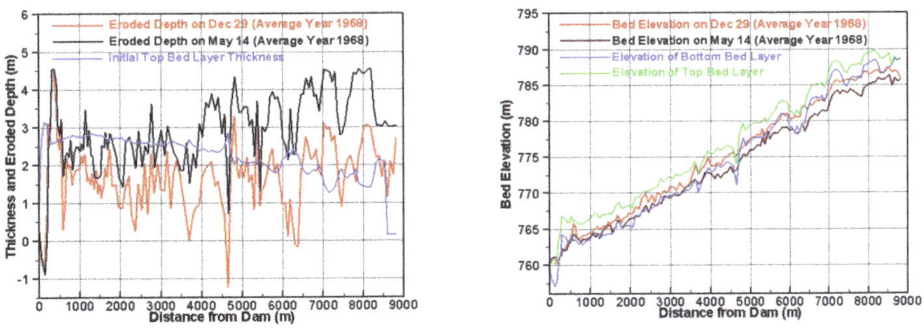

(**a**) (**b**)

Figure 12. (**a**) Net eroded depth; (**b**) bed elevation. Simulated net depth of erosion and deposition (**left**) and the predicted bed elevation (**right**) along the thalweg of the incised channel on two dates, compared with the initial thickness of the top bed layer deposit and the top and bottom bed layer elevations.

It has been commented that the deposition near the drawdown gate in zone 1 may be unrealistic given that (a) the model was depth-averaged but flow was highly 3D near the gate, and (b) the pressurized flow dominated near the gate. The inaccuracy of the erosion prediction near the gate, however, was expected to have negligible effect on the predicted erosion upstream.

5.3. Two-Dimensional Layer-Averaged Modeling of Turbidity Current at Shihmen Reservoir, Taiwan

A 2D model of turbidity current venting at Shihmen Reservoir, Taiwan, is presented to highlight the modeling process and model performance. Suspended sediment periodically moves into this reservoir in the form of a turbid undercurrent during large typhoons. Limited sediment venting capacity at the dam has led to the loss of reservoir capacity at a faster rate than the original design. The 2004 Typhoon Aere event is presented here as a case study, as the event was large and caused 11% loss of the reservoir storage capacity.

The 2D turbidity current model of [75] was used for the modeling and the results were compared with the 1-to-100 scale laboratory data. The numerical model covered the entire reservoir with about 15.5 km longitudinal length (Figure 13a), and a 2D mesh was developed consisting of 33,008 mesh cells. Initial reservoir bathymetry was derived from the measured data surveyed in 2003 (Figure 13b). At the reservoir inflow boundary, measured flow rate and suspended concentration were used as the boundary conditions. The median diameter of the turbidity current sediment was 5 μm. The simulation was carried out for the typhoon period of 02:00, 24 August to 21:00, 26 August 2004—a 67 h duration in the prototype.

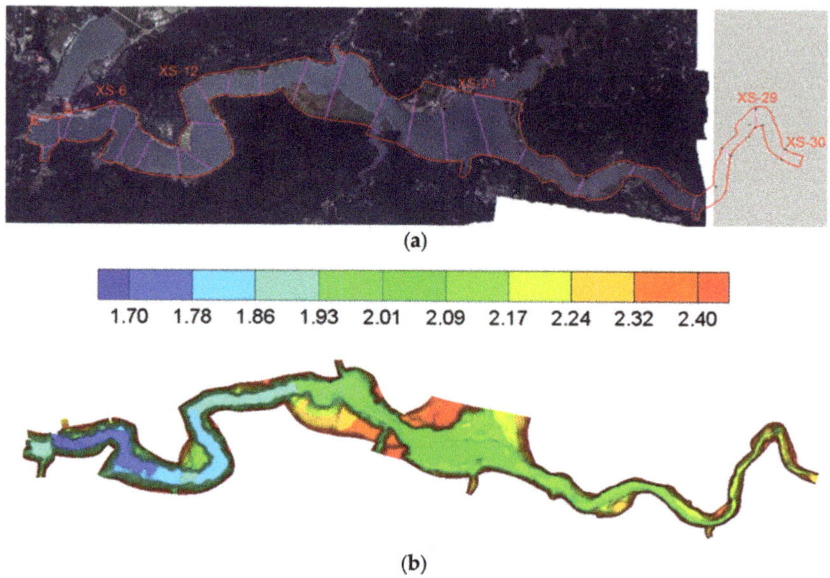

Figure 13. (**a**) Study domain and key cross sections; (**b**) reservoir bathymetry. The model domain and selected cross-section (XS) locations for the Shihmen Reservoir modeling along with the bathymetry (in meters).

Downstream boundary conditions were complex and demonstrated the need for 2D modeling. There were five outlets distributed along the dam and located at different depths. They included the spillway, the flood diversion tunnel, the powerhouse intake, the permanent river outlet, and the Shihmen intake. The spillway had an invert of 235 m, width of 107 m, and discharge capacity of 5800 m^3/s at the prototype scale. The flood diversion tunnel had an invert of 220 m, height of 8 m and a full-capacity discharge of 1800 m^3/s. The powerhouse outlet vented the highest amount of sediment, as it was located at the lowest depth. It had an invert of 171 m, a height of 5 m and a full-capacity discharge of 380 m^3/s. The permanent channel outlet was relatively small and less important in sediment venting. It had the same invert as the powerhouse but a full capacity of 30 m^3/s. The Shihmen intake was also small in its venting capacity; it had an invert of 193.6 m, height of 2.4 m, and full capacity of 13 m^3/s. The key model input was the drag coefficient; a value of 0.055 was adopted (calibrated). The initial reservoir was assumed to be clear (i.e., zero suspended sediment concentration without the undercurrent); the turbidity current then entered the model domain and moved towards the dam.

The predicted arrival times of the turbidity current at various cross-sections along the reservoir are compared with the measured data in Figure 14. Good agreement was obtained. The percentage error was less than 14% at all measured cross-sections. The numerical model predicted that it would take 0.925 h for the front of the current to reach the powerhouse from the upstream boundary (9.25 h in the prototype), which was comparable to 0.930 h of the physical model. The turbidity current front was predicted to move at a nearly constant speed of 0.0450 m/s; it was compared with the measured speed of 0.0448 m/s (0.448 m/s in the prototype). The results showed that the numerical model was capable of predicting the current movement through the reservoir well. Such predictions were important as they could help inform how the outlets might be operated to optimize the sediment venting.

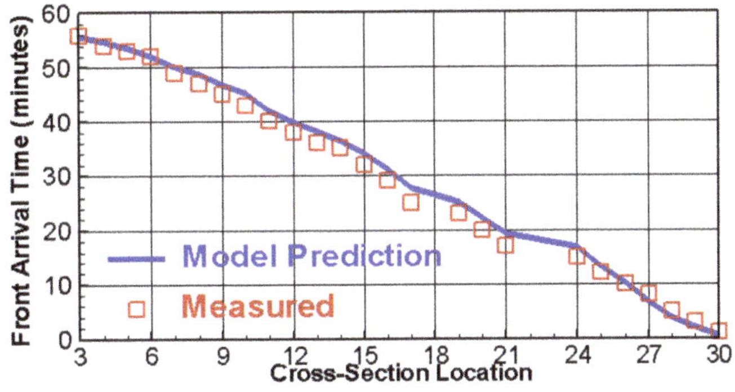

Figure 14. Comparison of model-predicted and measured current arrival times at various cross-sections along the reservoir during Typhoon Aere (time in the physical model scale).

Of particular interest was the total amount of sediment that might be vented out of the available outlets; this would produce the venting efficiency important for determining the reservoir storage loss and its sustained life. The sediment volumes delivered into and vented out of the reservoir through all outlets are compared in Table 1 between the numerical and physical models. It was shown that the total sediment volume vented out of the reservoir was 55.9% and 45.1%, respectively, for the numerical model and physical model. The higher venting efficiency predicted by the numerical model was due primarily to the higher predicted sediment through the spillway and the diversion tunnel—the two higher-level outlets whose venting ability depended heavily on the vertical distribution of the turbidity current, which was not predicted by the 2D model and obtained through an empirical relation.

Table 1. Summary of total sediment volumes moved into and out of the reservoir during the Typhoon Aere event from both the numerical and physical models.

	Numerical Model	Physical Model
Total sediment volume into reservoir (million-m^3)	10.93	10.97
Volume through Power House (million-m^3)	3.31 (30.3%)	3.18 (29.0%)
Volume through Spillway (million-m^3)	1.72 (15.7%)	1.02 (9.30%)
Volume through Flood Diversion (million-m^3)	0.754 (6.9%)	0.420 (3.92%)
Volume through Permanent Channel (million-m^3)	0.265 (2.42%)	0.259 (2.36%)
Volume through Shihmen Intake (million-m^3)	0.0569 (0.52%)	0.0593 (0.54%)

A sensitivity study was also carried out with several model input parameters. It was found that the drag coefficient and erosional rate were important, while the entrainment rate was not. The drag coefficient impacted the current front speed significantly but had negligible effect on the outlet sediment rates. For example, the times needed for the current front to reach the powerhouse outlet were 8.20, 9.25, and 10.03 h, respectively, with drag coefficients of 0.035, 0.055, and 0.075. The erosional rate coefficient was important for predicting the erosion and deposition characteristics during the venting event as well as the current movement speed. An increased erosional rate coefficient led to increased sediment volume out of the reservoir and faster current movement speed.

5.4. Three-Dimensional CFD Modeling of Pressure Flushing at Cherry Creek Reservoir, Colorado

A 3D CFD modeling study is presented for pressure flushing at the Cherry Creek Reservoir, Denver, Colorado. Three-dimensional modeling is often required for pressure flushing, as the above guidelines recommended.

A small horizontal domain near the outlets was chosen for the 3D modeling according to the guidelines discussed above and shown in Figure 15a. The far field flows, indeed, were found to be small and insignificant for the present modeling, based on the results of an entire reservoir simulation (results not shown here). The flushing outlets were located within an intake tower whose dimensions are marked in Figures 15b and 16. The outlet gate geometry needed to be represented as accurately as possible for adequate 3D modeling (see Figure 16). As can be seen, the five bottom gates were used for pressure flushing; they were extended out in the numerical model so that water release amounts might be implemented properly by the numerical model. The 3D mesh had a horizontal size of 9855 mixed quadrilateral and triangle cells (Figure 15a) and 47 vertical points. Note that the 3D model used solved the NS equations, and the k-ε turbulence model was adopted.

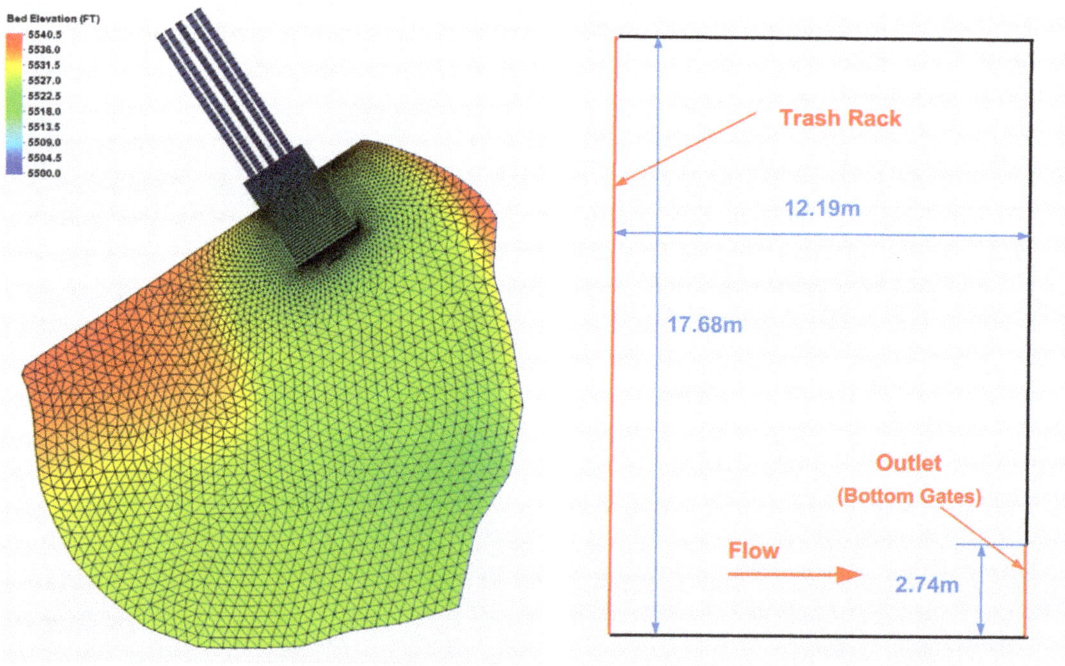

(a) (b)

Figure 15. (**a**) Model domain and horizontal mesh; (**b**) side view of the intake tower. Model domain, horizontal mesh, and a side view of the intake tower.

Two sets of modeling were carried out, corresponding to two pressure flushing operations at the Cheery Creek Reservoir: 2017 low-discharge and 2018 high-discharge flushing. The flushing release rate was an input in the model as the boundary condition, while other model inputs (to be discussed below) were kept the same for all modeling runs. Only the 2018 flushing results are discussed below.

Pressure flushing was conducted in the field on 23 May 2018 with a nominal discharge of 36.8 m^3/s and the reservoir pool maintained at the constant elevation of 45.71 m. The actual flow release rate through all intake gates is shown in Figure 17. The release was achieved by opening one gate at a time of the five gates, and following the sequence of gates 3, 1, 2, 4, and 5 (gates are numbered from right to left looking towards the intake in Figure 16).

The reservoir sediment consisted of clay, silt, and sand and the properties were measured by [90] and used by the model. The fractions for the clay, silt, and sand were 45%, 50%, and 5%, respectively. The bed was treated as cohesive. The erodibility properties of the cohesive sediment are important for modeling. The critical shear stress and erodibility were measured by [90]: the critical shear stress was 0.62 Pa and the erodibility was 3.55×10^{-4} m/s-Pa.

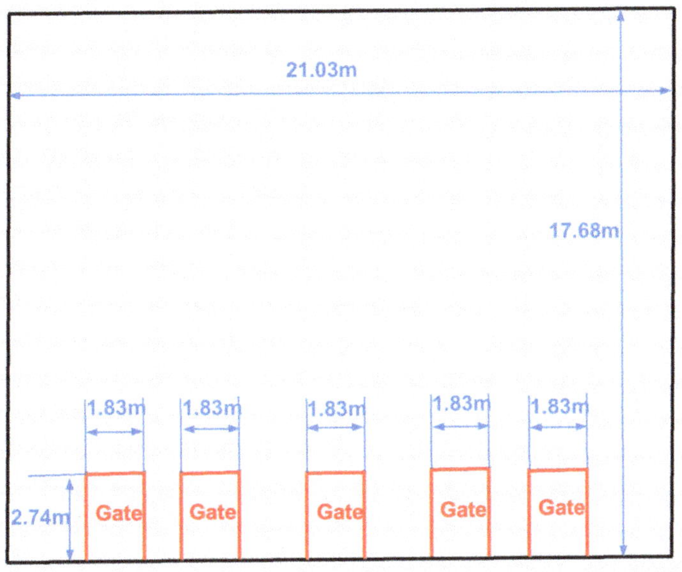

Figure 16. Front view of the intake tower with five bottom gates for water release.

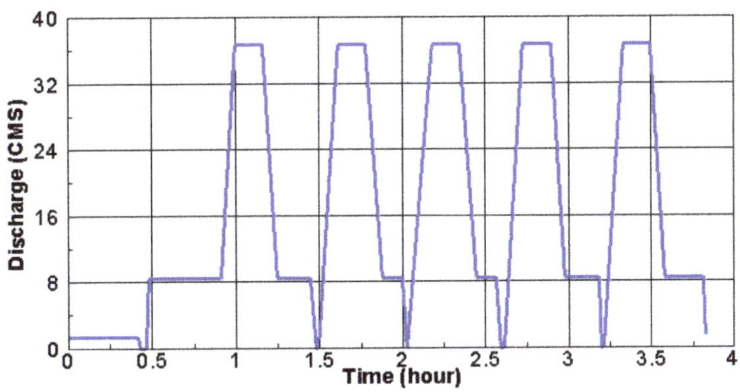

Figure 17. The actual flow release during the 2018 pressure flushing operation.

First, the predicted sediment concentration out of the reservoir was compared with the field data in Figure 18. The agreement was found to be good, although the concentration was under-predicted over the first two gate opening periods and over-predicted in the next three gate periods. Note that the predicted concentration was obtained right after the gates (within the outlet works), while the measured value was within the Cherry Creek, about 0.4 km downstream of the dam outlet. This may explain some of the above discrepancies. The measured high concentration during the first gate was partially contributed by the sediments stored downstream of the release outlet, and not entirely due to the reservoir release.

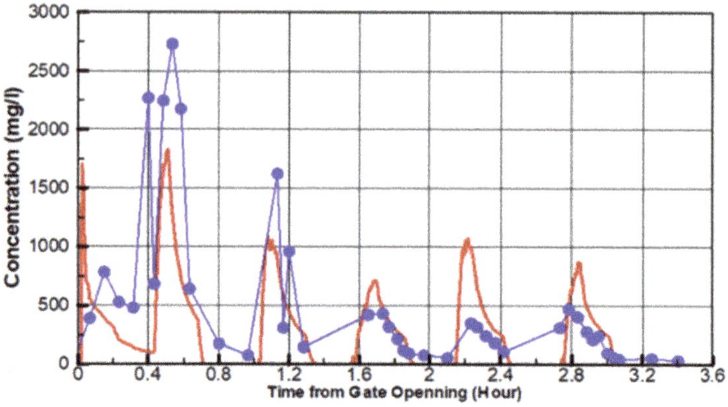

Figure 18. Baseline model prediction and field-measured sediment concentration downstream of the release gate during the 2018 pressure flushing (Red line: numerical model; blue symbol and line: measurement).

The predicted erosion pattern (scour zone) produced by the pressure flushing is shown in Figure 19. The results showed that the scour zone was limited to within the intake, which was confirmed by the field survey at the site. A quantitative comparison of the erosion pattern was not possible as the field measurement was not able to reach inside the intake.

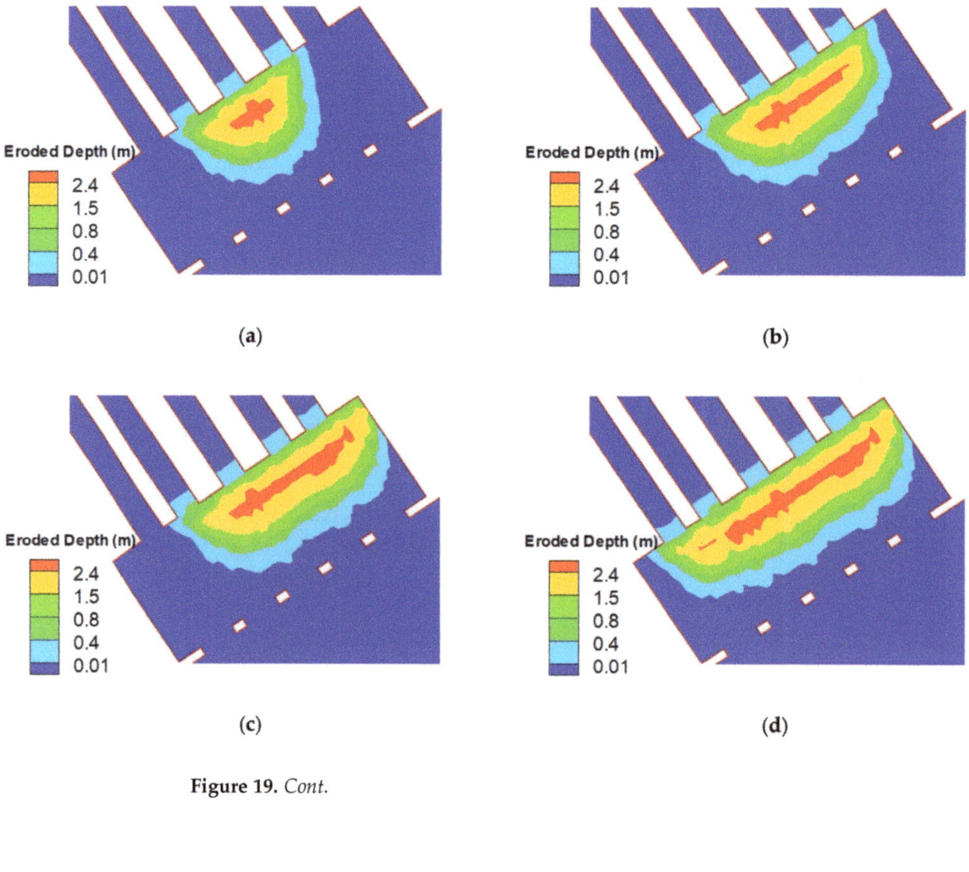

Figure 19. *Cont.*

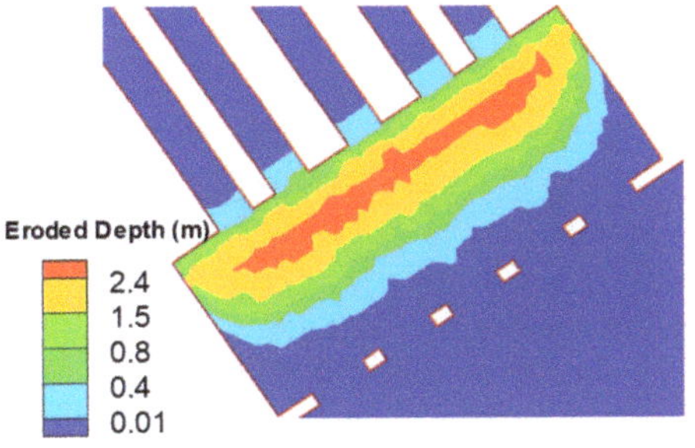

(**e**)

Figure 19. (**a**) Time = 1.0 h (Gate 3 complete); (**b**) time = 1.5 h (Gate 2 complete); (**c**) time = 2.0 h (Gate 1 complete); (**d**) time = 2.5 h (Gate 4 complete); (**e**) time = 3.5 h (end of flushing). Predicted scour development with time during the 2018 pressure flushing (contours represent the eroded depth in meters).

This case study shows that the 3D CFD model performed well in simulating the pressure flushing process, at least for the study site. For example, the sediment concentration released downstream was well predicted, pointing to its potential to be used for future pressure flushing modeling applications. It showed also that 3D numerical models may be useful in developing an effective flushing strategy. Based on the above results, for example, a three-gate release—gates 3, 1, and 5—would be more efficient than the current five-gate schedule. Further, the flushing efficiency might be maximized by flushing every other year or the flushing duration may be much reduced to save water release, as high erosion occurs primarily during the early stage of gate opening.

6. Concluding Remarks

A comprehensive review has been provided in this article with regard to the numerical modeling of hydraulic flushing to manage reservoir sedimentation. Three sediment removal types were covered: drawdown flushing, pressure flushing, and turbidity current venting. The review also provides the empirical/analytical methods that may be used for a quick assessment of the three types of hydraulic flushing, while the more comprehensive and general 1D, 2D, and 3D models were reviewed extensively. In particular, guidelines and best practices were presented for the three categories of the numerical models: 1D cross-sectionally averaged, 2D depth-averaged or layer-averaged, and 3D CFD models. Case studies were presented for each hydraulic flushing type and using each category of the numerical models. These cases illustrate the guidelines and discuss how study questions may be addressed. In addition, the case studies also discuss how a model is selected given a specific reservoir, the determination of the model domain and mesh, model inputs, model performance, and a results comparison. Important findings include the following: (a) the empirical/analytical method may be used for planning studies; (b) 1D, 2D, and 2D models are recommended for design and alternatives assessment studies; (c) 1D models are appropriate for narrow reservoirs or long-term simulations; (d) 2D models are recommended for wide reservoirs or where later changes are important; (e) 3D models are needed for pressure flushing modeling or any near-field processes for which vertical variation is important (1D and 2D models are not adequate); and (f) any numerical models need to be calibrated and/or validated for the study site before the results may be used for predictions.

Author Contributions: Conceptualization, Y.G.L. and B.P.G.; methodology, Y.G.L. and J.H.; software, Y.G.L. and J.H.; validation, Y.G.L. and J.H.; writing—original draft preparation, Y.G.L. All authors have read and agreed to the published version of the manuscript.

Funding: This research was funded by the Science and Technology Office, U.S. Bureau of Reclamation.

Data Availability Statement: Data sharing is not applicable.

Conflicts of Interest: The authors declare no conflicts of interest. B.P.G. is employed by the company Stantec Inc. The remaining authors declare that the research was conducted in the absence of any commercial or financial relationships that could be construed as a potential conflict of interest.

References

1. Annandale, G.W.; Morris, G.; Karki, P. Extending the life of reservoirs: Sustainable sediment management for dams and run-of-river hydropower. In *Directions in Development*; World Bank Group: Washington, DC, USA, 2016.
2. Moriasi, D.N.; Steiner, J.L.; Duke, S.E.; Starks, P.J.; Verser, A.J. Reservoir sedimentation rates in the Little Washita River experimental watershed, Oklahoma: Measurement and controlling factors. *J. Am. Water Resour. Assoc.* **2018**, *54*, 1011–1023. [CrossRef]
3. Shelley, J.; Kenney, M.; Layzell, A.; Brown, T. *Contribution of Two Eroding Banks to Multipurpose Pool Sedimentation at a Midwestern Reservoir*; Rep. No. ERDC/TN RSM-20-6; US Army Engineer Research and Development Center: Vicksburg, MS, USA, 2020.
4. Shelley, J.; Boywer, M.; Granet, J.; Williams, A. *Environmental Benefits of Restoring Sediment Continuity to the Kansas River*; Rep. No. ERDC/CHL CHETN-XIV-50; US Army Engineer Research and Development Center: Vicksburg, MS, USA, 2016.
5. Morris, G.L.; Fan, J. *Reservoir Sedimentation Handbook: Design and Management of Dams, Reservoirs, and Watersheds for Sustainable Use*; McGraw Hill Professional: New York, NY, USA, 1998.
6. Annandale, G.W. Reservoir conservation and sediment management. In *Water Week*; Water Bank Group: Washington, DC, USA, 2001.
7. White, R. *Evacuation of Sediments from Reservoirs*; Thomas Telford: London, UK, 2001.
8. Basson, G.R. Management of siltation in existing and new reservoirs. General Report. In Proceedings of the 23rd Congress of the International Commission on Large Dams, Brasilia, Brazil, 25–29 May 2009.
9. Wisser, D.; Frolking, S.; Hagen, S.; Bierkens, M.F.P. Beyond peak reservoir storage? A global estimate of declining water storage capacity in large reservoirs. *Water Resour. Res.* **2013**, *49*, 5732–5739. [CrossRef]
10. Huang, J.; Greimann, B.; Kimbrel, S. *Numerical Modeling and Analysis of Sediment Transport at Paonia Reservoir: SRH-1D Model Update, Calibration, Short-Term Prediction, and Long-Term Prediction*; Report No. SRH-2016-30; Sedimentation and River Hydraulics Group, Technical Service Center, U.S. Bureau of Reclamation: Denver, CO, USA, 2016.
11. Huang, J.; Greimann, B.; Kimbrel, S. Simulation of Sediment Flushing in Paonia Reservoir of Colorado. *J. Hydraul. Eng.* **2019**, *145*, 06019015. [CrossRef]
12. Shelley, J.; Hotchkiss, R.H.; Boyd, P.; Gibson, S. Discharging Sediment Downstream: Case Studies in Cost Effective, Environmentally Acceptable Reservoir Sediment Management in the United States. *J. Water Resour. Plann. Manag.* **2022**, *148*, 05021028. [CrossRef]
13. Shen, H.W. Flushing sediment through reservoirs. *J. Hydraul. Res.* **2010**, *37*, 743–757. [CrossRef]
14. Randle, T.J.; Bountry, J.A.; Ritchie, A.; Wille, K. Large-scale dam removal on the Elwha River, Washington, USA: Erosion of reservoir sediment. *Geomorphology* **2015**, *246*, 709–728. [CrossRef]
15. Madadi, M.R.; Rahimpour, M.; Qaderi, K. Improving the Pressurized Flushing Efficiency in Reservoirs: An Experimental Study. *Water Resour. Manag.* **2017**, *31*, 4633–4647. [CrossRef]
16. Atkinson, E. *The Feasibility of Flushing Sediment from Reservoirs*; Rep. No. OD137; HR Wallingford: Wallingford, UK, 1996.
17. Chaudhry, M.A. Worldwide experience of sediment flushing through reservoirs. *Mehran Univ. Res. J. Eng. Technol.* **2012**, *31*, 395–408.
18. Batalla, R.J.; Vericat, D. Hydrological and sediment transport dynamics of flushing flows: Implications for management in large Mediterranean rivers. *River Res. Appl.* **2009**, *25*, 297–314. [CrossRef]
19. Kantoush, S.A.; Sumi, T.; Suzuki, T.; Murasaki, M. Impacts of sediment flushing on channel evolution and morphological processes: Case study of the Kurobe River, Japan. In *River Flow, Proceedings of the 5th River Flow Conference, Braunschweig, Germany, 8–10 June 2010*; Dittrich, A., Koll, K., Aberle, J., Geisenhainer, P., Eds.; Bundesanstalt für Wasserbau: Karlsruhe, Germany, 2020; pp. 1165–1173.
20. Antoine, G.; Camenen, B.; Jodeau, M.; Nemery, J.; Esteves, M. Downstream erosion and deposition dynamics of fine suspended sediments due to dam flushing. *J. Hydrol.* **2020**, *585*, 124763. [CrossRef]
21. Sumi, T. Evaluation of Efficiency of Reservoir Sediment Flushing in Kurobe River. In Proceedings of the 4th International Conference on Scour and Erosion (ICSE-4), Tokyo, Japan, 5–7 November 2008; pp. 608–613.
22. Kondolf, G.M.; Gao, Y.; Annandale, G.W.; Morris, G.L.; Jiang, E.; Zhang, J.; Cao, Y.; Carling, P.; Fu, K.; Guo, Q. Sustainable sediment management in reservoirs and regulated rivers: Experiences from five continents. *Earth's Future* **2014**, *2*, 256–280. [CrossRef]
23. Dahl, T.A.; Ramos-Villanueva, M. *Overview of Historical Reservoir Flushing Events and Screening Guidance*; Coastal and Hydraulics Laboratory, U.S. Army Engineer Waterways Experiment Station: Vicksburg, MS, USA, 2019.
24. Anari, R.; Hotchkiss, R.H.; Langendoen, E.J. Elements for the Successful Computer Simulation of Sediment Management Strategies for Reservoirs. *Water* **2020**, *12*, 714. [CrossRef]

25. Lai, J.S.; Shen, H.W. Flushing sediment through reservoirs. *J. Hydraul. Res.* **1996**, *34*, 237–255. [CrossRef]
26. Shahmirzadi, M.M.; Dehghani, A.A.; Meftahh, M.; Mosaedi, A. Experimental investigation of pressure flushing technique in reservoir storages. *Water Geosci.* **2010**, *1*, 132–137.
27. Chamoun, S.; De Cesare, G.; Schleiss, A.J. Managing reservoir sedimentation by venting turbidity currents: A review. *Int. J. Sediment Res.* **2016**, *31*, 195–204. [CrossRef]
28. Annandale, G.W. *Developments in Water Science-Reservoir Sedimentation*; BV Rand Afrikaans University: Johannesburg, South Africa; Elsevier Science Publishers: Amsterdam, The Netherlands, 1987.
29. Ackers, P.; Thompson, G. Reservoir sedimentation and influence of flushing. In *Sediment Transport in Gravel-Bed Rivers*; John Wiley and Sons: New York, NY, USA, 1987.
30. Fan, J.; Morris, G.L. Reservoir sedimentation. II: Reservoir desiltation and long-term storage capacity. *J. Hydraul. Eng.* **1992**, *118*, 370–384. [CrossRef]
31. Kantoush, S.A. Experimental Study on the Influence of the Geometry of Shallow Reservoirs on Flow Patterns and Sedimentation by Suspended Sediments. Ph.D. Thesis, EPFL, Lausanne, Switzerland, 2008.
32. Lai, Y.G.; Greimann, B.P. *Modeling of Cherry Creek Reservoir Pressure Flush*; Report No. ENV-2020-83; Technical Service Center, U.S. Bureau of Reclamation: Denver, CO, USA, 2020.
33. Shammaa, Y.; Zhu, D.Z.; Rajaratnam, N. Flow upstream of orifices and sluice gates. *J. Hydraul. Eng.* **2005**, *131*, 127–133. [CrossRef]
34. Powell, D.N.; Khan, A. Scour upstream of a circular orifice under constant head. *J. Hydraul. Res.* **2012**, *50*, 28–34. [CrossRef]
35. Powell, D.N.; Khan, A. Flow field upstream of an orifice under fixed bed and equilibrium scour conditions. *J. Hydraul. Eng.* **2015**, *141*, 04014076. [CrossRef]
36. Fathi-Moghadam, M.; Emamgholizadeh, S.; Bina, M.; Ghomeshi, M. Physical modelling of pressure flushing for desilting of non-cohesive sediment. *J. Hydraul. Res.* **2010**, *48*, 509–514. [CrossRef]
37. Kamble, S.A.; Kunjeer, P.S.; Sureshkumar, B.; Isaac, N. Hydraulic model studies for estimating scour cone development during pressure flushing of reservoirs. *ISH J. Hydraul. Eng.* **2017**, *24*, 337–344. [CrossRef]
38. Emamgholizadeh, S.; Fathi-Moghadam, M. Pressure flushing of cohesive sediment in large dam reservoirs. *J. Hydrol. Eng.* **2014**, *19*, 674–681. [CrossRef]
39. Hajikandi, H.; Vosoughi, H.; Jamali, S. Comparing the Scour Upstream of Circular and Square Orifices. *Int. J. Civ. Eng.* **2018**, *16*, 1145–1156. [CrossRef]
40. Ford, D.; Johnson, M. *An Assessment of Reservoir Density Currents and Inflow Processes*; Technical Report E-83-7; U.S. Army Engineer Waterways Experiment Station: Vicksburg, MS, USA, 1983.
41. Imberger, J.; Patterson, J.; Hebbert, B.; Loh, I. Dynamics of reservoir of medium size. *J. Hydraul. Div.* **1978**, *104*, 725–743. [CrossRef]
42. Simoes, F.J.M. *User's Manual for DCURL—A Model for Routing Density Currents in Reservoirs and Lakes*; Technical Service Center, Bureau of Reclamation: Denver, CO, USA, 1999.
43. Akiyama, J.; Stefan, H.G. Prediction of turbidity currents in reservoirs and coastal regions. In Proceedings of the Third Symposium River Sedimentation, Jackson, MS, USA, 31 March–4 April 1986; pp. 1295–1305.
44. Singh, B.; Shah, C.R. Plunging phenomenon of density currents in reservoirs. *Houille Blanche* **1971**, *26*, 341–352. [CrossRef]
45. Jirka, G.H.; Watanabe, M. Thermal Structure of Cooling Ponds. *J. Hydraul. Div.* **1980**, *106*, 701–715. [CrossRef]
46. Fleenor, W.E. Effects and Control of Plunging Inflows on Reservoir Hydrodynamics and Downstream Releases. Ph.D. Thesis, Civil and Environmental Engineering, University of California, Davis, CA, USA, 2001.
47. Parker, G.; Fukushima, Y.; Pantin, H.M. Self-accelerating turbidity currents. *J. Fluid Mech.* **1986**, *171*, 145–181. [CrossRef]
48. Ashida, K.; Egashira, S. Hydraulic characteristics of thermocline in reservoirs. In Proceedings of the XVII Congress of the IAHR, Baden, Germany, 15–19 August 1977; Volume 2, pp. 33–40.
49. Imberger, J.; Patterson, J.C. A Dynamic Reservoir Simulation Model–DYRESM: 5. In *Transport Models for Inland and Coastal Waters*; Fischer, H.B., Ed.; Academic Press: Berkeley, CA, USA, 1981; pp. 310–361.
50. Teal, M.J.; Bountry, J.; Pridal, D. *Modeling Sediment Movement in Reservoirs*; U.S. Society on Dams: Denver, CO, USA, 2015; p. 25.
51. Morris, G.L.; Hu, G. HEC-6 modeling of sediment management in Loíza. Hydraulic Engineering: Saving A Threatened Resource. In *Search of Solutions, Proceedings of the Hydraulic Engineering Sessions at Water Forum, Baltimore, Maryland, 2–6 August 1992*; ASCE: Reston, VA, USA, 1992; pp. 630–635.
52. Chang, H.H.; Harrison, L.L.; Lee, W.; Tu, S. Numerical modeling for sediment-pass-through reservoirs. *J. Hydraul. Eng.* **1996**, *122*, 381–388. [CrossRef]
53. Liu, J.; Minami, S.; Otsuki, H.; Liu, B.; Ashida, K. Prediction of concerted sediment flushing. *J. Hydraul. Eng.* **2004**, *130*, 1089–1096. [CrossRef]
54. Ahn, J.; Yang, C.T.; Pridal, D.B.; Remus, J.I. Numerical modeling of sediment flushing from Lewis and Clark Lake. *Int. J. Sediment Res.* **2013**, *28*, 182–193. [CrossRef]
55. Guertault, L.; Camenen, B.; Paquier, A.; Peteuil, C. 1D Modelling of fine sediments dynamics in a dam reservoir during a flushing event. In Proceedings of the 7th International Conference on Fluvial Hydraulics, Lausanne, Switzerland, 3–5 September 2014; CRC Press: Lausanne, Switzerland, 2014; pp. 147–154.
56. Boyd, P.M.; Gibson, S.A. *Applying 1D Sediment Models to Reservoir Flushing Studies: Measuring, Monitoring, and Modeling the Spencer Dam Sediment Flush with HEC-RAS*; Coastal and Hydraulics Laboratory, U.S. Army Engineer Waterways Experiment Station: Vicksburg, MS, USA, 2016.

57. Gibson, S.; Crain, J. *Modeling Sediment Concentrations during a Drawdown Reservoir Flush: Simulating the Fall Creek Operations with HEC-RAS*; ERDC/TN RSM-19-7; U.S. Army Engineer Research and Development Center: Vicksburg, MS, USA, 2019. [CrossRef]
58. Brignoli, M.L. Improving Sustainability of Sediment Management in Alpine Reservoirs: Control of Sediment Flushing Operations to Mitigate Downstream Environmental Impacts. Ph.D. Thesis, Università degli Studi dell'Insubria, Como, Italiy, 2017.
59. Olsen, N.R.B. Two-dimensional numerical modelling of flushing processes in water reservoirs. *J. Hydraul. Res.* **1999**, *37*, 3–16. [CrossRef]
60. Scheuerlein, H.; Tritthart, M.; Nunez-Gonzalez, F. Numerical and physical modeling concerning the removal of sediment deposits from reservoirs. In *Hydraulics of Dams and River Structures, Proceedings of the International Conference on Hydraulics of Dams and River Structures, Tehran, Iran, 26–28 April 2004*; Yazdandoost, F., Attari, J., Eds.; CRC Press: London, UK, 2004; pp. 245–254. [CrossRef]
61. Dewals, B.; Erpicum, S.; Archambeau, P.; Detrembleur, S.; Fraikin, C.; Pirotton, M. Large scale 2D numerical modelling of reservoirs sedimentation and flushing operations. In Proceedings of the 9th International Symposium on River Sedimentation, Yichang, China, 18–21 October 2004.
62. Boeriu, P.; Roelvink, D.; Mulatu, C.A.; Thilakasiri, C.N.; Moldovanu, A.; Margaritescu, M. Modeling the flushing process of reservoirs. *Rom. Rev. Precis. Mech.* **2011**, *40*, 54.
63. Chen, C.-N.; Tsai, C.-H. Estimating sediment flushing efficiency of a shaft spillway pipe and bed evolution in a reservoir. *Water* **2017**, *9*, 924. [CrossRef]
64. Iqbal, M.; Ghumman, A.R.; Haider, S.; Hashmi, H.N.; Khan, M.A. Application of Godunov type 2D model for simulating sediment flushing in a reservoir. *Arab. J. Sci. Eng.* **2019**, *44*, 4289–4307. [CrossRef]
65. Lai, J.S.; Shen, H.W. An experiment study on reservoir drawdown flushing. *Int. J. Sediment Res.* **1995**, *10*, 19–36.
66. Chaudhary, H.P.; Isaac, N.; Tayade, S.B.; Bhosekar, V.V. Integrated 1D and 2D numerical model simulations for flushing of sediment from reservoirs. *ISH J. Hydraul. Eng.* **2019**, *25*, 19–27. [CrossRef]
67. Stillwater Sciences. *Matilija Dam Removal 65% Design Subtask 2.9: Hydraulic and Sediment Transport Modeling in SRH-2D*; Prepared by Stillwater Sciences, Berkeley, California; AECOM Technical Services, Inc.: Los Angeles, CA, USA, 2021.
68. Wang, H.-W.; Tsai, B.-S.; Hwang, C.; Chen, G.W.; Kuo, W.-C. Efficiency of the Drawdown Flushing and Partition Desilting of a Reservoir in Taiwan. *Water* **2020**, *12*, 2166. [CrossRef]
69. Wells, S.A.; Gordon, J.A. A Three-Dimensional Field Evaluation and Analysis of Water Quality–Implications of the Third Dimension. In *Proceedings of the Symposium on Surface Water Impoundments, Minneapolis, MN, USA, 2–5 June 1980*; Stefan, A.G., Ed.; ASCE: Reston, VA, USA, 1980; pp. 644–653.
70. Cole, T.M.; Buchak, E.M. *CE-QUAL-W2: A Two Dimensional, Laterally Averaged, Hydrodynamic and Water-Quality Model, Version 2.0, User's Manual*; Instruction Rep. No. EL-95-1; U.S. Army Engineer Waterways Experiment Station: Vicksburg, MS, USA, 1995.
71. Ahlfeld, D.; Joaquin, A.; Tobiason, J. Case study: Impact of reservoir stratification on interflow travel time. *J. Hydraul. Eng.* **2003**, *129*, 966–975. [CrossRef]
72. Bender, M.D.; Kubitschek, J.P.; Vermeyen, T.B. *Temperature Modeling of Folsom Lake, Lake Natoma, and the Lower American River*; Project Report; Technical Service Center, Bureau of Reclamation: Denver, CO, USA, 2007.
73. Bradford, S.F.; Katopodes, N.D. Hydrodynamics of turbidity underflows. I: Formulation and numerical analysis. *J. Hydraul. Eng.* **1999**, *125*, 1006–1015. [CrossRef]
74. Groenenberg, R.; Sloff, K.; Weltje, G.J. A high-resolution 2-DH numerical scheme for process-based modeling of 3-D turbidite fan stratigraphy. *Comput. Geosci.* **2009**, *35*, 1686–1700. [CrossRef]
75. Lai, Y.G.; Huang, J.; Wu, K. Reservoir Turbidity Current Modeling with a Two-Dimensional Layer-Averaged Model. *J. Hydraul. Eng.* **2015**, *141*, 04015029. [CrossRef]
76. Lai, Y.G.; Wu, K. A numerical modeling study of sediment bypass tunnels at Shihmen reservoir, Taiwan. *Int. J. Hydrol.* **2018**, *2*, 00056. [CrossRef]
77. Lai, Y.G.; Wu, K. A Three-Dimensional Flow and Sediment Transport Model for Free-Surface Open Channel Flows on Unstructured Flexible Meshes. *Fluids* **2019**, *4*, 18. [CrossRef]
78. Ghoreishi, S.H.; Tabatabai, M.M.R. Model study reservoir flushing. *J. Water Sci. Res. JWSR* **2010**, *2*, 1–8.
79. Haun, S.; Olsen, N.R.B. Three-dimensional numerical modelling of reservoir flushing in a prototype scale. *Int. J. River Basin Manag.* **2012**, *10*, 341–349. [CrossRef]
80. Olsen, N.R.B.; Haun, S. *Numerical Modelling of Bank Failures during Reservoir Draw-Down*; EDP Sciences: Les Ulis, France, 2018.
81. Saam, L.; Mouris, K.; Wieprecht, S.; Haun, S. Three-dimensional numerical modelling of reservoir flushing to obtain long-term sediment equilibrium. In Proceedings of the 38th IAHR World Congress, Panama City, Panama, 1–6 September 2019.
82. Esmaeili, T.; Sumi, T.; Kantoush, S.A.; Kubota, Y.; Haun, S. Numerical study on flushing channel evolution, case study of Dashidaira reservoir, Kurobe river. *J. Jpn. Soc. Civ. Eng.* **2015**, *71*, 115–120. [CrossRef] [PubMed]
83. Esmaeili, T.; Sumi, T.; Kantoush, S.A.; Kubota, Y.; Haun, S.; Rüther, N. Three-dimensional numerical study of free-flow sediment flushing to increase the flushing efficiency: A case-study reservoir in Japan. *Water* **2017**, *9*, 900. [CrossRef]
84. Ermilov, A.A.; Baranya, S.; Rüther, N. *Numerical Simulation of Sediment Flushing in Reservoirs with TELEMAC*; Norwegian University of Science and Technology: Trondheim, Norway, 2018.
85. Castillo, L.; Carrillo, J.; Álvarez, M.A. Complementary methods for determining the sedimentation and flushing in a reservoir. *J. Hydraul. Eng.* **2015**, *141*, 05015001–05015010. [CrossRef]

86. Lai, Y.G.; Liu, X.; Bombardelli, F.A.; Song, Y. Three-Dimensional Numerical Modeling of Local Scour: A State-of-the-Art Review and Perspective. *J. Hydraul. Eng.* **2022**, *148*, 03122002. [CrossRef]
87. USBR (U.S. Bureau of Reclamation). *Hydrology, Hydraulics and Sediment Transport Studies for the Secretary's Determination on Klamath River Dam Removal and Basin Restoration*; Technical Report No. SRH-2011-02, Prepared for Mid-Pacific Region; US Bureau of Reclamation, Technical Service Center: Denver, CO, USA, 2011.
88. GEC (Gathard Engineering Consulting). *Klamath River Dam and Sediment Investigation*; Technical Report; GEC (Gathard Engineering Consulting): Seattle, WA, USA, 2006; 96p.
89. PWA (Phillip Williams and Associates, Ltd.). *A River Once More: Restoring the Klamath River Following Removal of the Iron Gate, Copco, and J. C. Boyle Dams*; California State Coastal Conservancy and California Department of Fish and Game: Oakland, CA, USA, 2009.
90. Armstrong, B. *Cherry Creek Reservoir Sediment Erosion Testing Results*; Technical Memorandum No. 8530-2017-22; US Bureau of Reclamation: Denver, CO, USA, 2017.

fluids

Article

Comparison of Two Hydrological Models, HEC-HMS and SWAT in Runoff Estimation: Application to Huai Bang Sai Tropical Watershed, Thailand

Imiya M. Chathuranika [1], Miyuru B. Gunathilake [2,3], Pavithra K. Baddewela [4], Erandi Sachinthanie [5], Mukand S. Babel [6], Sangam Shrestha [6], Manoj K. Jha [7] and Upaka S. Rathnayake [1,*]

1 Department of Civil Engineering, Faculty of Engineering, Sri Lanka Institute of Technology, Malabe 10115, Sri Lanka
2 Hydrology and Aquatic Environment, Division of Environmental and Natural Resources, Norwegian Institute of Bioeconomy and Research, 1431 Ås, Norway
3 Water, Energy and Environmental Engineering Research Unit, Faculty of Technology, University of Oulu, 90014 Oulu, Finland
4 Department of Transport and Logistics Management, Faculty of Engineering, University of Moratuwa, Katubedda 10400, Sri Lanka
5 Department of Physics, Faculty of Science, University of Peradeniya, Peradeniya, Kandy 20400, Sri Lanka
6 Water Engineering and Management, School of Engineering and Technology, Asian Institute of Technology, Pathum Thani 12120, Thailand
7 Department of Civil, Architectural and Environmental Engineering, North Carolina Agricultural and Technical State University, Greensboro, NC 27411, USA
* Correspondence: upaka.r@sliit.lk or upakasanjeewa@gmail.com; Tel.: +94-71-988-3318

Citation: Chathuranika, I.M.; Gunathilake, M.B.; Baddewela, P.K.; Sachinthanie, E.; Babel, M.S.; Shrestha, S.; Jha, M.K.; Rathnayake, U.S. Comparison of Two Hydrological Models, HEC-HMS and SWAT in Runoff Estimation: Application to Huai Bang Sai Tropical Watershed, Thailand. *Fluids* **2022**, *7*, 267. https://doi.org/10.3390/fluids7080267

Academic Editors: Jaan H. Pu, Prashanth Reddy Hanmaiahgari, Manish Pandey and Mohammad Amir Khan

Received: 25 May 2022
Accepted: 30 July 2022
Published: 4 August 2022

Abstract: In the present study, the streamflow simulation capacities between the Soil and Water Assessment Tool (SWAT) and the Hydrologic Engineering Centre-Hydrologic Modelling System (HEC-HMS) were compared for the Huai Bang Sai (HBS) watershed in northeastern Thailand. During calibration (2007–2010) and validation (2011–2014), the SWAT model demonstrated a Coefficient of Determination (R^2) and a Nash Sutcliffe Efficiency (NSE) of 0.83 and 0.82, and 0.78 and 0.77, respectively. During the same periods, the HEC-HMS model demonstrated values of 0.80 and 0.79, and 0.84 and 0.82. The exceedance probabilities at 10%, 40%, and 90% were 144.5, 14.5, and 0.9 mm in the flow duration curves (FDCs) obtained for observed flow. From the HEC-HMS and SWAT models, these indices yielded 109.0, 15.0, and 0.02 mm, and 123.5, 16.95, and 0.02 mm. These results inferred those high flows were captured well by the SWAT model, while medium flows were captured well by the HEC-HMS model. It is noteworthy that the low flows were accurately simulated by both models. Furthermore, dry and wet seasonal flows were simulated reasonably well by the SWAT model with slight under-predictions of 2.12% and 13.52% compared to the observed values. The HEC-HMS model under-predicted the dry and wet seasonal flows by 10.76% and 18.54% compared to observed flows. The results of the present study will provide valuable recommendations for the stakeholders of the HBS watershed to improve water usage policies. In addition, the present study will be helpful to select the most appropriate hydrologic model for humid tropical watersheds in Thailand and elsewhere in the world.

Keywords: Hydrologic Engineering Center-Hydrologic Modeling System (HEC-HMS); Huai Bang Sai (HBS) watershed; streamflow; Soil and Water Assessment Tool (SWAT)

1. Introduction

Water resource management and operational hydrology require reliable predictions of water balance components including runoff, evapotranspiration, infiltration, and groundwater flow. Hydrologic models are used for the planning of water resources [1–3], for flood predictions [4–6], to understand the hydrology due to changes in land use and climate [7,8], for water quality monitoring [9], to formulate aquifer recharge management strategies [10], to design hydraulic infrastructure [11], for ecological restoration design [12], etc. The evolvement of computer technology and programming has benefited researchers, academia, and commercial-based companies to develop different software to simulate watershed processes. The Soil and Water Assessment Tool (SWAT) [13], the Hydrologic Engineering Center-Hydrologic Modeling System (HEC-HMS) [14], the Hydrologiska Byråns Vattenbalansavdelning (HBV light) model [15], the J2000 model [16], the GR4J model [17], the Hydrologic Simulation Program FORTRAN (HSPF) model [18], and the MIKE-SHE model [19] are some of the widely used hydrologic models used in different regions of the world today [20].

In this study, the widely used SWAT and the HEC-HMS hydrologic models were used to compare the streamflow simulation capacities in the Huai Bang Sai (HBS) watershed, which flows into the greater Mekong River. These two hydrologic models have been frequently used in tropical regions by many researchers [21,22]. The HEC-HMS model is a lump-based model, while the SWAT model is a semi-distributed model. Lump-based models consider the total basin as a "single homogeneous element". On the other hand, the semi-distributed models discretize the drainage basin into homogeneous units of landform, soil, and topography of the watershed [23]. Hence, in the present study, the hypothesis tested was whether the spatial discretization of the watershed through different hydrologic models had an impact on the response to the streamflow simulation.

It is noteworthy that hydrologic models that were developed in environments were later applied in watersheds that had different climatic and watershed characteristics than they were originally developed for. For instance, Gunathilake et al. [24] and Chathuranika et al. [25] used the HEC-HMS model to simulate streamflow in the Seethawaka and Nilwala watersheds in Sri Lanka. However, the HEC-HMS was developed for the temperate climatic conditions of the USA. This model is widely used and was developed by the United States Army Corps of Engineers (USACE). In another study, the SWAT model developed by the Agricultural Research Services of the United States Department of Agriculture, (originally developed for the temperate climatic conditions in the USA) was applied in the sub-humid tropical Indian region by Shekar and Vinay [26]. Phomcha et al. [27] applied the SWAT model to the Lam Sonthi tropical watershed in Thailand and found that the model was able to simulate sediments accurately. Phomcha et al. [27] stated that although the SWAT model can be simulated with a small amount of input data, some of its model algorithms are inefficient for tropical watersheds. Supakosol and Boonrawd [28] and Rossi et al. [29] used the SWAT model to simulate streamflow in the Nong Han lake basin and Mae Nam Chi in Yasothon in northeastern Thailand. Furthermore, the HEC-HMS model was used by Kuntiyawichai et al. [30] to simulate flow into the Ubol Ratana reservoir in northeastern Thailand. These results showcased that both these models are capable of simulating streamflow in watersheds of northeastern Thailand.

Shekar and Vinay [26] conducted a comparison study for HEC-HMS and SWAT for a river basin in India and demonstrated that the SWAT model outperformed the HEC-HMS model. On the contrary, Aliye et al. [31] showcased that the HEC-HMS model performed better than the SWAT model for the Ethiopian Rift valley lake basin. A similar comparison study was carried out in the Srepok river basin in Vietnam by Khoi [32]. The results of this study inferred that the SWAT model outperformed the HEC-HMS. However, Ismail et al. [33] reported that the HEC-HMS model outperformed the SWAT model in the tropical Bernam river basin in Malaysia. The above studies indicate that the hydrological performance of models should be determined for the individual watershed for its suitability.

As it was stated in the preceding paragraphs, streamflow computation is highly important to major streams and rivers in a watershed. Even though the field measurements are highly accurate, the logistic difficulties might devalue the field measurements. Hourly instrument usage is not an easy task for a major catchment. Therefore, computational modeling is convenient in this situation. However, the streamflow calculations from different computational models can have some mismatches to the ground-measured streamflows. Therefore, identification of the best suited hydrologic model for a particular watershed is important in the context of streamflows.

This study presents the comparative analysis of SWAT and HEC-HMS hydrologic models in the HBS of northeastern Thailand. The overall idea of the comparative analysis is to investigate the capabilities of streamflow calculations by the two widely used computational models. Both models were calibrated between 2007 and 2010 (for 4 years) and validated from 2011 to 2014 (for 4 years) at the same discharge station. The hydrologic model statistical performances for both models were examined using the Coefficient of Determination (R^2) and the Nash Sutcliffe Efficiency (NSE). Error in Peak flow, Percentage Error in Volume (PEV) were calculated to compare the performances of both models. These parameters were investigated to determine the streamflow computational capabilities. In the present study, the hypothesis tested was whether the spatial discretization of the watershed through different hydrologic models had an impact on the response to the streamflow simulation. The results of the present study will essentially provide valuable recommendations and insights to select the most appropriate hydrologic model simulating rainfall–runoff processes of the tropical humid northeastern part of Thailand. The research findings would also be useful for the policymakers to take necessary actions to achieve sustainable water resource management.

2. The Study Area and Data Required

The HBS watershed is a sub-catchment of the greater Mekong River watershed, which is in northeastern Thailand. This region is situated at the Thailand-Lao PDR border (neighboring the Mekong River) on the eastern side. The HBS drains an area of 1340 km^2 before joining the Mekong River near Mukdahan town (refer to Figure 1). The HBS watershed lies between 16°35′ N–16°55′ N and 104°02′ E–104°44′ E. The altitude of the HBS watershed varies between 140 and 640 m above mean sea level (AMSL). This area receives a mean annual precipitation of 1200 mm with the majority of this received during the southwest monsoon season from May to September. Dense deciduous forests are the primary land use type of the area, and it covers nearly 68% of the catchment. Other main land use types in the area are cassava, sugarcane, and rubber plantations (refer to Figure 2a). Forest cover deterioration due to the cultivation of cash crops and the reduction in soil and water quality are identified as the key environmental problems during the last 30 years in the region. This phenomenon has happened due to poor soil and water conservation practices [34]. Hang Chat is the primary soil type in the catchment. It has a loamy sand texture and belongs to the hydrologic soil group C (loamy sand nature) of the "United States Department of Agriculture" (USDA)'s soil classification. Figure 2b represents the soil distribution map of the HBS watershed.

Figure 1 shows the locations of the rain gauges and the flow gauge in the study area. Daily rainfall data were obtained from the "Royal Irrigation Department" (RID), Thailand for Dong Luang, Wan Yai, Tong Khop, Mukdahan, and Huai Ta Poe stations for 8 years (from 2007 to 2014). Daily temperature records were obtained for a similar period from "Thai Meteorological Department" (TMD) from the nearby meteorological stations. Similarly, daily streamflow data for station kh.92 (Ban Kan Luang Dong) were also collected from RID. The land use types for the year 2015 were obtained from the Land Development Department (LDD) of Thailand with a 500 m resolution, and in a 1:50,000 scale. Soil data for 2015 were also acquired from the LDD of Thailand with a 1 km resolution, and in a 1:100,000 scale. "Digital Elevation Model" (DEM) of 30 m resolution was downloaded from the "United States Geological Survey" (USGS) website; https://earthexplorer.usgs.gov/ (accessed on 31 June 2022).

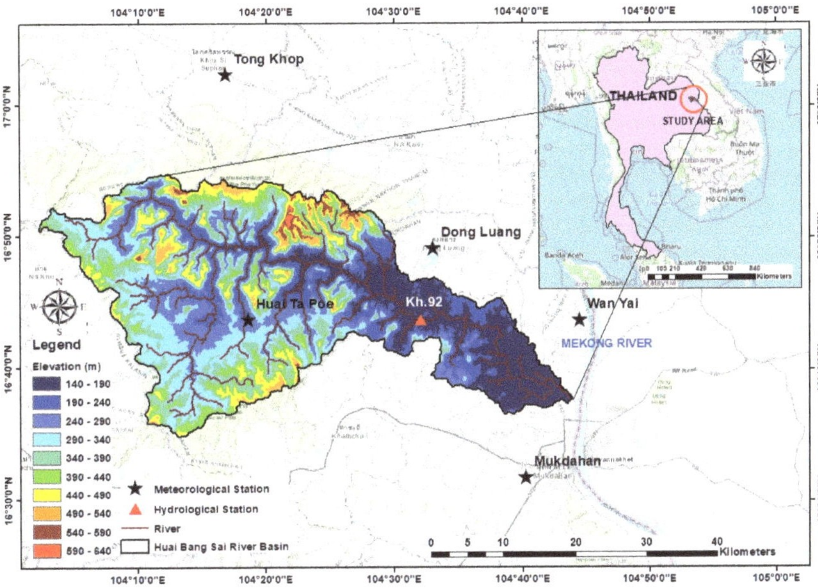

Figure 1. Location and topography map with hydrometeorological stations in Huai Bang Sai watershed in northeastern Thailand.

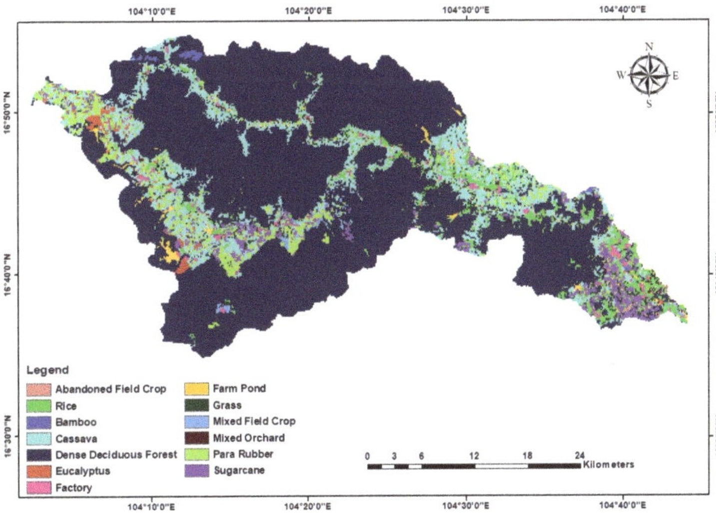

(a)

Figure 2. *Cont.*

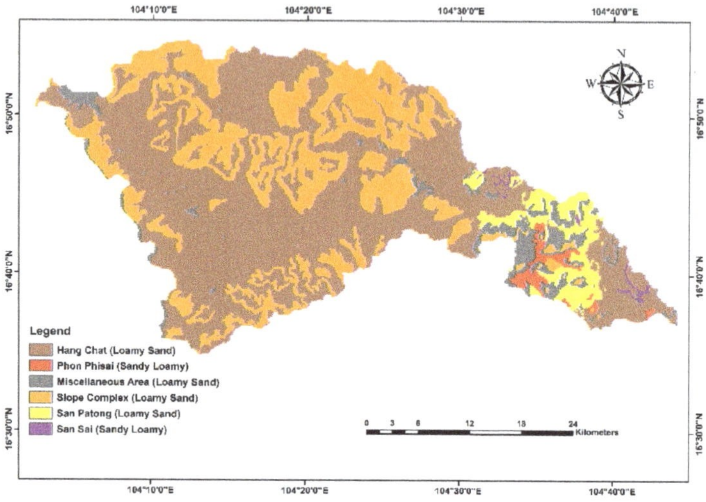

(**b**)

Figure 2. (**a**) Land-use map and (**b**) soil distribution map for the HBS watershed.

3. Rainfall–Runoff Modelling of the HBS Watershed

This study compared the performance of HEC-HMS and SWAT hydrological models to simulate streamflow at the kh.92 hydrological station from 2007 to 2014. For the present study, a HEC-HMS model was developed, and the model was compared with the SWAT model, which was developed previously by Babel et al. [21].

Both the HEC-HMS and SWAT models are capable of continuous simulations [35]. These models simplify water resource systems for ease of understanding of the model behavior. Understanding the components of the hydrological cycle including surface runoff, infiltration, evaporation, transpiration, and precipitation is of prime importance in hydrologic modeling studies.

In the case of the HEC-HMS model, the HEC-GeoHMS and Archydro tools were used to delineate and calculate the physical and drainage characteristics of the watershed. They were used in the HEC-HMS model as inputs for the initial simulation. The whole watershed was delineated into 12 sub-watersheds considering nearly equal surface areas for each sub-watershed. HEC-HMS model development process includes four main components namely, basin model, input data (time series, paired and gridded data), meteorological model, and control specifications [36,37]. The basin model connects sub-watersheds, reaches, junctions, diversions, reservoirs, etc., to create a drainage system [38]. The time interval for a simulation is controlled by control specifications [39]. In this study, climate data were added to the meteorological model to distribute them spatially and temporally over the watershed through the Thiessen polygon method. Time series data of precipitation, temperature, and streamflow data in a daily step were included in the model. The HEC-HMS model was calibrated from 1 January 2007 to 31 December 2010 and validated between 1 January 2011 and 31 December 2014. A sensitivity analysis was conducted by considering the changes in the percentage error in runoff volume (PEV). The normalized objective function (NOF), the Nash Sutcliffe Efficiency (NSE), the percentage of bias (PBIAS), and the Coefficient of Determination (R^2) were used to determine the statistical performance of the HEC-HMS model on a daily and monthly basis. If the simulated values exactly match with the observed values, NOF, NSE, PBIAS, and R^2 would be equal to zero, one, zero percent, and one, respectively. These skill matrices were calculated using the following equations from (1) to (4) [40,41].

$$NOF = \frac{1}{\bar{O}}\sqrt{\frac{1}{n}\sum_{i=1}^{n}(O_i - S_i)^2} \tag{1}$$

$$NSE = 1 - \frac{\sum_{I=1}^{n}(S_i - O_i)^2}{\sum_{I=1}^{n}(O_i - \bar{O})^2} \quad (2)$$

$$PBIAS = \frac{\sum_{i=1}^{n}(S_i - O_i)}{\sum_{i-1}^{n} O_i} \times 100\% \quad (3)$$

$$R^2 = \frac{n\sum O_i \cdot S_i - \sum O_i \cdot S_i}{\left(\sqrt{n(\sum O_i{}^2) - (\sum O_i)^2}\right) \times \left(\sqrt{n(\sum S_i{}^2) - (\sum S_i)^2}\right)} \quad (4)$$

where O_i = observed discharge, S_i = simulated discharge, n = number of observed or simulated data points, and $\bar{O}$ = mean of the observed discharge.

4. Development of Hydrologic Models

4.1. SWAT Model Development

The SWAT model, which operates in daily time steps was developed by the Agricultural Research Services (ARS) division of the USDA [42]. This model is efficient in assessing hydrological processes and non-point source pollution at different spatial scales. The SWAT model divides a watershed into multiple watersheds, and then further discretizes into hydrologic response units (HRUs) consisting of a combination of similar land use, soil, and slope characteristics [43]. HRUs also represent sub-basin area percentages and they are not recognized spatially within a model simulation [44]. The water balance equation (refer to Equation (5)) is the governing equation in the SWAT model [45].

$$SW_t = SW_0 + \sum_{i=1}^{t}\left(R_{day} - Q_{surf} - ET_a - W_{seep} - Q_{gw}\right) \quad (5)$$

where SW_t = final soil water content (mm), SW_0 = initial soil water content (mm), t = time (days), R_{day} = amount of precipitation (mm), Q_{surf} = amount of surface runoff (mm), ET_a = amount of evapotranspiration (mm), W_{seep} = amount of water entering the vadose zone from the soil profile (mm), and Q_{gw} = amount of return flow (mm).

Babel et al. [21] developed the SWAT model for the HBS river basin using the SWAT 2012 version. In this previous study, firstly, the entire watershed was delineated into 7 sub-watersheds in the model setup process. Then, these sub-basins were subdivided into 797 HRUs. Afterward, the model was calibrated from 1 January 2007 to 31 December 2010 (4 years) and validated from 1 January 2011 to 31 December 2014 (4 years). A warm-up period of 3 years (1 January 2004 to 31 December 2006) was considered to equilibrate between various water storages in the model. Streamflows observed at the kh.92 hydrological station were used for hydrologic model development. A sensitivity analysis was conducted through the manual calibration process. SWAT Calibration and Uncertainty Procedures (SWAT-CUP) were initially used to identify the most sensitive parameters to streamflow. In the SWAT model developed through the previous study, surface runoff was predicted by the Soil Conservation Service Curve Number (SCS-CN) method. Moreover, the potential evapotranspiration was calculated by the Hargreaves method.

Initially, the Digital Elevation Model (DEM) of the HBS watershed was delineated into sub watersheds by the watershed delineation tool available in SWAT. Thereafter, the reclassified land use and soil maps were used as input in the SWAT model. Then, the weather data were inserted to run the model. Finally, the observed streamflow data at kh.92 was used to calibrate the model.

4.2. HEC-HMS Model Development

A well-calibrated model depends upon the technical abilities of the hydrological model as well as the quality of the input data. The HEC-HMS model for the HBS watershed was manually calibrated for daily streamflow predictions for the period 2007–2010 by comparing the observed streamflow for the peaks, timing, and runoff volumes.

In order to determine the critical parameters affecting the calibration of the rainfall–runoff model, a sensitivity analysis was carried out. One parameter was varied at a time between −50% and 50% within the augmentations of 10% [4]. This was performed while keeping other hydrological parameters constant until the best agreement between observed streamflow and simulated streamflow was achieved. The soil percolation (mm/hr), impervious percentage (%), soil storage (mm), and groundwater 1 storage (mm) are found to be the most sensitive parameters.

5. Results and Discussion

5.1. Results from the HEC-HMS Model

Figure 3 depicts the "percentage error in runoff volume" (PEV) for the calibration period (2007–2010).

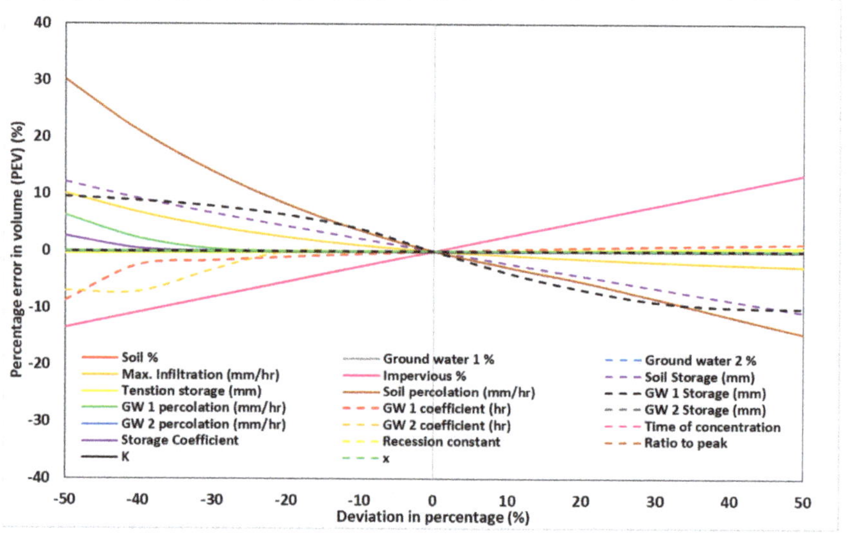

Figure 3. Sensitivity analysis of the HEC-HMS model for the calibration period (2007-2010).

The HEC-HMS model produced a reasonable agreement between observed and simulated discharges on daily and monthly time scales. Figure 4 depicts the daily hydrograph comparison of simulated and observed discharges at kh.92 hydrological station for the calibration (January 2007–December 2010) and validation (January 2011–December 2014) periods. The HEC-HMS model underestimated observed streamflow in certain time periods. During calibration and validation, the model underestimated the water volume by 14.43% and 16.62%, respectively. Optimized values for the HBS watershed are provided in Table 1. The model validation results proved that there was an acceptable agreement between observed and simulated hydrographs for the period between January 2011 and December 2014. Among the different loss methods available in the HEC-HMS model for the present study, the soil and moisture accounting method, which is capable of simulating continuous events, was used. The direct runoff was simulated by the Clark Unit Hydrograph. The recession method was used to simulate baseflow, while flow routing was carried out by the Muskingum method.

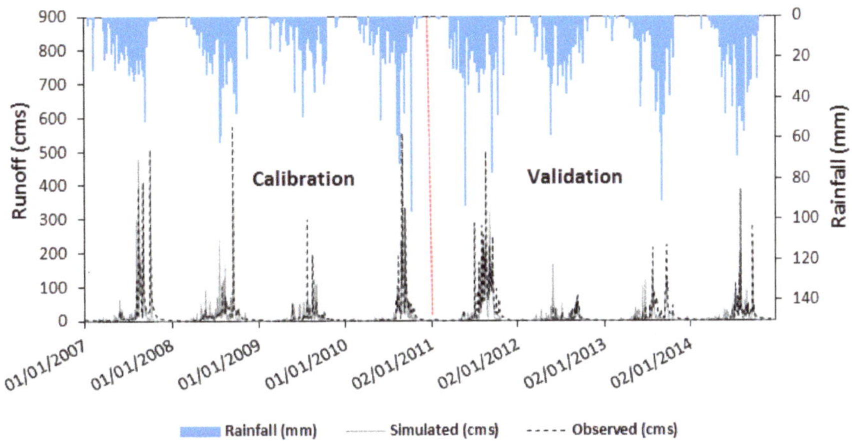

Figure 4. Comparison of daily observed and simulated streamflows at kh.92 for the 2007–2014 period from the HEC-HMS model.

Table 1. Optimized parameters of HEC-HMS model.

Method	Parameter	Unit	Optimized Value
Soil Moisture Accounting	Soil Percentage	%	70
	Groundwater 1	%	45
	Groundwater 2	%	82
	Max. Infiltration	mm/hr	4.5
	Impervious Percentage	%	16.2
	Soil Storage	mm	276
	Tension Storage	mm	30
	Soil Percolation	mm/hr	0.42
	GW 1 Storage	mm	9
	GW 1 Percolation	mm/hr	0.675
	GW 1 Coefficient	hr	120
	GW 2 Storage	mm	100
	GW 2 Percolation	mm/hr	1
	GW 2 Coefficient	hr	100
Clark Unit Hydrograph	Time of Concentration	hr	10
	Storage Coefficient	hr	42
Recession	Initial Discharge	m^3/s	0.1
	Recession Constant		0.35
	Ratio to Peak		0.4
Muskingum	K	hr	0.02
	x		0.3

The statistics obtained during the calibration and validation of the HEC-HMS model are given in Table 2 below. The statistical indicators including NOF, NSE, PBIAS, and R^2 were calculated to evaluate the model performance. Skill metrics for calibration and validation on a daily basis for different metrics including NOFs of 1.52 and 1.58, NSEs of 0.70 and 0.60, PBIAS of 14.44% and 16.63%, and R^2 of 0.70 and 0.55, respectively. Monthly skill performance for calibration and validation presented better performance compared to the daily performance, which was calculated for NOF as 0.71 and 0.80, NSE as 0.79 and 0.82, and R^2 as 0.80 and 0.84. PBIAS demonstrated an underestimation in both calibration and validation periods. Moreover, daily and monthly R^2 values are found to be higher than 0.5, while NSE values were found to be greater than 0.60. The NOF was closer to zero. The obtained results demonstrated that the model is acceptable for use in hydrological studies.

Table 2. Statistical performance criterions during calibration and validation for the HEC-HMS model.

Cluster	Daily				Monthly			
	NOF	NSE	PBIAS	R^2	NOF	NSE	PBIAS	R^2
Calibration (2007–2010)	1.52	0.70	−14.45	0.70	0.71	0.79	−14.45	0.80
Validation (2011–2014)	1.58	0.60	−16.63	0.55	0.80	0.82	−16.63	0.84

5.2. Streamflow Prediction Capacities between the HEC-HMS and SWAT Models

The streamflow predictions of HEC-HMS were compared with that of SWAT [21] for the HBS watershed.

The monthly observed and simulated streamflows for the period 2007–2014 for the HBS watershed using the HEC-HMS and SWAT models are depicted in Figure 5. Both models performed fairly well although with few discrepancies. The times to peak of simulated discharges from the two hydrological models were comparable. The highest observed monthly discharge at kh.92 during the 2007–2014 period happened in August 2011 and this peak discharge was underestimated by the HEC-HMS and SWAT models by 26.62% and 15.98%. Moreover, a reasonable amount of flood peaks were captured from the SWAT model compared to the HEC-HMS model. The total water volume from the 2007–2014 period has been underestimated through the HEC-HMS and SWAT simulations by 17.76% and 12.37%. In addition, the monthly statistical performances for R^2 and NSE obtained by Babel et al. [21] were 0.83 and 0.82 during the calibration (2007–2010) and 0.78 and 0.77 for validation (2011–2014) from the SWAT model. During the same periods, the values obtained for the same metrics from the HEC-HMS model were 0.80 and 0.79 and 0.84 and 0.82. Therefore, the SWAT model statistically performed well during the calibration period (2007–2010).

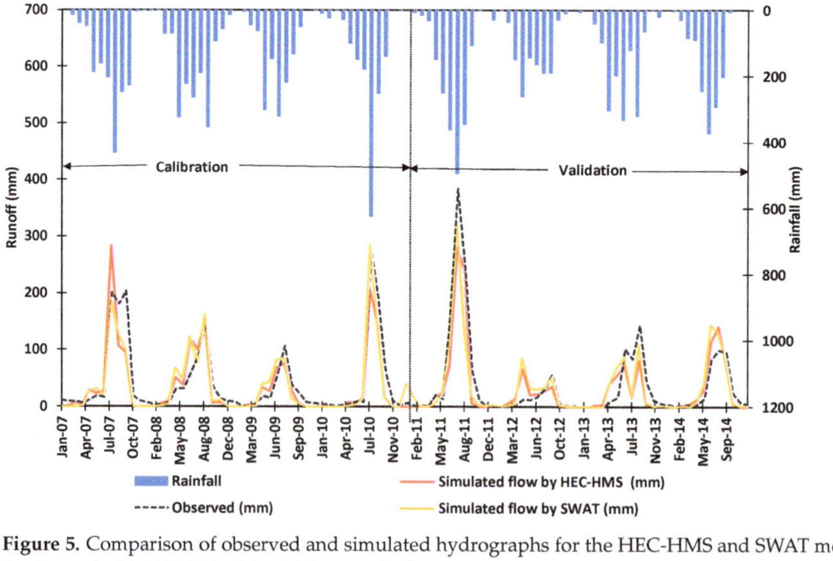

Figure 5. Comparison of observed and simulated hydrographs for the HEC-HMS and SWAT models for the period 2007–2014 of the HBS watershed.

Figure 6 illustrates scatter plots for the HEC-HMS and SWAT models after the optimization for the kh.92 hydrological station considering monthly simulated and observed streamflows under calibration (2007–2010) and validation (2011–2014) periods. Simulated and observed values are well distributed along the uphill and downhill compared to the 1:1 line for both models. Some points can be seen along the 1:1 lines as well. Therefore, predictive capacities for both models show satisfactory agreement during calibration and validation periods. Linear graphs for the calibration period (2007–2010) are very close to the 1:1 line for both models and HEC-HMS shows a better performance compared to the SWAT model if both hydrological models are exactly matched, then the green color linear curve in Figure 6c should be on the 1:1 line. This gap represents the strengths and weaknesses of both models. The HEC-HMS and SWAT models performed similarly while providing a few months of contradictory results. Through visual inspection, it is clear that the SWAT model was able to catch higher monthly flows compared to the HEC-HMS model.

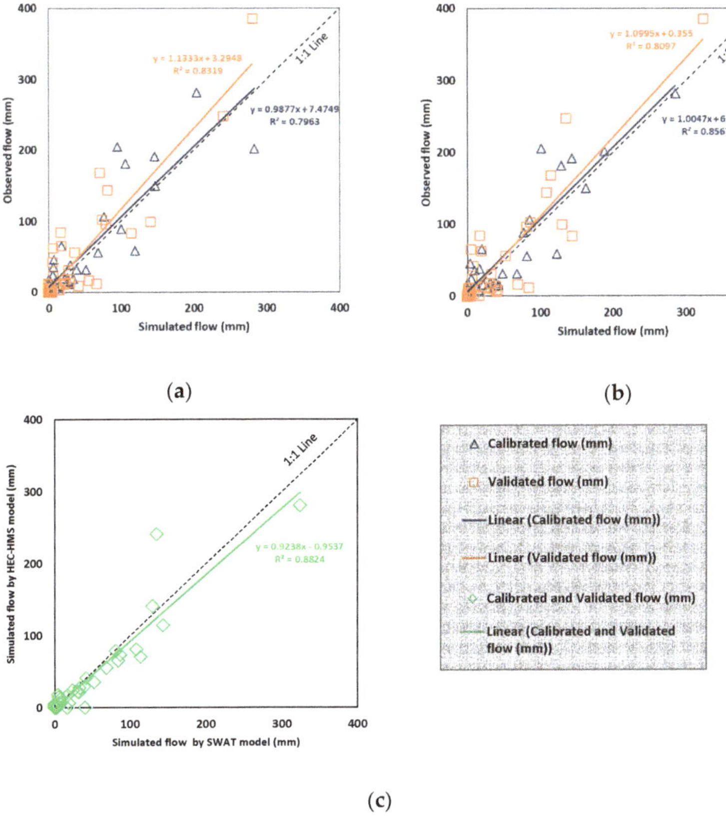

Figure 6. Scatter plots for monthly data comparing between: (**a**) simulated streamflow by HEC-HMS model versus observed streamflow in mm for calibration and validation; (**b**) simulated streamflow by SWAT model versus observed streamflow in mm for calibration and validation; (**c**) simulated streamflow by SWAT model versus simulated streamflow by HEC-HMS during the 2007–2014 period.

Flow duration curves (FDCs) were created to compare high flows (10% exceedance), medium flows (40% exceedance), and low flows (90% exceedance) for simulated monthly discharges through the HEC-HMS and SWAT models and observed monthly discharges at the kh.92 hydrological station in the HBS catchment during the 2007–2014 period. Figure 7 illustrates the magnitude of observed monthly streamflows for 10%, 40%, and 90% exceedance percentages as 144.5, 14.5, and 0.9 mm. Simulated monthly discharges show exceedance probabilities of 10%, 40%, and 90% as 109.0, 15.0, and 0.02 mm for the HEC-HMS model and 123.5, 16.95, and 0.02 mm for the SWAT model, respectively. This infers that the SWAT model can capture high flows compared to the HEC-HMS model for the 2007–2014 period. Medium flows can be obtained through the HEC-HMS model more accurately compared to the SWAT model. Both models provide similar performance for the generation of low flows within the considered duration.

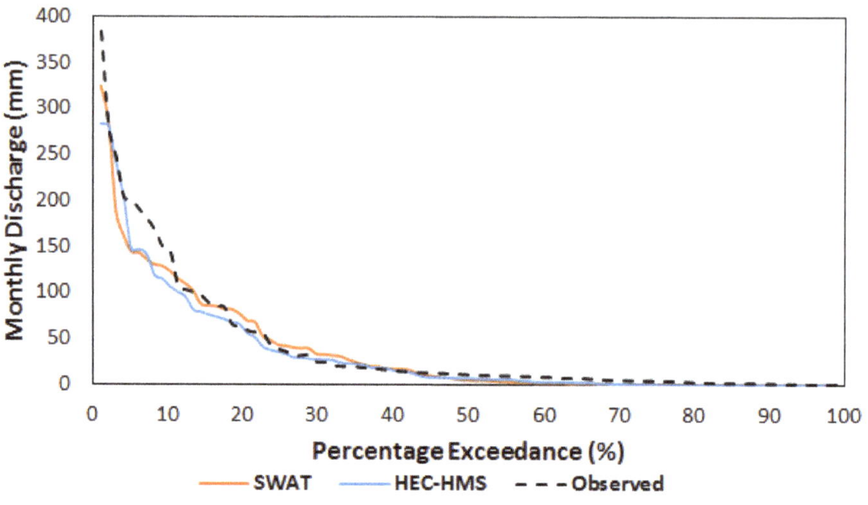

Figure 7. FDCs for monthly simulated streamflows by the HEC-HMS and SWAT models compared to the observed monthly streamflow within the period 2007–2014.

Figure 8 shows the mean seasonal discharge for the HEC-HMS and SWAT models with the observed mean seasonal discharge at kh.92 hydrologic station for the 2007–2014 period. Northeastern Thailand has two major rainfall seasons, namely the dry (November to May) and wet (June to October) seasons. The SWAT and HEC-HMS models performed with under-predictions compared to the observed for dry and wet seasons as 2.12% and 13.52%, and 10.76% and 18.54%, respectively. Therefore, the SWAT model performed moderately well compared with the HEC-HMS model on a seasonal basis for the HBS watershed, Thailand. Lacombe et al. [46] stated that northeastern Thailand receives approximately 80–90% of annual precipitation from May to October and above.

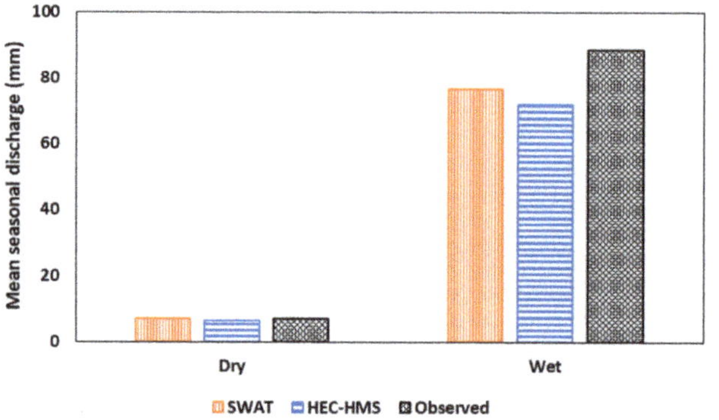

Figure 8. Mean seasonal discharge for observed as well as the HEC-HMS and SWAT models for the period 2007–2014.

In fact, in the SWAT model, the precipitation for a specific sub-basin is derived from the nearest weather station. However, in the HEC-HMS model, a Thiessen Polygon weighted rainfall is assigned for different sub-basins. The Thiessen Polygon weights are user given. Since rainfall is the main input in a rainfall–runoff model this process might have an impact on streamflow simulations as observed through the two hydrologic models. The HEC-HMS uses the Clark Unit Hydrograph method to simulate streamflow, while the SCS-CN method is adopted in the SWAT model. The SCS-CN method accounts for soil, land use, and slopes, while the Clark Unit Hydrograph accounts for the basin shape, watershed storage, and timing. The above-mentioned reasons can be some of the contributing factors to the variations in streamflow results obtained.

The above results also focus on parameter uncertainty and it is clear that better results can be obtained using different parameter sets in different hydrological models. Additionally, errors in the observed datasets make it difficult to perfectly determine the accuracy of runoff predictions through hydrological modelling. The results of this study can be further improved if observed data for evapotranspiration, wind speed, radiation, etc., are available for comparison purposes. Moreover, the robustness of the hydrological model changes according to the chosen time scale. It is recommended to determine suitable hydrological models using alternatives and also to determine hydrology using ensembles of several model structures for tropical catchments.

6. Conclusions

A tropical watershed in northeastern Thailand, the Huai Bang Sai (HBS) was selected in this research to compare the hydrologic performances of two widely used hydrologic models the HEC-HMS and the SWAT. The current work was carried out for the period 2007–2014. For both models, the calibration was carried out at the kh.92 (Ban Kan Luang Dong) hydrological station. The HEC-HMS model developed for the HBS watershed was calibrated and validated and then compared with the earlier developed SWAT model. The R^2 and NSE obtained during the calibration process were 0.80 and 0.79, and 0.83 and 0.82 during validation in the HEC-HMS model. For the SWAT model, during calibration, these indices yielded 0.83 and 0.82 and for validation, they were 0.78 and 0.77. The performance of both models is deemed to be satisfactory. The SWAT model was able to capture high flows compared to the HEC-HMS model more accurately for the 2007–2014 period, whereas medium flows were captured through the HEC-HMS model more accurately. Low flows were obtained with good accuracy by both models. In seasonal scales, the SWAT models outperformed the HEC-HMS model. Hence, the SWAT model can be attractive for both wet and dry seasonal flow simulations. The study results demonstrated that the spatial discretization of the HBS watershed through the SWAT and HEC-HMS models did not have a significant impact on response to streamflow simulations. The differences in equations used to compute hydrologic processes did not demonstrate large deviations in reproducing streamflow. Hence, both the SWAT and HEC-HMS are recommended to be used in the tropical humid conditions in Thailand and elsewhere in the world.

Author Contributions: Conceptualization, I.M.C., M.B.G. and U.S.R.; methodology, I.M.C.; software, I.M.C.; formal analysis, I.M.C.; writing—original draft preparation, I.M.C., M.B.G., P.K.B. and E.S.; writing—review and editing, I.M.C., M.B.G., M.K.J., M.S.B., S.S. and U.S.R. supervision, M.B.G., M.K.J., M.S.B., S.S. and U.S.R. All authors have read and agreed to the published version of the manuscript.

Funding: This research received no external funding.

Institutional Review Board Statement: Not applicable.

Informed Consent Statement: Not applicable.

Data Availability Statement: The climatic data used in this research study are available upon request for research purposes.

Acknowledgments: The authors of this manuscript are grateful to all data provision government agencies of Thailand for allowing us to carry out the work successfully. They are also thankful to Sri Lanka Institute of Information Technology, Sri Lanka for allowing the first author to carry out her research work.

Conflicts of Interest: The authors declare no conflict of interest.

References

1. Gunathilake, M.B.; Amaratunga, Y.V.; Perera, A.; Chathuranika, I.M.; Gunathilake, A.S.; Rathnayake, U. Evaluation of Future Climate and Potential Impact on Streamflow in the Upper Nan River Basin of Northern Thailand. *Adv. Meteorol.* **2020**, *2020*, 8881118. [CrossRef]
2. Li, T.; Duan, Y.; Guo, S.; Meng, L.; Nametso, M. Study on Applicability of Distributed Hydrological Model under Different Terrain Conditions. *Sustainability* **2020**, *12*, 9684. [CrossRef]
3. Flores, N.; Rodríguez, R.; Yépez, S.; Osores, V.; Rau, P.; Rivera, D.; Balocchi, F. Comparison of Three Daily Rainfall-Runoff Hydrological Models Using Four Evapotranspiration Models in Four Small Forested Watersheds with Different Land Cover in South-Central Chile. *Water* **2021**, *13*, 3191. [CrossRef]
4. Gunathilake, M.B.; Panditharathne, P.; Gunathilake, A.S.; Warakagoda, N. Application of a HEC-HMS model on event-based simulations in a tropical watershed. *Eng. Appl. Sci. Res.* **2020**, *47*, 349–360.
5. Yang, S.-C.; Yang, T.-H.; Chang, Y.-C.; Chen, C.-H.; Lin, M.-Y.; Ho, J.-Y.; Lee, K.T. Development of a Hydrological Ensemble Prediction System to Assist with Decision-Making for Floods during Typhoons. *Sustainability* **2020**, *12*, 4258. [CrossRef]
6. Hanif, A.; Dhanasekar, A.; Keene, A.; Li, H.; Carlson, K. Flood risk assessment methodology for planning under climate change scenarios and the corresponding change in land cover. *J. Water Clim. Chang.* **2019**, *11*, 1370–1382. [CrossRef]
7. Bhatta, B.; Shrestha, S.; Shrestha, P.K.; Talchabhadel, R. Evaluation and application of a SWAT model to assess the climate change impact on the hydrology of the Himalayan River Basin. *CATENA* **2019**, *181*, 104082. [CrossRef]
8. Chen, Q.; Chen, H.; Wang, J.; Zhao, Y.; Chen, J.; Xu, C. Impacts of Climate Change and Land-Use Change on Hydrological Extremes in the Jinsha River Basin. *Water* **2019**, *11*, 1398. [CrossRef]
9. Zhou, Y.; Cui, Z.; Lin, K.; Sheng, S.; Chen, H.; Guo, S.; Xu, C.-Y. Short-term flood probability density forecasting using a conceptual hydrological model with machine learning techniques. *J. Hydrol.* **2021**, *604*, 127255. [CrossRef]
10. Mvandaba, V.; Hughes, D.; Kapangaziwiri, E.; Kahinda, J.M.; Hobbs, P.; Madonsela, S.; Oosthuizen, N. The delineation of alluvial aquifers towards a better understanding of channel transmission losses in the Limpopo River Basin. *Phys. Chem. Earth Parts A/B/C* **2018**, *108*, 60–73. [CrossRef]
11. Aureli, F.; Mignosa, P.; Prost, F.; Dazzi, S. Hydrological and Hydraulic Flood Hazard Modeling in Poorly Gauged Catchments: An Analysis in Northern Italy. *Hydrology* **2021**, *8*, 149. [CrossRef]
12. Peng, H.; Jia, Y.; Tague, C.; Slaughter, P. An Eco-Hydrological Model-Based Assessment of the Impacts of Soil and Water Conservation Management in the Jinghe River Basin, China. *Water* **2015**, *7*, 6301–6320. [CrossRef]
13. Arnold, J.G.; Muttiah, R.S.; Williams, J.R. Large-area hydrologic modeling and assessment: Part I. *J. American Water Resour. Assoc.* **1998**, *34*, 73–89. [CrossRef]
14. Feldman, A. *Hydrological Modeling System HEC-HMS, Technical Reference Manual*; Hydrologic Engineering Center HEC; US Army Corps of Engineers: Davis, CA, USA, 2020.
15. Bergström, S. *The HBV Model-Its Structure and Applications*; Swedish Meteorological and Hydrological Institute: Norrköping, Sweden, 1992.
16. Krause, P. *Das hydrologische Modellsystem J2000-Beschreibung und Anwendung in großen Flußgebieten (No. PreJuSER-37462)*; Programmgruppe Systemforschung und Technologische Entwicklung: Jülich, Germany, 2001; ISBN 3-89336-283-5.
17. Perrin, C.; Michel, C.; Andréassian, V. Improvement of a parsimonious model for streamflow simulation. *J. Hydrol.* **2003**, *279*, 275–289. [CrossRef]
18. Bicknell, B.R.; Imhoff, J.C.; Kittle, J.L.; Donigian, A.S.; Johanson, R.C. *Athens, Greece: Hydrological Simulation Program—FORTRAN*; User's Manual for Version 11; Rep. No. EPA/600/R-97/080; U.S. EPA: Washington, DC, USA, 1997.
19. Abbott, M.B.; Bathurst, J.C.; Cunge, J.A.; O'Connell, P.E.; Rasmussen, J. An introduction to the European Hydro-logical System-Système Hydrologique Européen, "SHE", 1: History and philosophy of a physically-based, distributed modelling system. *J. Hydrol.* **1986**, *87*, 45–59. [CrossRef]
20. Đukić, V.; Erić, R. SHETRAN and HEC HMS Model Evaluation for Runoff and Soil Moisture Simulation in the Jičinka River Catchment (Czech Republic). *Water* **2021**, *13*, 872. [CrossRef]
21. Babel, M.S.; Gunathilake, M.B.; Jha, M.K. Evaluation of Ecosystem-Based Adaptation Measures for Sediment Yield in a Tropical Watershed in Thailand. *Water* **2021**, *13*, 2767. [CrossRef]
22. Gunathilake, M.B.; Amaratunga, Y.V.; Perera, A.; Karunanayake, C.; Gunathilake, A.S.; Rathnayake, U. Statistical evaluation and hydrologic simulation capacity of different satellite-based precipitation products (SbPPs) in the Upper Nan River Basin, Northern Thailand. *J. Hydrol. Reg. Stud.* **2020**, *32*, 100743. [CrossRef]
23. Fathy, I.; Negm, A.M.; El-Fiky, M.; Nassar, M.; Al-Sayed, E. Runoff hydrograph modeling for arid regions (Case study: Wadi Sudr-Sinai). *Int. Water Technol. J.* **2015**, *5*, 58–68.
24. Gunathilake, M.B.; Karunanayake, C.; Gunathilake, A.S.; Marasingha, N.; Samarasinghe, J.T.; Bandara, I.M.; Rathnayake, U. Hydrological Models and Artificial Neural Networks (ANNs) to Simulate Streamflow in a Tropical Catchment of Sri Lanka. *Appl. Comput. Intell. Soft Comput.* **2021**, *2021*, 6683389. [CrossRef]
25. Chathuranika, I.M.; Gunathilake, M.B.; Azamathulla, H.; Rathnayake, U. Evaluation of Future Streamflow in the Upper Part of the Nilwala River Basin (Sri Lanka) under Climate Change. *Hydrology* **2022**, *9*, 48. [CrossRef]
26. Shekar, N.C.S.; Vinay, D.C. Performance of HEC-HMS and SWAT to simulate streamflow in the sub-humid tropical Hemavathi catchment. *J. Water Clim. Chang.* **2021**, *12*, 3005–3017. [CrossRef]

27. Phomcha, P.; Wirojanagud, P.; Vangpaisal, T.; Thaveevouthti, T. Predicting sediment discharge in an agricultural watershed: A case study of the Lam Sonthi watershed, Thailand. *Sci. Asia* **2011**, *37*, 43–50. [CrossRef]
28. Supakosol, J.; Boonrawd, K. Hydrologic evaluation and effects of climate change on the Nong Han Lake Basin, northeastern Thailand. *J. Water Clim. Chang.* **2019**, *11*, 992–1000. [CrossRef]
29. Rossi, C.G.; Srinivasan, K.; Jirayoot, T.; Le Due, P.; Souvannabouth, N.B.; Gassman, P.W. Hydrologic evaluation of the lower mekong river basin with the soil and water assessment tool model. *Int. Agric. Eng. J.* **2009**, *18*, 13.
30. Kuntiyawichai, K.; Sri-Amporn, W.; Wongsasri, S.; Chindaprasirt, P. Anticipating of Potential Climate and Land Use Change Impacts on Floods: A Case Study of the Lower Nam Phong River Basin. *Water* **2020**, *12*, 1158. [CrossRef]
31. Aliye, M.A.; Aga, A.O.; Tadesse, T.; Yohannes, P. Evaluating the Performance of HEC-HMS and SWAT Hydrological Models in Simulating the Rainfall-Runoff Process for Data Scarce Region of Ethiopian Rift Valley Lake Basin. *Open J. Mod. Hydrol.* **2020**, *10*, 105–122. [CrossRef]
32. Khoi, D.N. Comparison of the HEC-HMS and SWAT hydrological models in simulating the streamflow. *J. Sci. Technol.* **2016**, *53*, 189–195.
33. Ismail, H.; Kamal, M.R.; Hin, L.S.; Abdullah, A.F. Performance of HEC-HMS and ArcSWAT models for assessing climate change impacts on streamflow at Bernam River Basin in Malaysia. *Sci. Technol.* **2020**, *28*, 1027–1048.
34. Kheereemangkla, Y.; Shrestha, R.P.; Shrestha, S.; Jourdain, D. Modeling hydrologic responses to land management scenarios for the Chi River Sub-basin Part II, Northeast Thailand. *Environ. Earth Sci.* **2016**, *75*, 793. [CrossRef]
35. Bekele, W.T.; Haile, A.T.; Rientjes, T. Impact of climate change on the streamflow of the Arjo-Didessa catchment under RCP scenarios. *J. Water Clim. Chang.* **2021**, *12*, 2325–2337. [CrossRef]
36. Bhuiyan, H.A.K.M.; McNairn, H.; Powers, J.; Merzouki, A. Application of HEC-HMS in a Cold Region Watershed and Use of RADARSAT-2 Soil Moisture in Initializing the Model. *Hydrology* **2017**, *4*, 9. [CrossRef]
37. Haque, S.; Ali, M.; Islam, A.K.M.S.; Khan, J.U. Changes in flow and sediment load of poorly gauged Brahmaputra river basin under an extreme climate scenario. *J. Water Clim. Chang.* **2020**, *12*, 937–954. [CrossRef]
38. Hamdan, A.; Almuktar, S.; Scholz, M. Rainfall-Runoff Modeling Using the HEC-HMS Model for the Al-Adhaim River Catchment, Northern Iraq. *Hydrology* **2021**, *8*, 58. [CrossRef]
39. Tassew, B.G.; Belete, M.A.; Miegel, K. Application of HEC-HMS Model for Flow Simulation in the Lake Tana Basin: The Case of Gilgel Abay Catchment, Upper Blue Nile Basin, Ethiopia. *Hydrology* **2019**, *6*, 21. [CrossRef]
40. Ouédraogo, W.A.A.; Raude, J.M.; Gathenya, J.M. Continuous Modeling of the Mkurumudzi River Catchment in Kenya Using the HEC-HMS Conceptual Model: Calibration, Validation, Model Performance Evaluation and Sensitivity Analysis. *Hydrology* **2018**, *5*, 44. [CrossRef]
41. Chiang, S.; Chang, C.-H.; Chen, W.-B. Comparison of Rainfall-Runoff Simulation between Support Vector Regression and HEC-HMS for a Rural Watershed in Taiwan. *Water* **2022**, *14*, 191. [CrossRef]
42. Melesse, A.M.; McClain, M.; Wang, X.; Abira, M.; Mutayoba, W. Modeling the Impact of Land-Cover and Rainfall Regime Change Scenarios on the Flow of Mara River, Kenya. In Proceedings of the World Environmental and Water Resources Congress 2008, Honolulu, HI, USA, 12–16 May 2008; pp. 1–10. [CrossRef]
43. Akoko, G.; Le, T.; Gomi, T.; Kato, T. A Review of SWAT Model Application in Africa. *Water* **2021**, *13*, 1313. [CrossRef]
44. Pignotti, G.; Rathjens, H.; Cibin, R.; Chaubey, I.; Crawford, M. Comparative Analysis of HRU and Grid-Based SWAT Models. *Water* **2017**, *9*, 272. [CrossRef]
45. Marahatta, S.; Devkota, L.; Aryal, D. Application of SWAT in Hydrological Simulation of Complex Mountainous River Basin (Part I: Model Development). *Water* **2021**, *13*, 1546. [CrossRef]
46. Lacombe, G.; Trébuil, G.; Polthanee, A. Long-term change in rainfall distribution in Northeast Thailand: Will cropping systems be able to adapt? *Cah. Agric.* **2017**, *26*, 25001. [CrossRef]

fluids

Article

Numerical Investigation of Critical Hydraulic Parameters Using FLOW-3D: A Case Study of Taunsa Barrage, Pakistan

Muhammad Waqas Zaffar [1,*], Ishtiaq Haasan [1] and Abdul Razzaq Ghumman [2]

1 Department of Civil Engineering, Capital University of Science and Technology (CUST), Expressway, Road Zone-V Sihala, Kahuta 44000, Islamabad Capital Territory, Pakistan; eishtiaq@cust.edu.pk
2 Department of Civil Engineering, College of Engineering, Qassim University, Buraydah 51452, Saudi Arabia; abdul.razzaq@qec.edu.sa
* Correspondence: dce171001@cust.pk

Abstract: Hydraulic structures, such as barrages, play an important role in the sustainable development of several regions worldwide. The aim of this novel study is to identify the critical hydraulic parameters (CHPs) of Taunsa Barrage, built on the Indus River. These CHPs, including free surface profiles, flow depths, Froude number, velocity profiles, energy dissipation and turbulence kinetic energy, were investigated using simulation via FLOW-3D numerical models. Incompressible Reynolds-averaged Navier–Stokes (RANS) equations on each computational cell were solved using the numerical methods available in FLOW-3D. The simulation results indicated that the locations of hydraulic jumps (HJs) were lower than that were reported in the previous one-dimensional study. Similarly, the distances of the HJs from the downstream toe of the glacis were reached at 2.97 m and 6 m at 129.10 m and 130.30 m tailwater levels, respectively, which deviated from the previous studies. In higher tailwater, the sequent depth ratio also deviated from the previous data. The maximum turbulent kinetic energies were observed in the developing regions of HJs, which were found to be decreased as the distance from the HJ was increased. The results of this research will be highly useful for engineers working in the field of design of hydraulic structures.

Keywords: hydraulic parameters; numerical methods; simulation; stilling basin; Taunsa barrage; energy dissipation; efficiency

Citation: Zaffar, M.W.; Haasan, I.; Ghumman, A.R. Numerical Investigation of Critical Hydraulic Parameters Using FLOW-3D: A Case Study of Taunsa Barrage, Pakistan. *Fluids* **2023**, *8*, 310. https://doi.org/10.3390/fluids8120310

Academic Editors: D. Andrew S. Rees, Prashanth Reddy Hanmaiahgari, Manish Pandey, Mohammad Amir Khan and Jaan H. Pu

Received: 3 October 2023
Revised: 17 November 2023
Accepted: 17 November 2023
Published: 28 November 2023

1. Introduction

1.1. Significance of Hydraulic Jumps (HJs)

Due to the sufficient head upstream of hydraulic structures, the outflow water has immense kinetic energy that can damage the downstream structures. Researchers and hydraulic engineers have devised many measures such as baffle blocks, friction blocks, chute blocks, and end sills to control the above-mentioned issues. These devices stabilize the hydraulic jumps (a vital energy dissipation process) formed at the location. Hydraulic jumps occur in natural systems and artificial channels such as rivers, streams, spillways, sluice gates, barrages, and weirs. At the different locations of these hydraulic structures, the flow passes through different conditions, i.e., subcritical (Froude number ((Fr) < 1), critical (Fr = 1), and supercritical (Fr > 1)). In the subcritical condition, the actual flow depth is higher than the critical, whereas in the supercritical condition the flow depth is always found to be less than critical. In the hydraulic jumps, the flow changes suddenly from supercritical to subcritical conditions and vice versa. During this chaotic phenomenon, a rapid rise in the free surface occurs that dissipates a large amount of energy due to the turbulent mixing [1–4]. A hydraulic jump (HJ hereafter) occurs in gravity-driven flows when Fr crosses unity and is defined as the ratio of inertial to gravitational forces.

Numerous researchers have conducted experiments on HJs by employing various geometries and hydraulic conditions, such as vegetated bed (Bai et al. [5]), close conduits

(Li et al. [6]), stilling basins of open-channel hydraulic structures [7–13]; and horizontal smooth and rough beds [14]. In addition, Balachandar et al. [15] investigated the effects of tailwater variation on HJs and studied downstream bed profiles, whereas [16–20] investigated submerged HJs for different beds and flow conditions. Furthermore, a few studies have investigated the transition region of an HJ (Zobeyer et al. [21]; Abbaspour et al. [22]) and the weak HJ (Mignot and Cienfuegos [23]) on rough beds and in small open channels. On the whole, these studies tested the basin's appurtenances to increase the energy dissipation and mainly investigated the gate openings, Fr, Reynolds number, roller lengths of HJs, sequent depths, tailwater, momentum and velocity decay, and turbulent kinetic energy. However, the literature did not reveal any detailed experimental study on the flow patterns downstream of the Taunsa barrage, which is an important research area because of the significant importance of the barrage.

1.2. Role of Stilling Basins

Stilling basins are accepted for the dissipation of the surplus kinetic energy of an HJ downstream of spillways, dams, barrages, and pipe outlets. The hydraulic performance of the stilling basin is very much dependent on its shape and size, which affect the flow patterns [24,25]. Furthermore, a substantial amount of energy is still to be dissipated downstream of the stilling basin, for which flexible aprons are provided. The risk of scouring is also safeguarded by the flexible aprons, which alleviate the uplift pressure that is left behind. Ali and Mohamed [26] and Mishra [27] conducted experiments to study the effects of stilling basins shapes downstream of radial gates. Alikhani et al. [28] conducted experiments on a single vertical sill in a stilling basin for a forced HJ. Elsaeed et al. [29] investigated the effects of end steps on the length of submerged HJs and measured the energy losses and velocity profile along the stilling basin. Tiwari et al. (a, b) [30,31] investigated the stilling basin shape for pipe outlets and used different shapes of intermediate sills with heights equivalent to the diameter of the pipes. Hager and Li [32] and Herrera-Granados and Kostecki [33] conducted experiments to investigate the effects of different energy dissipators on the characteristics of HJ. Ali and Kaleem [34] investigated the energy dissipation of Taunsa Barrage stilling basins before and after remodeling. The results showed that, due to the remodeling of the basin, the stone apron was launched during the 2010 flood and the river course also shifted towards the left side. Chaudary and Sarwar [35] and Chaudhry [36] analyzed the tailwater effects on different stilling basins of Taunsa Barrage. The results indicated that the existing tailwater levels downstream of the prototype barrage were appropriate for the formation of an HJ.

1.3. Computational Fluid Dynamic and Hydraulic Modelling

The previous sections (Sections 1.1 and 1.2) discussed the methods and the significance of physical and experimental investigations of HJs and basin appurtenances, which can be assisted by three dimensional (3D) numerical codes. Additionally, experimental investigations and on-site measurements are usually expensive and time-consuming and due to hindrance of flow devices and scaling effects, the output results are found to deviate from the prototypes. Therefore, the use of numerical modelling to investigate the hydraulic characteristics of grade-control structures is becoming popular. Such modeling tools are helpful, especially when the basic fundamental equations are not adequate to give solutions such as multifaceted geometries [33]. Recently, with the development in computer technology, the problems of complicated hydraulic structures can be studied by improved via numerical methods and turbulence models, which further harmonize physical modeling in the design stages. These numerical models also test various feasible flow phenomena, with little change in the input to obtain further data for the computed domains [14,37,38]. Many numerical investigations are found on HJs and energy dissipation using various numerical codes; however, Table 1 highlights a few of the most relevant studies dealing with HJs and energy dissipators.

Table 1. Numerical simulations on HJs and energy dissipation basins.

Reference	Numerical Codes	Modelling Approach	Turbulence Scheme
Chaudhry [36]	HEC-RAS	Energy equation	1-D model
Aydogdu et al. [39]	ANSYS-FLUENT	RANS	RNG K-ε
Mukha et al. [40]	Open Foam	Large Eddy Simulation (LES)	LES-VOF
Abd El Azim et al. [41]	FLOW-3D	RANS	RNG K-ε
Kosaj et al. [42]	FLOW-3D	RANS	RNG K-ε
Mirzaei and Tootoonchi [43]	FLOW-3D	RANS	Standard K-ε, LES, RNG K-ε,
Macián-Pérez et al. [44]	FLOW-3D	RANS	Standard K-ε, K-ω, RNG K-ε,
Daneshfaraz and Ghaderi [45]	FLUENT	RANS	RNG K-ε
Anjum et al. [46]	FLUENT	RANS	Reynolds Stress Model (RSM)
Dargahi [47]	FLUENT	RANS	RSM, Standard K-ε, RNG K-ε
Karim and Ali [48]	FLUENT	RANS	RSM, Standard K-ε, RNG K-ε
Liu and García [49]	Open Foam	RANS	Standard K-ε
Bayon et al. [50]	Open Foam, FLOW-3D	RANS	RNG K-ε
Bayon-Barrachina et al. [51]	Open Foam	RANS	Standard K-ε, SST K-ω, RNG K-ε
Nguyen et al. [52] and Riad et al. [53]	Finite Element Model (FEM)	RANS	Standard K-ε, Prandtl Mixing Length
Chatila and Tabbara [54]	ADINA-F	RANS	Standard K-ε
Cassan and Belaud [55]	FLUENT	RANS	RSM, Standard K-ε, RNG K-ε
Carvalho et al. [56]	FLOW-3D	RANS	RNG K-ε

From Table 1, some studies, e.g., [39,40,43,44,47,50,51,56], have focused HJs and their associated parameters such as velocity, free surface profile, sequent depths, roller lengths, and turbulent kinetics energy, whereas other researchers have investigated energy dissipation [41,45,46,48,53–55] and scour [42,49] downstream of different hydraulic structures. However, except for the 1D Hydrologic Engineering Center River Analysis System (HEC-RAS) study on Taunsa barrage, the literature did not reveal any study that studied the effects of tailwater on the HJ and flow pattern in the barrage's basin.

1.4. Research Motives and Problem Statement

Soon after the operation of the Taunsa barrage, due to the retrogression of the downstream riverbed, the tailwater levels were lowered, which consequently damaged the basin's floor; during 1959–1962 these regular issues were resolved. Additionally, due to the structural flaws, some of the baffle blocks were also found to be uprooted [55]. After so many years of partial repairs, it was declared that the barrage was to be remodeled; it was reported that due to the lowering of the tailwater levels, the HJ was sweeping on the floor. On the basis of a model study report, the basin was remodeled; however, even after remodeling, the data from the 2010 flood revealed damage downstream of barrage [57,58]. To understand the tailwater levels and HJ locations, Chaudhry [36] carried out a one-dimensional HEC-RAS study.

In the year 2008, Taunsa barrage was remodeled based on a sectional model study on the rigid bed. Except tailwater, no other hydraulic parameters were thoroughly investigated. Furthermore, these kinds of physical studies are generally associated with scaling effects such as roughness and length. However, due to the advent of computer technology and advanced turbulence models, as compared with HEC-RAS, more dedicated hydraulic modeling tools are available to investigate hydraulic issues such as those found downstream of the Taunsa barrage. Therefore, in this study, firstly, using the frequency of occurrence and relative importance index (RII), the critical hydraulic parameters (CHPs) are identified from the literature. The prime objective of this study is to investigate the tailwater and HJ locations at 44 m^3/s discharge downstream of the old basin of Taunsa barrage using FLOW-3D. This study further investigates the identified CHPs on the basin before its remodeling and mainly focuses on free surface profiles, flow depths, Froude number, velocity profiles, and turbulent kinetic energy. Additionally, the results of CHPs are also compared with the available data from prototypes, HEC-RAS, and studies from the literature. The specific research goals for the present numerical study are as below:

- To calibrate and validate a 3D model under the field conditions of a full-scale hydraulic structure using data from Taunsa Barrage.
- To identify the critical hydraulic parameters (CHPs) that play a crucial role in the design of graded control structures.
- To investigate the flow patterns of CHPs for stilling basins at various tailwater levels using the data from the old Taunsa Barrage.
- To study the effects of different tailwater levels on the locations of the HJs and compare them with relevant field and numerical data.
- To provide an overview of HJ locations in various basins of a barrage for different tailwater and discharge levels (a case study for Taunsa Barrage).

2. Study Area

Pakistan is an agricultural country; the major source of economy and livelihood depends on the agriculture sector. This sector provides about 25% of GDP and engages 50% of labor from rural areas [56]. More than 18 million hectares (ha) of land is irrigated by the Indus River and its branches. In the Indus Basin, barrages are essential components that divert water into the canals and also serve as roads, bridges and power transmission lines [56]. Taunsa Barrage was completed in 1958 on the Indus River. The barrage diverts water to Dera Ghazi Khan Division through the Muzaffargarh and Taunsa-Panjnad link canals and irrigates about 809,371 ha of land. The barrage was designed for a discharge capacity of 28,313 m^3/s. The total length of the barrage is 1325 m, whereas the clear width for flow passage is 1171 m. The maximum upstream and downstream flood levels are 136.94 m and 135.33 m, respectively, whereas 136.24 m is the normal pond level for the operation of the canals. The upstream and downstream floor levels are designed at 128.31 m and 126.79 m, respectively. The weir's crest is located at 130.44 m, whereas the waterfall is kept at 3.66 m. It should also be noted that all the elevations given herein are from the mean sea level. Taunsa barrage's location and its typical cross section is shown in Figure 1.

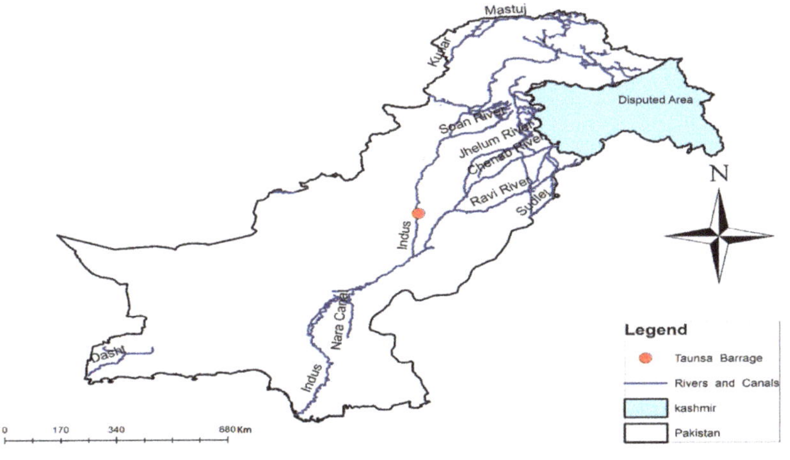

Figure 1. Location of the study area.

The stilling basin of the barrage is of a modified form similar to the United States Bureau of Reclamation's (USBR's) stilling basin type-III. Within the basin, two rows of baffle and friction blocks facilitate energy dissipation and stabilize the HJ even in cases of minimum tailwater requirements. However, soon after the barrage construction, multiple problems such as the oblique right-sided river approach, heavy siltation in one of the major canals, uprooting of impact baffle blocks, damage to the stilling basin floor, and retrogression of water levels appeared in its downstream areas. During 1959–1962, 1966, and 1973, repairs works were carried out to cater to the problems mentioned above; however, the problems remained persistent. In the reports by Zaidi et al. [58] and the World Bank [57], sweeping of the HJ was believed to be the main reason for the issues highlighted above. To resolve these issues, the Punjab Government constituted committees of experts; however, no specific measures were taken and the issues continued to be aggravated. The typical cross section of the Taunsa barrage stilling basin is shown in Figure 2.

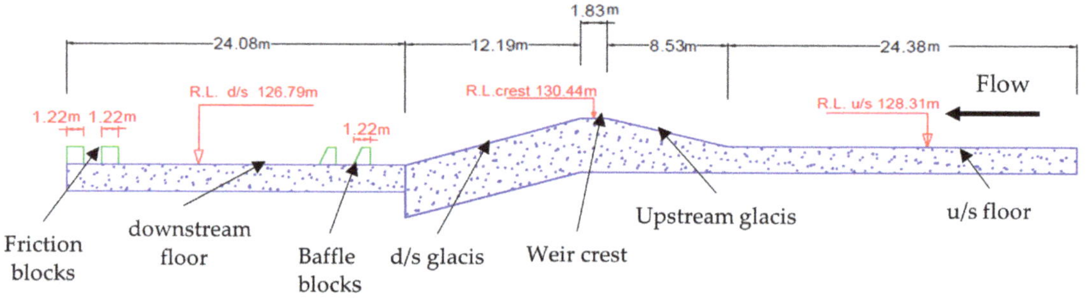

Figure 2. Typical cross section of the barrage.

3. Material and Methods

The methodology for the present study has two phases, as shown in Figure 3.

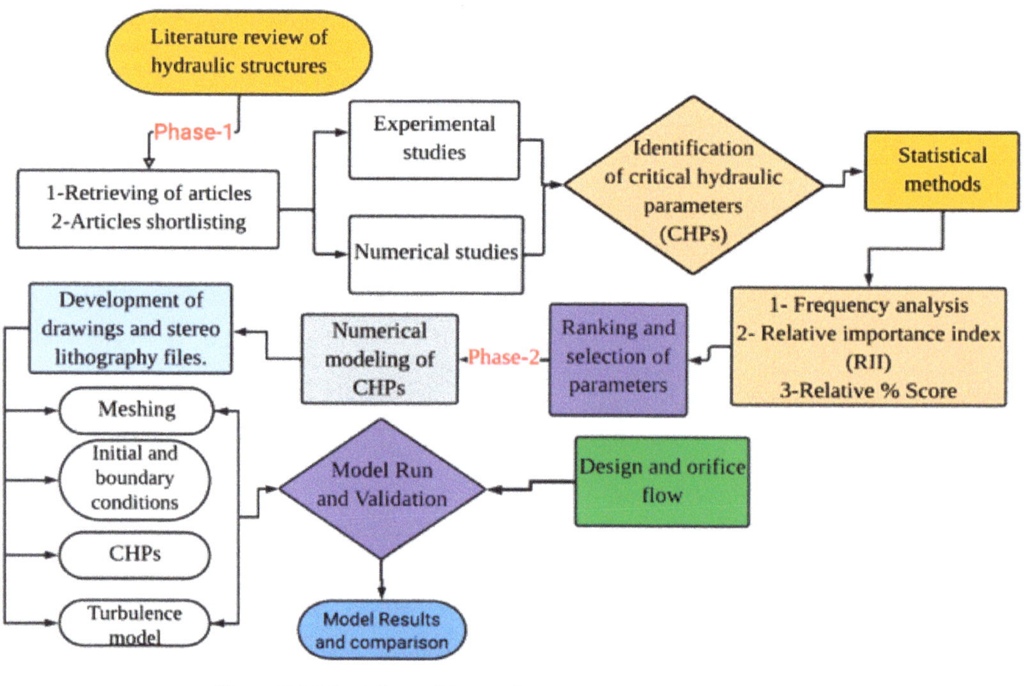

Figure 3. Methodology of the study.

3.1. Phase 1: Identification of the Critical Hydraulic Parameters

In the first phase, the critical hydraulic parameters (CHPs) were identified from in-depth review of the published experimental and numerical data, whereas in the second phase the identified CHPs from the first phase were numerically investigated using FLOW-3D within the basin of Taunsa Barrage. For the first phase, a systematic review of the previous studies published from 2000 to 2020 was carried out and all those articles were retrieved that addressed the hydraulic parameters on downstream sides of any hydraulic structures such as spillways, barrages, sluice gates, weirs, and falls. The retrieved articles were divided into two categories, one was experimental studies and the other was numerical studies. By doing so, eighty research articles were retrieved, whereas the unrelated articles were discarded. A detailed review of the retrieved article was carried out and, skimming through these papers, forty-two articles on experimental studies and twenty-four articles on numerical studies were selected for further analysis to identify the CHPs. The statistical methods used for the identification of parameters are explained below.

The ranking of CHPs was performed on the basis of the relative percentage score, which was calculated by Formula (1) [59–61].

$$\text{Percent Score} = R_f \times \text{RPA} \tag{1}$$

where R_f and RPA are the relative frequency and relative portion of the party affected, respectively, which have been assigned to each parameter. In the present investigation, articles from the experimental and numerical categories were selected as parties [60–65].

The relative importance of the parameters can be calculated using the relative importance index (RII). A similar approach was used in the present study. The RII for each parameter was calculated using Formula (2):

$$\text{RII} = \sum\nolimits_{i=0}^{n} \frac{\text{Wi}}{\text{A} \times \text{N}} \tag{2}$$

where RII is the relative importance index and W_i and A are the weight and highest weight given to each parameter. In Formula (2), N is the total number of research articles from where these hydraulic parameters were taken. The RII value ranges from 0 to 1; the higher the RII, the more critical is the parameter.

3.2. Phase 2: Numerical Model Implementation

Choosing the most suitable CFD codes from many available options is crucial [63–65]. Still, it is tedious as the investigative parameters [66–69] are strongly case dependent (Bennett et al. [70]). However, according to Chen et al. [14] and Babaali et al. [24], FLOW-3D has been the most widely used modelling tool for hydraulic investigations. Bayon et al. [50] recommended FLOW-3D for HJ characteristics after comparing its results with similar numerical models. Based on the recommendations of studies mentioned above, the present study implemented FLOW-3D numerical models to investigate the identified CHPs within the stilling basin of the studied barrage. FLOW-3D software (https://www.flow3d.com/) is considered as one of the most potent computer tools for performing three-dimensional (3D) flow analyses and uses multiple techniques to investigate the issue of multi-fluid by solving incompressible Reynolds-averaged Navier–Stokes (RANS) equations on each computational cell. FLOW-3D subdivides the flow and solid domains into structured grid blocks to resolve these domains for obtaining solutions of hydraulic issues. These structured rectangular grids are easy to develop and store essential information on cell faces and nodes. However, non-uniform mesh grid facilitates users to create meshes for complex geometries. Each cell within the grid is identified with a specific number in three dimensions: i in the x-direction, j in the y-direction, and k in the z-direction. The fundamentals of finite difference and finite volume methods were formerly developed on such meshes. These methods are central for the development of FLOW-3D. This 3D-modeling tool applies the finite volume method (FVM) derived directly from the conservation law to hold fluid properties. In FLOW-3D, the fluid–solid interface is tracked using the FAVOR method, whereas the generalized minimum residual method (GMRES) is implemented to solve issues of pressure velocity coupling. The proceeding sections show the equations used for the present models.

FLOW-3D discretizes the governing equations such as continuity and momentum equations. The general form of the mass continuity is described by Equation (3). For incompressible flow simulation, considering ρ as constant, Equation (3) is transformed for the incompressibility conditions, as provided in the following Equation (4):

$$\frac{V_F \partial \rho}{\partial t} + \frac{\partial}{\partial x}(\rho u Ax) + R\frac{\partial}{\partial y}(\rho v Ay) + \frac{\partial}{\partial z}(\rho w Az) + \xi\frac{\rho u Ax}{x} = Rsor + R_{DIF} \tag{3}$$

$$\frac{\partial}{\partial x}(uAx) + R\frac{\partial}{\partial y}(vAy) + \frac{\partial}{\partial z}(wAz) + \xi\frac{\partial uAx}{x}(uAx) = \frac{R_{SOR}}{\rho} \tag{4}$$

In Equations (3) and (4), V_F is the partial volume of flow, ρ is the fluid density, R_{SOR} is the mass source, and R_{DIF} is the diffusion term of turbulence. In case of Cartesian coordinates, R is equal to unity and ξ is set as zero. The fluid velocity's components in three dimensions are computed using the following Reynolds-averaged Navier–Stokes (RANS) Equation (5):

$$\frac{\partial u}{\partial t} + \frac{1}{V_F}\left[uA_x\frac{\partial u}{\partial_x} + vA_y\frac{\partial u}{\partial_y} + wA_{zy}\frac{\partial u}{\partial_z}\right] = \frac{1}{\rho}\frac{\partial p}{\partial_x} + Gx + fx \tag{5a}$$

$$\frac{\partial v}{\partial t} + \frac{1}{V_F}\left[uA_x\frac{\partial u}{\partial_x} + vA_y\frac{\partial u}{\partial_y} + wA_{zy}\frac{\partial u}{\partial_z}\right] = \frac{1}{\rho}\frac{\partial p}{\partial_y} + Gy + fy \tag{5b}$$

$$\frac{\partial w}{\partial t} + \frac{1}{V_F}\left[uA_x\frac{\partial u}{\partial_x} + vA_y\frac{\partial u}{\partial_y} + wA_{zy}\frac{\partial u}{\partial_z}\right] = \frac{1}{\rho}\frac{\partial p}{\partial_z} + Gz + fz \tag{5c}$$

where u, v, and w are the velocity components, A_x, A_y, and A_z are the flow areas, G_x, G_y, and G_z are body accelerations, f_x, f_y, and f_z are viscous accelerations, and ρ is the fluid density.

3.2.1. Model Meshing and the Initial and Boundary Conditions

The solid geometry of the model was prepared in AutoCAD and converted into a stereolithographic file (Stl.). Before importing the geometry into FLOW-3D, it was checked by Netfab software (https://inno-venture.com/netfabb/) to remove errors, holes, and facets. The model geometry and simulation domain were resolved using structured hexahedral mesh.

A single mesh block of 55.47 m long, 20.42 m wide, and 10.06 m high was implemented. Initially, a coarse mesh of 0.50 m cell size was applied to resolve the geometry; however, the stilling basin appurtenances were not fully resolved. Gradually reducing the cell size to 0.16 m, the geometry displayed a more suitable resolution to run the simulations. In total, 2,890,443 cubic mesh cells were used for models. To reduce the simulation time, a domain-removing component was added on the downstream side to deactivate the empty cells. The cell deactivation region of the domain-removing component was defined from the gate to the end of the stilling basin, and it was ensured that the region of domain-removing component did not contain the fluid. Figure 4 shows meshing applied to the models.

Routinely, barrage gates are not opened to the same levels. The openings are set according to the flows in the river. For reproducing the similar conditions for 44 m^3/s of flow, the models were set for a constant elevation of 136.24 m for the pond levels, whereas five different tailwater levels ranging from 129.10 m to 130.30 m with an equal increment of 0.30 m were implemented on the downstream side. The flux surface (porosity = 1) was set at the end of the stilling basin and a movable probe was assigned to measure the free surface profile and other essential parameters in the stilling basin. The volume flow rate of single bay was used for the discharge calculation of 64 bays of the barrage, thereby the actual conditions for 44 m^3/s flow were generated in the models. For all the models, the initial boundary conditions of discharge (44 m^3/s), upstream pond level (136.24 m), and turbulence model (RNG-K-ε) were kept constant, whereas the tailwater level was changed from 129.10 m to 130.30 m. The total simulation length of the model was 55.47 m, of which 38.10 m comprised the downstream area. Table 2 shows the initial conditions for gated- and free-flow analysis.

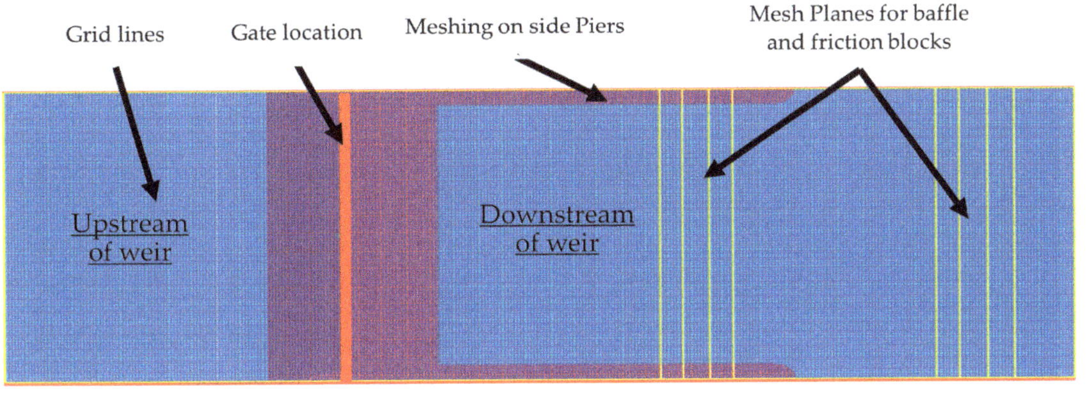

Figure 4. Meshing setup for the simulation domain.

Table 2. Initial conditions for the gated and free flow.

Discharge (m^3/s)	Minimum Tailwater Required for HJ (m)	Maximum Tailwater Required for HJ (m)	Upstream Water Level Maintained (m)	Turbulence Model	Models Operation
44	129.10	132.28	136.24	RNG K-ε	Gated Flow
444	133.8	-----	135.93	RNG K-ε	Free Designed Flow

Figure 5 shows that the pressure (P) boundaries were set for upstream (X_{min}) and downstream (X_{max}), whereas the lateral sides and bed were set as the rigid boundaries (W) and no-slip conditions were imposed which were expressed as zero tangential and normal velocity (u = v = w = 0) on the wall. u, v, and w are the velocity in the x, y, and z directions, respectively. For all variables (except pressure (P), which was set to zero), the upper boundaries (Z_{max}) were set as atmospheric pressure to allow water to null von Neumann.

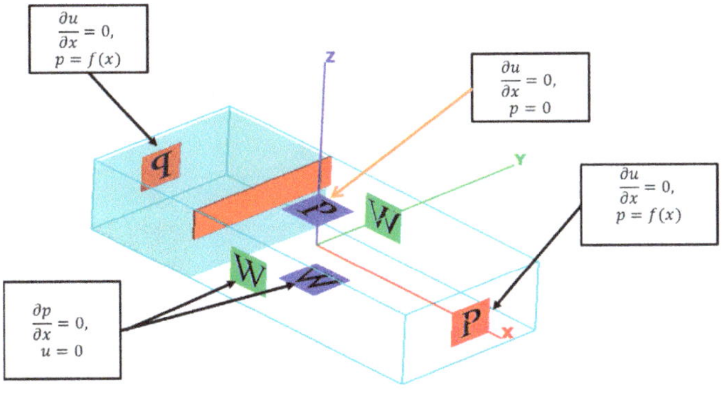

Figure 5. Boundary conditions governing the models.

3.2.2. Turbulence Modelling and Free Surface Tracking

One of the important aspects of computational fluid dynamic (CFD) models is turbulence closure. These numerical models implement Reynolds-averaged Navier–Stokes equations (RANS) to find the turbulence closure and solve high Reynolds numbers that develop flow instabilities. These models solve Reynolds stresses terms in the Navier–Stokes equation and calculate solutions for the additional equations of the turbulent viscosity and transport variable. Out of the six turbulence models in FLOW-3D, the two equation turbulence models (K-ε), standard K-ε and renormalization group (RNG K-ε), are the most widely used in hydraulic investigations. The above-mentioned turbulence models were applied by Bayon-Barrachina and Lopez-Jimenez [4], Macián-Pérez [9], and Bayon et al. [50] to investigate the HJ characteristics and the results showed that RNG K-ε produced reasonable accuracy for the efficiency of the HJ, sequent depths, roller lengths, and turbulent kinetic energy. Similarly, Nikmehr and Aminpour [71] also used RNG K-ε to investigate the free surface profile, flow rate, and Fr_1 on the corrugated bed and the model results were well in agreement with the compared experiments. Furthermore, studies such as those by Carvalho et al. [56], Savage and Johnson [72], and Johnson and Savage [73] investigated HJ characteristics within the stilling basins of spillways and indicated that the RNG K-ε turbulence model produced free surface, velocity, pressure profiles, and turbulent kinetic energy that were well in agreement with the experimental results. Based on the published data from similar studies, the present studies implemented the RNG K-ε turbulence model within the stilling basin of the studied barrage. In the RNG K-ε turbulence model [73–81], Equations (6) and (7) were applied to model the turbulent kinetic energy (k) and its dissipation (ε), respectively.

$$\frac{\partial}{\partial t}(\rho k) + \frac{\partial}{\partial xi}(\rho kui) = \frac{\partial}{\partial xj}\left[\mu + \frac{\mu t}{\sigma k}(\rho k)\frac{\partial k}{\partial xj}\right] + Pk + \rho\varepsilon \tag{6}$$

$$\frac{\partial}{\partial t}(\rho \varepsilon) + \frac{\partial}{\partial xi}(\rho \varepsilon ui) = \frac{\partial}{\partial xj}\left[\mu + \frac{\mu t}{\sigma_\varepsilon}(\rho k)\frac{\partial \varepsilon}{\partial xj}\right] + C1_\varepsilon\frac{\varepsilon}{k}Pk - C2_\varepsilon\frac{\varepsilon^2}{k} \tag{7}$$

In Equations (6) and (7), x_i, μ, μ_t, k, ε, ρ, and P_k are the coordinate in x direction, dynamic viscosity, turbulent dynamic viscosity, turbulent kinetic energy (TKE), turbulent dissipation, fluid density, and produced TKE, respectively. Finally, the terms σk, σ_ε, $C_{1\varepsilon}$, and $C_{2\varepsilon}$ are model parameters whose values are given in the study by Yakhot et al. [67]. The volume of fluid (VOF) method was used to track the free surface in which the fraction of the fluid (F) was implemented to find the fractional volume (i.e., water or air). To track the free surface within the simulation domain, Equation (8) was used.

$$\frac{\partial F}{\partial t} + \nabla x(\bar{u}F) = 0 \tag{8}$$

where, in the modelling domain, the fluid fraction (F) is represented by the below three possibilities.

1. If F approaches 0, the cell is considered as empty;
2. When F reaches 1, the cell is believed to be occupied by fluid;
3. If $0 < F < 1$, the cell represents a surface between the two fluids.

Presently, one fluid (water) with free surface is considered, whereas other advection schemes are selected by the models.

3.3. Model Verification and Validation

For validation of free designed flow analysis, $h_e/H_d = 0.998$ was implemented as previously used in [72,73,82,83], whereas h_e and H_d were the effective head and designed heads, respectively, as shown in Figure 6. For the designed flow analysis to run the simulations, pond and tailwater levels of 135.93 m and 133.8 m were used, respectively.

Figure 6. Operating conditions for gated/orifice flow.

On the other hand, for gated flow, Formula (9) was used to calculate the discharge through the orifice. For 44 m^3/s discharge, D = 0.280 m and H_d = 136.24 m were used to operate the models, where D is the gate opening and Hd is the design head for orifice discharge.

$$Q = \frac{2}{3}Cd * A * \sqrt{2gh_c} \quad (9)$$

where Q, A, and g are the volume flow rate, area of orifice, and acceleration due to gravity, respectively, and are measured in m^3/s, m^2, and m/s^2, respectively. However, in Equation (9), h_c is the centerline head, which is calculated using the gate openings and pond levels. For gated flow, a 0.816 value for the coefficients of discharge (C_d) was used, whereas a C_d value of 0.819 was obtained from the models.

Courant number stability criteria [66,67] were adopted to control the time steps and varied from 0.06 to 0.0023 and 0.015 to 0.0025 for free and gated flow, respectively. The volume flow rates at inlet and outlet boundaries were monitored to check the steady state of the models [83–85]. For modelling the discharges of 44 m^3/s and 444 m^3/s, a simulation finish time of (T = 80 s) was selected, as can be seen in Figure 7; the models achieved steady state at T = 60 s and T = 75 s for free designed and gated flow, respectively.

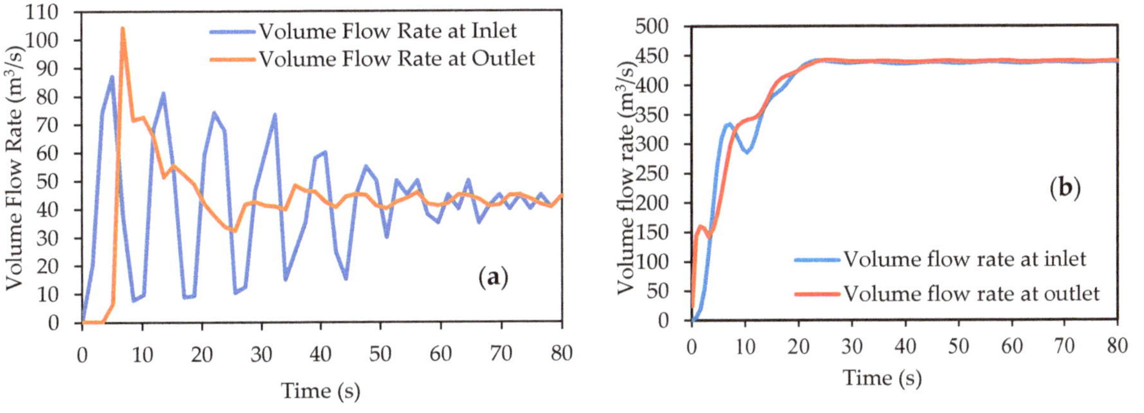

Figure 7. Time rate of change of flow at inlet and outlet boundaries. (**a**) Gated flow. (**b**) Free designed flow.

Analysis of free designed and flow analysis showed that at the beginning the free surface on the upstream and downstream of weir was found to be fluctuating; this became stable when the models reached the steady state, as shown in Figure 7. Free stable HJs were observed at T = 60 s and T = 75 s for free and gated flow, respectively. In free flow, the model underestimated the flow and the maximum error reached −5%, whereas a 1.14% error was observed in gated flow. From the gated and free flow analysis, it was found that the present models produced acceptable discharge accuracy, which allowed us to study the CHPs downstream of the studied barrage as described in Section 4.2.

4. Results

4.1. Identification of Critical Hydraulic Parameters (CHPs)

Appendices A and B show the sources of hydraulic parameters extracted from the numerical and experimental studies. The parameters in the appendixes were based on their frequency of occurrence. From the initial analysis, twenty-four and twenty-three hydraulic parameters were found from the numerical and experimental studies, respectively, as provided in Tables 3 and 4, respectively. However, fifteen parameters were found to be common to both types of studies (numerical and experimental). By taking the union of both sources (numerical and experimental), a total of thirty-three parameters were identified from the published data; these are listed in Table 5.

Table 3. Frequency of occurrence, RII, and ranking of parameters from numerical studies.

Parameters	Frequency	RII	Rank	Parameters	Frequency	RII	Rank
VP	16	0.155	1st	WSS	2	0.019	10th
FSP	15	0.146	2nd	BSS	2	0.019	
PP	11	0.107	3rd	TWL	1	0.010	11th
TKE	8	0.078	4th	CC	1	0.010	
AV	7	0.068	5th	HL	1	0.010	
DM	6	0.058	6th	FF	1	0.010	
Fr_1	6	0.058		MC	1	0.010	
η	5	0.049	7th	EC	1	0.010	
SS	4	0.039	8th	DC	1	0.010	
SDHJ	4	0.039		SP	1	0.010	
RSS	3	0.029	9th	BP	1	0.010	
LHJ	3	0.029		ED	1	0.010	

Table 4. Frequency of occurrence, RII, and ranking of parameters from experimental studies.

Parameters	Frequency	RII	Rank	Parameters	Frequency	RII	Rank
VP	24	0.125	1st	TI	5	0.026	11th
Fr_1	24	0.125		SM	4	0.021	12th
TWL	20	0.104	2nd	LHJ	4	0.021	
FSP	19	0.099	3rd	RG	3	0.016	13th
SS	17	0.089	4th	η	3	0.016	
BP	14	0.073	5th	MF	2	0.010	14th
SP	13	0.068	6th	VF	2	0.010	
ED	9	0.047	7th	SDHJ	2	0.010	
BSS	7	0.036	8th	TP	1	0.005	15h
TKE	6	0.031	9th	PS	1	0.005	
PP	6	0.031		$\Delta E/E1$	1	0.005	
RSS	5	0.026	10th				

In Table 3, by applying the relative importance index (RII), the velocity profile (VP), free surface profile (FSP), pressure profile (PP), turbulence kinetic energy (TKE), air volume value (AV), discharge measurement (DM), shape of stilling basin (SS), Fr_1, and HJ efficiency (η) were ranked as the most CHPs in numerical studies that were conducted from 2000 to 2020.

Table 4 shows the ranking of the parameters that were computed for experimental studies. The RII showed that VP, Fr_1, TWL, SS, FSP, BP, SP, and ED were the most CHPs, upon which several studies have been performed during recent years. After adding up the parameters from the experimental and numerical studies based on their RII, the VP, FSP, Fr_1, SS, TWL, PP, BP, TKE, SP, and ED were found to be the most important CHPs, as shown in Table 5. In Table 6, the relative percentage score of the 33 parameters was calculated; to do so, the Rf and RPA for each parameter were computed and results were presented in three different ranks.

Table 5. RII and ranking of parameters by combining numerical and experimental studies.

Parameters	Frequency	RII	Rank	Parameters	Frequency	RII	Rank
VP	40	0.140	1st	SDHJ	6	0.021	14th
FSP	31	0.110	2nd	TI	5	0.017	15th
Fr_1	30	0.100	3rd	SM	4	0.014	16th
SS	21	0.080	4th	RG	3	0.010	17th
TWL	21	0.070	5th	MF	2	0.007	
PP	17	0.050	6th	VF	2	0.007	18th
BP	15	0.060		WSS	2	0.007	
TKE	14	0.060	7th	CC	1	0.003	
SP	14	0.060		HL	1	0.003	
ED	10	0.040	8th	FF	1	0.003	
BSS	9	0.030	9th	MC	1	0.003	
RSS	8	0.030		EC	1	0.003	19th
η_L	8	0.020	10th	DC	1	0.003	
AV	7	0.030	11th	ES	1	0.003	
LHJ	7	0.010	12th	TP	1	0.003	
DM	6	0.020	13th	PS	1	0.003	
				(ΔE/E1)	1	0.003	

Table 6. Relative % score and ranking based on frequency of occurrence of the parameters in the literature [62,64,65,69].

Parameters	Frequency	Relative Frequency	PA	RPA	% Score	R % Score	Parameters	Frequency	Relative Frequency	PA	RPA	% Score	R % Score
VP	40	0.137	2	1	0.137	14.71%	TI	5	0.017	1	0.34	0.006	0.63%
FSP	31	0.106	2	1	0.106	11.40%	SM	4	0.014	1	0.34	0.005	0.50%
Fr_1	30	0.103	2	1	0.103	11.04%	RG	3	0.010	1	0.34	0.003	0.38%
SS	21	0.072	2	1	0.072	7.72%	MF	2	0.007	1	0.34	0.002	0.25%
TWL	21	0.072	2	1	0.072	7.72%	VF	2	0.007	1	0.34	0.002	0.25%
PP	17	0.058	2	1	0.058	6.25%	WSS	2	0.007	1	0.66	0.005	0.49%
BP	15	0.051	2	1	0.051	5.52%	CC	1	0.003	1	0.66	0.002	0.24%
TKE	14	0.048	2	1	0.048	5.15%	HL	1	0.003	1	0.66	0.002	0.24%
SP	14	0.048	2	1	0.048	5.15%	FF	1	0.003	1	0.66	0.002	0.24%
ED	10	0.034	2	1	0.034	3.68%	MC	1	0.003	1	0.66	0.002	0.24%
BSS	9	0.031	2	1	0.031	3.31%	EC	1	0.003	1	0.66	0.002	0.24%
RSS	8	0.027	2	1	0.027	2.94%	DC	1	0.003	1	0.66	0.002	0.24%
η_L	8	0.027	2	1	0.027	2.94%	ES	1	0.003	1	0.66	0.002	0.24%
AV	7	0.024	1	0.66	0.016	1.70%	TP	1	0.003	1	0.34	0.001	0.13%
LHJ	7	0.024	2	1	0.024	2.57%	PS	1	0.003	1	0.34	0.001	0.13%
DM	6	0.021	1	0.66	0.014	1.46%	ΔE/E1	1	0.003	1	0.34	0.001	0.13%
SDHJ	6	0.021	2	1	0.021	2.21%							

Table 7 shows the overall relative position of the CHPs based on frequency analysis, RII, and relative % score. It can be seen from Table 7 that, except for VP, all the other parameters changed their position when employing different statistical methods. From Table 7, it is found that VP, Fr, FSP, SS, TKE, and TWL are the most significant parameters that have been widely investigated in the literature. Hence, for the present study, these hydraulic parameters are focused on and investigated downstream of the studied barrage.

Table 7. Relative position of CHPs in the literature with different statistical methods.

Parameters	Ranking Extracted from Table 2	Ranking Extracted from Table 3	Ranking Extracted from Table 4	Ranking Extracted from Table 5	Overall Occurrence
(1)	(2)	(3)	(4)	(5)	(6) = (2) + (3) + (4) + (5)
VP	1st	1th	1th	1th	4
Fr_1	6th	1th	3rd	3rd	4
FSP	2nd	NA	2nd	2nd	3
SS	NA *	4th	4th	4th	3
TKE	4th	NA	7th	8th	3
TWL	NA	2nd	5th	5th	3

* NA = Did not appear.

4.2. Phase 2: Results for Critical Hydraulic Parameters (CHPs) Using Numerical Models

4.2.1. Free Surface Profiles

Figure 8 compares the free surface profiles at five different TWLs. The free surface data were taken from the downstream of the gate to the end of the stilling basin. The total upstream ponding length was 16.45 m, which ranged from X_{min} to the gate. At start, with the time rate of change, the free surface profiles were fluctuating [86–89]; however, they became stable when the models achieved a steady state. For the investigated tailwater levels, HJs were located at the glacis. However, the locations and lengths of the HJs were found to be different as the tailwater levels were changed; the shapes of the HJs were also changed.

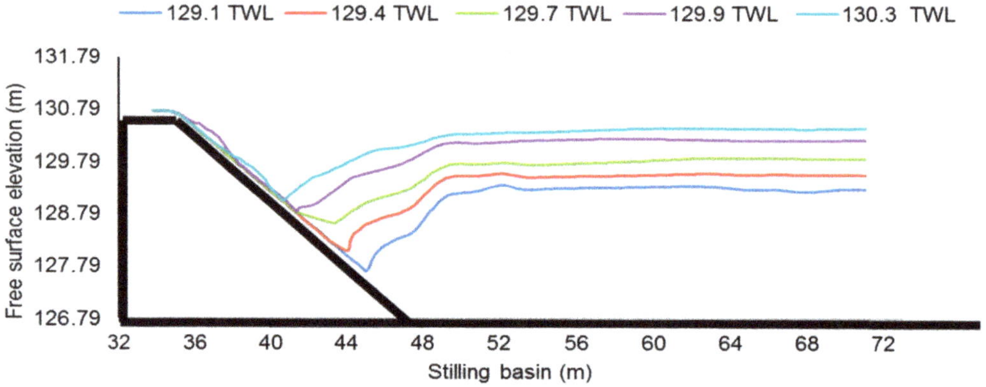

Figure 8. Free surface profiles and location of HJs at various tailwater levels.

The results further indicated that, at low TWLs, the undulating free water surface was found in the HJ, which continued to the end of the stilling basin, as shown in Figure 8 in 129.10 m and 129.40 m TWLs. At higher TWLs, the HJ was shifted to the upstream side of the glacis and different free surface profiles were noticed compared with the lower tailwater levels, as shown in Figure 8.

At 129.10 m tailwater, the results showed a 127.90 m elevation for the HJ; this is in good agreement with [82,90], in which the HJ elevation was observed at 128 m. Furthermore, the results of the present models are also compared with the designed and prototype data for the year 2010. In these reports, the distance of the HJ at 130.30 m was 1.22 m from the toe of the glacis, as shown in Table 8. However, the HJ locations were found to be missing on the lower tailwater [91–94]. Upon comparison with the available data point, the present model showed a five times higher HJ distance from the toe of the downstream glacis. In addition, even at a lower TWL of 129.10 m, the distance of the HJ was found to be 2.5 times higher than that observed at the barrage site during 2010. A detailed comparison of the HJ elevation and its location is provided in Table 8.

Table 8. Comparison of the results of the hydraulic jump with HEC-RAS and field data.

Q (m^3/s)	TWLs (m)	Present Study 3D Models		Chaudhry [36] HEC-RAS		Zaidi et al. [95]	
		HJ Elevation (m)	HJ Distance from Glacis Toe (m)	HJ Elevation (m)	HJ Distance from Glacis Toe (m)	Designed HJ Location(m)	HJ Location Obserevd at Prototype (m)
44	129.10	127.91	2.97	128	3.20	....	....
	129.40	128.33	4.17	....	....	....	....
	129.70	128.60	5.92	....	....	....	....
	129.9	128.87	6.15	....	....	....	....
	130.30	129.03	6.03	129.5	6.5	1.22	1.22

From Table 8, it can be realized that compared with the present models, the HEC-RAS models in [36] overestimated the length and location of the HJs. On the contrary, even after the remodeling of basin, the location and distances of HJs from the downstream toe of the glacis were found to be less than in the old basin, i.e., studied presently.

4.2.2. Froude Number

Figure 9 shows Froude number variations in the stilling basin. In the tested models, large numbers of oscillations were observed from the gate opening to the jump initiating point. The maximum value for Fr_1 was found for 129.10 m tailwater, which reached to 5.87, whereas the minimum value for Fr_1 was 5.30 at 129.70 m.

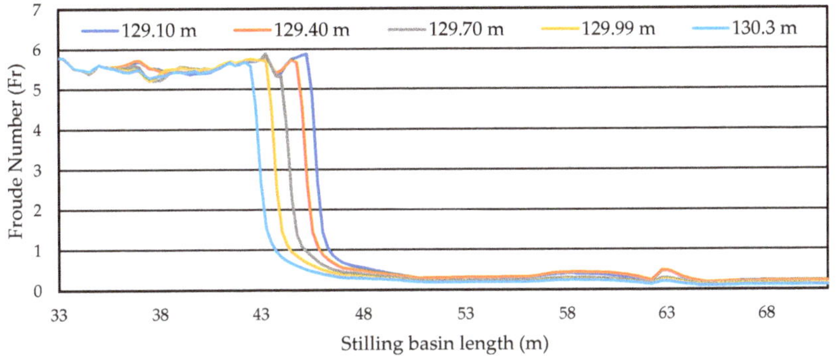

Figure 9. Variation in Froude number in the basin at different tailwater levels.

As compared with [36] for all the investigated tailwater levels, the present models showed higher values for Fr_1, whereas in the subcritical region, the results for the Froude number agreed with the study of Chaudhry [36]. A gradual decrease in the Fr_1 was observed when the TWLs increased. The minimum value for Fr_1 was observed at the maximum TWL. After the HJ, the flow changed into the subcritical state and the maximum value of Fr_2 was 0.22 at a 129.10 m TWL. At a constant discharge, pond level, and gate opening, the results of TWLs against Fr_1 showed a nonlinear trend. Two-dimensional illustrations of Fr in the stilling basin are shown in Figure 10a–e.

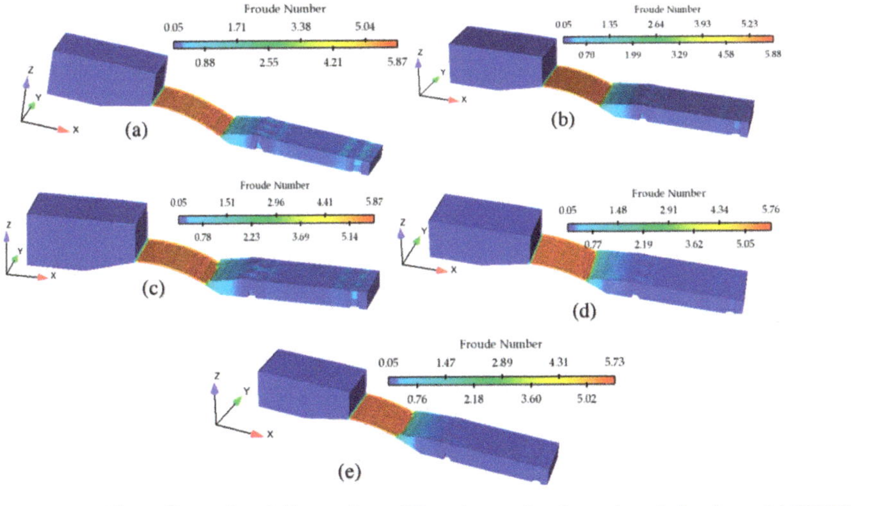

Figure 10. Three-dimensional illustration of Froude number in various tailwaters. (**a**) 129.10 m, (**b**) 129.40 m, (**c**) 129.70 m, (**d**) 129.99 m, and (**e**) 130.30 m.

4.2.3. Flow Depths

Figure 11 shows that at all the investigated tailwater levels, the flow depths up to the jump initiating point show identical behavior, whereas fluctuations in the flow depths were observed in the HJ region. Except at a 129.99 m TWL, the results indicate the smooth transition from supercritical (y_1) into subcritical flow depths (y_2), whereas at a 129.99 m TWL, large fluctuations were seen in the jump. At a 129.10 m TWL, due to the presence of friction blocks, the results indicate small oscillations in the flow depths at the stilling basin's end that deflected the flow towards the free surface.

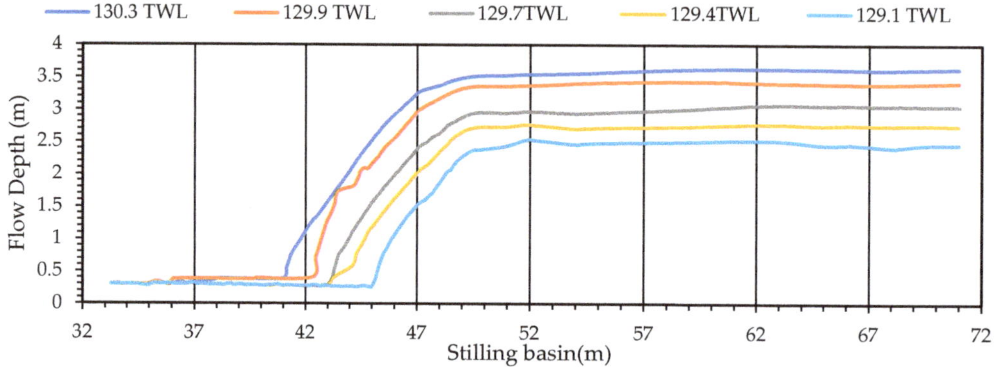

Figure 11. Flow depth variation at different TWLs in the stilling basin.

From Figure 12, it can be observed that as the tailwater was increased, the sequent depth ratio increased, and the Froude number values were found to be decreased. Figure 12 also compares the sequent depths of present study with the previous experimental and numerical studies. It can be seen from Figure 12 that at lower tailwater levels, i.e., 129.10 m, the sequent depths agreed well with the experimental data from Kucukali and Chanson [94] and numerical study by Bayon-Batrachia and Lope-Jimenez [4]. However, as the tailwater levels increased, the sequent depths showed deviation from the compared studies.

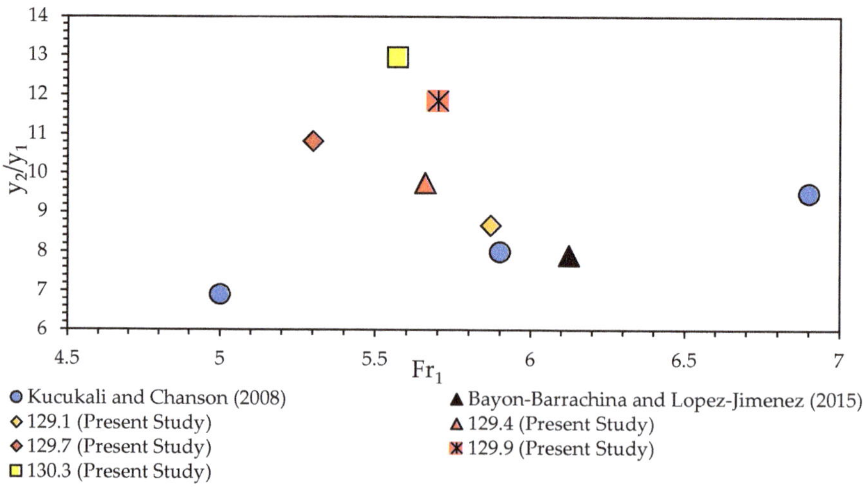

Figure 12. Comparison of sequent depth with the relevant literature studies [2,94] at various tailwater levels.

4.2.4. Free-Surface and Depth-Averaged Velocities

Figure 13a,b shows the free surface and depth-averaged velocities at various tailwater levels in the studied basin. The maximum values for longitudinal velocity were observed in the supercritical region before the HJ. Due to the recirculation and turbulence in the HJ region, negative velocity was observed. The maximum negative velocities were observed at lower tailwater levels, as shown in Figure 13a.

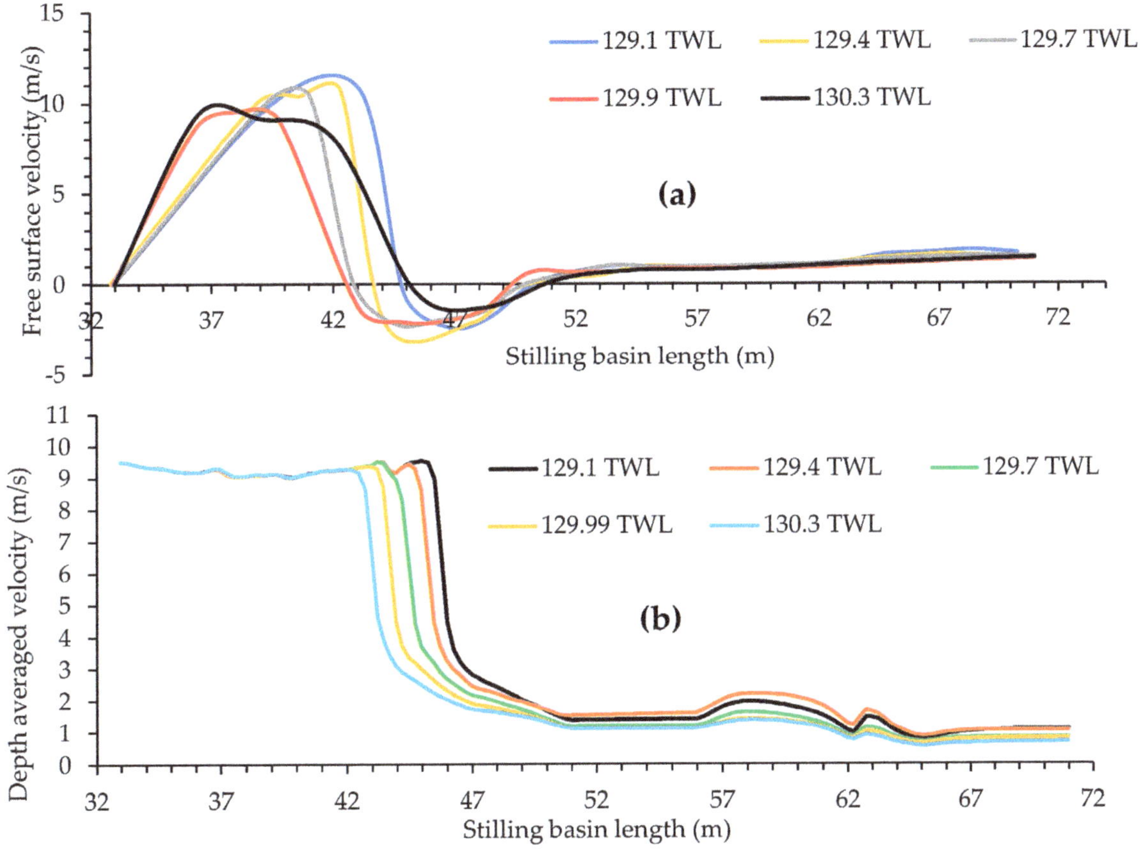

Figure 13. Velocity profiles in the stilling basin at the investigated TWLs. (**a**) Free−surface velocity profiles, (**b**) depth−averaged velocity.

The results showed that as the tailwater levels increased, the velocity values in the HJ and in the subcritical region decreased. Compared with lower tailwater levels, at a constant discharge, the velocity decay in the HJ region was found more at higher tailwater levels. Additionally, as compared with lower tailwater levels, at higher tailwater levels, the values of the velocities were small at the end of the end stilling basin. As compared with the numerical study by Chaudhry [36], at all the studied tailwater levels, the observed velocity values in the present numerical study were found to be higher.

Figure 14b shows depth-averaged velocity profiles at the studied TWLs. These velocity profiles are drawn from the centerline of the bay. At all the TWLs, the maximum velocity was found just before the initial locations of the HJs, which reached 9.49, 9.45, 9.30, 9.38, and 9.28 m/s at TWLs of 129.10, 129.40, 129.70, 129.99, and 130.30, respectively. In the transition regions of the HJ, due to the eddies and fluid recirculation, the velocity rapidly decreased and remained consistent in the baffle block region. After the baffle blocks, a slight increase in the velocity values was noticed, from X = 57 m to 62 m. From the jump initiating location to the start of friction blocks region, the maximum velocity was noticed at the 129.40 m TWL; however, after the friction blocks, the velocity values became equivalent to those that were observed at the 129.10 TWL. In addition, in and after the friction blocks region, the velocity values at the studied TWLs were further reduced. The results further showed that as the TWL was increased, the velocity values in the stilling basin were found to be decreased due to the higher tailwater depths. At the basin's end, the maximum and minimum velocity values reached 1.1 m/s and 0.70 m/s at the 129.10 and 130.30 TWLs, respectively.

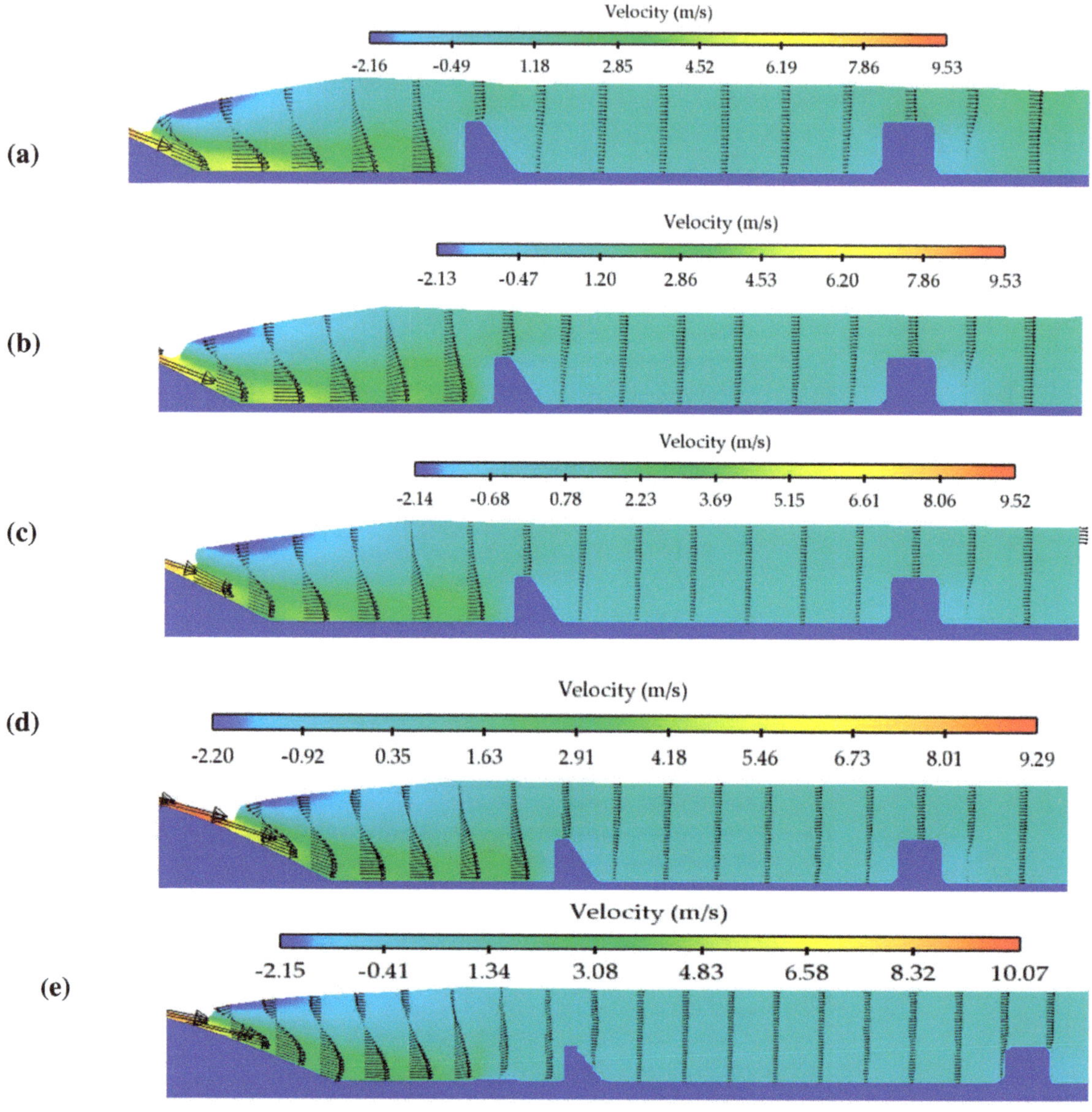

Figure 14. Velocity profiles at different tailwater levels. (**a**) 129.10 m, (**b**) 129.40 m, (**c**) 129.70 m, (**d**) 129.99 m, and (**e**) 130.30 m.

Figure 14 shows the behavior of velocity vectors in the longitudinal direction at various tailwater levels. Due to the supercritical velocity, the contracted flow jet was impinging at the toe of the glacis, which decreased the velocities at the upper fluid regions. Before the stilling basin appurtenances, reverse flows and eddies can be seen in the HJ at all the investigated TWLs. The flow behaviors of the upper fluid region of the HJ indicated typical backward velocity profiles, as described by Ead and Rajaratnam [11–13]. As the flow reached a steady state, these reverse fluid circulations stabilized, which showed the stagnation zones. The analysis showed that the recirculation region occurred in the HJ and between and after the baffle blocks, and the maximum backward velocity profiles were found in the developed areas. At all the investigated TWLs, after the HJ the flow recovering zone starts and the negative velocity profile becomes positive.

Additionally, the effect of energy dissipation devices shows that after the baffle blocks, the velocity profiles near the bed decreased and became positive on the free surface. In addition, at all the studied tailwater levels the vertical velocity profiles followed the trend of Ead and Rajaratnam [11–13]. The maximum velocities were found at the floor level before the baffle blocks and decreased as the flow moved forward to the end of the stilling basin.

4.2.5. Turbulent Kinetic Energies (TKEs)

The root mean square values of velocity fluctuations were used to calculate turbulent kinetic energies (TKE) at various locations. By considering the successive velocity values, the value of root mean square velocity (u_{rms}) can be obtained using Equation (10).

$$U_{rms} = \sqrt{\frac{\left(u_1^2 + u_2^2 + u_3^2 + \ldots + u_n^2\right)}{n}} \tag{10}$$

In the above Equation (10), u_1, u_2, and u_3 are the successive velocity values in the horizontal direction. Using the velocity values, the TKE can be calculated using Equation (11).

$$\text{TKE} = 1/2\left(u_{rms}^2 + v_{rms}^2 + w_{rms}^2\right) \tag{11}$$

whereas u_{rms}, v_{rms}, and w_{rms} are the root mean square velocities in the x, y, and z directions, respectively.

Figure 15a shows the variation in the TKE in 129.10 m tailwater. The TKE values were computed from HJ initiation to the termination points, such as at the supercritical flow region and within the HJ and subcritical regions. The maximum value for the TKE was found after the HJ and decreased afterward. At a 129.10 m tailwater level, the maximum value for the TKE reached 3.72 m^2/s^2 at X =10. At all the fluid depths at X = 52, after the HJ the TKE values were found to level off, as shown in Figure 15a. The result further showed that the TKEs were found to be maximal in the middle regions of the flow depths and that that minimum values were found at the free surface. In Figure 15b, as in the lower tailwater levels, the maximum value for the TKE was found after the HJ initiating location, and the value reached 3.12 m^2/s^2. However, as compared with lower tailwater levels, the values for TKEs in 129.40 m tailwater were found to be lower and the TKEs levelled off earlier within and after the HJ. In the HJ regions, following the similar trends for the 129.10 and 129.40 m tailwater levels, the TKEs in the 129.70, 129.99, and 130.30 m tailwater levels reached 3.20, 3.15, and 2.90 $m^{2/}s^2$, respectively, and their values decreased after the HJ, as illustrated in Figure 15c–e. Additionally, the results for the TKEs showed a trend that was noted in the numerical studies (Nikmehr and Aminpour [71]; Soori et al. [76]).

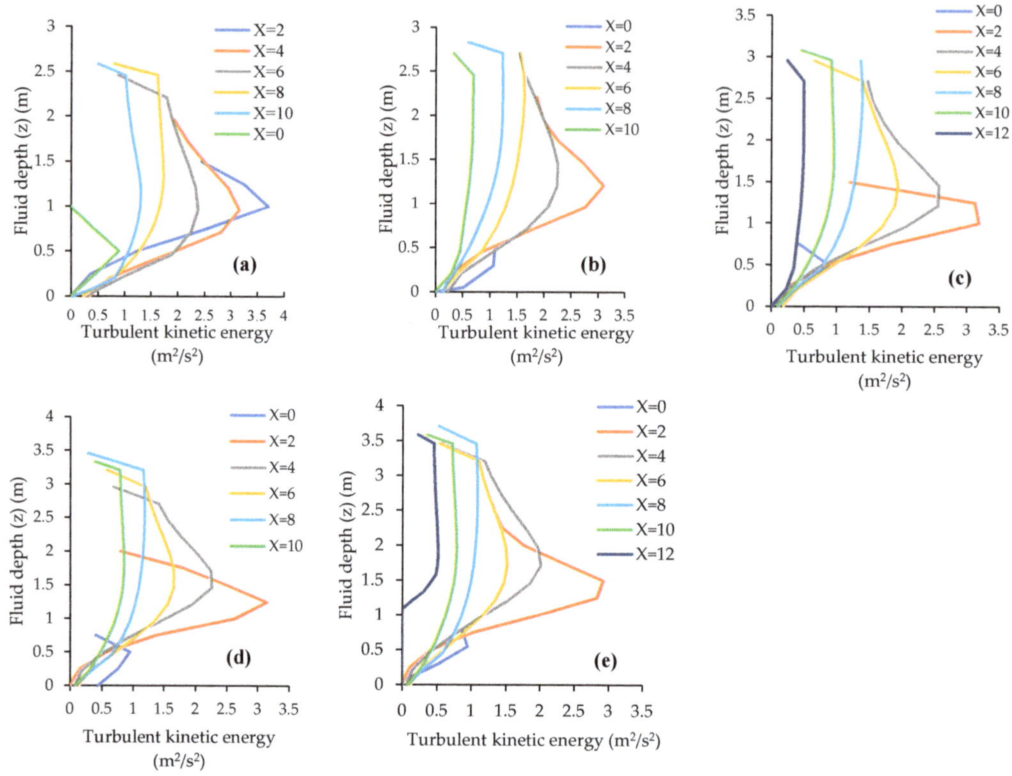

Figure 15. TKEs in the hydraulic jump region at different tailwater levels. (**a**) 129.10 m, (**b**) 129.40 m, (**c**) 129.70 m, (**d**) 129.99 m, and (**e**) 130.30 m.

Figure 16 shows 2D illustrations of turbulent kinetic energies captured using RNG-K-ε at five different tailwater levels. Figure 16a shows that in 129.10 m tailwater, the maximum turbulent kinetic energy was found before and within the HJ. The rest of the kinetic energy after the jump was dissipated by two rows of baffle blocks. After the baffle blocks, the flow became less turbulent and TKE was reduced and gradually leveled off at the end of stilling basin. Similarly, in the case of a 129.40 m tailwater level, the overall values for the TKE decreased compared with the 129.10 m tailwater. However, the maximum TKE is found before and within the HJ, as shown in Figure 16b. After the baffle blocks, TKEs gradually reduced up to the end of stilling basin. Following the same development, at the 129.70 m, 129.99 m, and 130.30 m TWLs, the maximum TKEs are found before and within the HJ; this can be seen in Figure 16c–e.

The 2D illustrations further indicate that as the tailwater increased downstream of the barrage, the TKEs within the HJ decreased but the turbulent behavior of all the fluid layers was observed differently. At the 130.30 m tailwater level, the TKEs within the upper and lower layers of the HJ decreased compared with the central region, as illustrated in Figure 16e. This different behavior within the HJ could be due to the greater flow depth available for the jump.

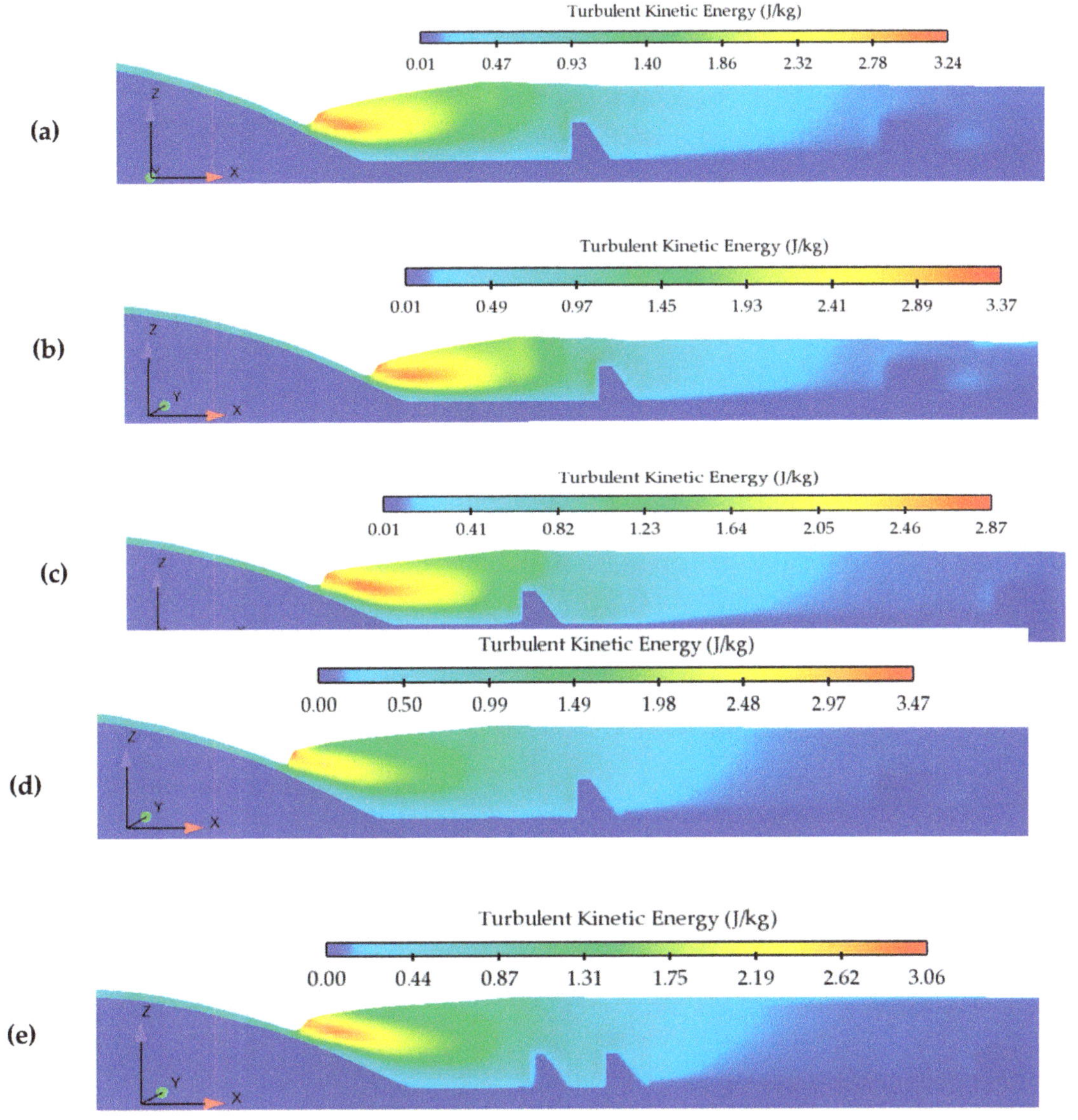

Figure 16. Two-dimensional representation of turbulent kinetic energy at a (**a**) 129.10 m TWL, (**b**) 129.40 m TWL, (**c**) 129.70 m TWL, (**d**) 129.99 m TWL, and (**e**) 130.30 m TWL.

5. Discussion and Real-World Implications

The preceding sections, Sections 4.1 and 4.2, focused on CHP identification and the effects of these parameters downstream of the Taunsa barrage, respectively. Frequency analysis, RII, and relative % score revealed different rankings for the hydraulic parameters, which were focused in many experimental and numerical studies. However, out of the CHPs, the velocity profile and Froude number were found to be the most investigated hydraulic parameters, as can be seen in Table 6. On the contrary, after extensive analysis of the literature, the investigation of the identified CHPs is found to be lacking on the studied barrage. The literature only revealed one study that employed a one-dimensional (1D) HEC-RAS model to investigate the limited hydraulic parameters downstream of the studied barrage [36]. Therefore, the present FLOW-3D model results are compared with 1D and the relevant studies from the literature, i.e., Bayon-Barrachina and Lope-Jimenez [4], Chaudhry [36], Ead and Rajaratnam [11–13], Nikmehr and Aminpour [71], Soori et al. [76], and Kucukali and Chanson [94]. After comparing with the results of the free surface profiles of HJs at the studied tailwater levels, it is revealed that the previously used 1D model underestimates the locations of the HJs; for which the errors reached 8%. Similarly, as compared with the present 3D models, the values for the Froude number in the supercritical region were found to be lower in 1D model, whereas the models' results for the Froude number in the subcritical region agree with previous data [36]. At all the tested tailwater levels, the analysis of sequent depths ratios showed agreement with the previous studies. In comparison with the 1D study [36], the results of velocity profiles within the stilling basin were found to be higher in the present models. However, at all the tested tailwater levels, the results of the vertical velocity profiles in the HJ region showed wall-jet-like profiles that showed agreement with Ead and Rajaratnam [11–13]. The maximum TKEs were found within the HJ region, and these declined as the distance from the HJ increased. In addition, the maximum amount of TKE was found within the central fluid depth; this started to decline towards the free surface and the basin's bed.

The findings of the present study will facilitate both hydraulic researchers and practitioners. Firstly, the results identified critical hydraulic parameters that should be given significant importance when a new hydraulic intervention is to be carried out, i.e., remodeling of hydraulic structures (Chaudary and Sarwar [35]; World Bank [57]; Zaidi et al. [58]). On the other hand, the use of numerical models is becoming prevalent in hydraulic investigations such as those carried out in the present study. The results showed that previous one-dimensional HEC-RAS studies are limited and unable to describe the flow characteristics spatially; this is why the FLOW-3D models' results are found to be different. In addition, the results from the FLOW-3D model also highlight that tailwater levels before the remodeling of the barrage were appropriate to hold the HJs at the glacis well above its toe. Therefore, it is believed that, in future, these modeling tools will also eliminate physical modeling, because conventionally scaled modeling is usually associated with the difficulties of terrain, concrete roughness, and flow measuring devices. The present study was limited to a single discharge value and turbulence model. Therefore, the model should also be tested and evaluated for higher values of discharge and other turbulence models.

6. Conclusions

The present study identified critical hydraulic parameters (CHPs) from the literature and studied these parameters downstream of Tuansa barrage using FLOW-3D numerical models. The study also investigated the effects of changes in tailwater levels on the HJ characteristics and compared the results with available previous data for the studied barrage. Following main conclusions drawn are:

- The literature review outlined thirty-three hydraulic parameters; out of those, the velocity profile, Froude number, free surface profile, shape of stilling basin, tailwater, and turbulent kinetic energy were the highly significant hydraulic parameters in the literature that were studied downstream of hydraulic structures.
- At all the investigated tailwater levels, no sweeping of the HJ was observed as reported in the previous studies. The location and elevation of HJs were observed to be different compared with a previous HEC-RAS one-dimensional hydraulic study. Upon comparison with the HJ results of the designed and downstream of a prototype barrage (i.e., remodeled basin), the distance of the HJ from the glacis toe was found to be higher, which further revealed the old basin (i.e., studied presently) was efficiently holding the HJ at the investigated discharge and TWLs.
- Non-linear trends for the Froude number and sequent depths were observed as the tailwater levels varied. On comparison with previous studies, the present models showed higher values for the Froude number and sequent depths, which showed deviation at higher tailwater levels.

- At lower tailwater levels, the vertical velocity profiles in the developing region of the HJ near the floor were found to be higher than the results at higher tailwater levels. At the investigated tailwater levels, jet-like velocity profiles were obtained in the HJ regions that levelled off as the distance from the HJ was increased.
- The maximum turbulent kinetic energy was found in the developing region of the HJ at the minimum tailwater level. After the impact and baffle blocks, the kinetic energy gradually reduced, and the minimum kinetic energy was observed at higher tailwater levels.

Based on the results of the numerical models, it can be said that the tailwater envelope for the studied discharge (44 m^3/s) for the Tuansa Barrage was within the acceptable limit, which holds the HJ on the downstream glacis. Using FLOW-3D numerical models, the study further confirmed that the previously used one-dimensional HEC-RAS models produced higher values for the hydraulic parameters. As the present study was limited to one discharge, more comprehensive studies for other discharges and turbulence models must be carried out on the stilling basin of the studied barrage. The study further suggests investigating the retrogression analysis downstream of the Taunsa barrage using three-dimensional numerical models.

Author Contributions: Conceptualization, M.W.Z. and I.H.; methodology, M.W.Z. and I.H.; software, i.e., FLOW-3D model, M.W.Z.; validation, M.W.Z. and A.R.G.; formal analysis, M.W.Z. and I.H.; investigation, M.W.Z. and A.R.G.; resources, M.W.Z. and I.H.; data curation, M.W.Z.; writing—original draft preparation, M.W.Z.; writing—review and editing, M.W.Z. and I.H.; visualization, M.W.Z. and A.R.G.; supervision, I.H.; project administration, I.H. All authors have read and agreed to the published version of the manuscript.

Funding: This research received no external funding except the publication fee expected to be granted by the Deanship of Scientific Research, Qassim University, Saudi Arabia.

Institutional Review Board Statement: Not applicable.

Informed Consent Statement: Not applicable.

Data Availability Statement: All relevant data are included in the paper or its Appendices A and B.

Acknowledgments: The researchers would like to thank the Deanship of Scientific Research, Qassim University, Saudi Arabia, for funding publication of this study.

Conflicts of Interest: The authors declare no conflict of interest.

Appendix A. Hydraulic Parameters from the Numerical Studies and Their Sources

Sr.No.	Hydraulic Parameters	Articles Cited
1	Velocity profile (VP)	[4,24,27,37,43,44,46,47,49,52,55,71,74–77]
2	Free surface profile (FSP)	[4,9,24,27,37,44,45,47,49,71,75–78]
3	Pressure profile (PP)	[4,24,44,45,49,50,72–74,76,78]
4	Turbulent kinetic energy (TKE)	[4,27,37,46,47,55,75,76]
5	Air volume value (AV)	[4,24,27,37,44,45,75]
6	Discharge measurement (DM)	[24,72,73,76–78]
7	Shape of stilling basin (SS)	[16,18,47,72]
8	Froude number (Fr_1)	[24,37,43,50,75,79]
9	HJ efficiency (η)	[24,44,50,74,76]
10	Reynolds shear stress (RSS)	[4,46,49]
11	Tailwater level (TWL)	[75]
12	Contraction coefficient (CC)	[55]
13	Head loss (HL)	[55]
14	Frictional forces (FF)	[55]
15	Momentum coefficient (MC)	[55]
16	Energy coefficient (EC)	[47]
17	Discharge coefficient (DC)	[47]
18	Wall shear stress (WSS)	[4,74]
19	Sequent depth of HJ (SDHJ)	[4,49,71,74]
20	Length of HJ (LHJ)	[49,71,74]
21	Scour profile (SP)	[9]
22	Bed profile (BP)	[37]
23	Energy dissipation (ED)	[47]
24	Bed shear stress (BSS)	[47,74]

Appendix B. Hydraulic Parameters from the Experimental Studies and Their Sources

Sr.No.	Hydraulic Parameters	Articles Cited
1	Velocity profile (VP)	[2,3,8–10,14,15,18,21–26,32,48,73,80–86]
2	Froude number (Fr_1)	[2,3,8,11,13,17–22,24,25,28,32,73,82–84,88,90,95]
3	Tailwater level (TWL)	[3,12,13,15,20,22,32,43,72–74,81,84–87,89–92]
4	Shape of stilling basin (SS)	[12,19,20,24–26,28,72,73,79,81,85,88,90,91,93]
5	Free surface profile (FSP)	[3,8,9,11,12,14,16,22–24,32,80,81,84,87,89,90,92,94]
6	Bed profile (BP)	[3,11,12,16,22–24,32,73,80,84,90,94,95]
7	Scour profile (SP)	[3,15,32,45,76,84,87,88,90,91,93,94]
8	Energy dissipation (ED)	[14,20,22–24,82,83,87,91]
9	Turbulence kinetic energy (TKE)	[14,20,22–24,73]
10	Pressure profile (PP)	[14,24,32,80,81,92]
11	Bed shear stress (BSS)	[2,11,17,22,48,73,85]
12	Reynolds shear stress (RSS)	[17,18,21,85,90]
13	Turbulence intensity (TI)	[17,18,21,81,85]
14	Submergence (SM)	[3,19,25,89]
15	Retrogression (RG)	[73,80,89]
16	HJ efficiency (η)	[22,23,29]
17	Momentum flux (MF)	[11,13]
18	Volume flux (VF)	[11,13]
19	Length of HJ (LHJ)	[8,26,86,92]
20	Turbulence production (TP)	[23]
21	Power spectra (PS)	[18]
22	Sequent depth of HJ (SDHJ)	[9,74]
23	Relative energy loss ($\Delta E/E1$)	[22]

References

1. Mahtabi, G.; Chaplot, B.; Azamathulla, H.M.; Pal, M. Classification of hydraulic jump in rough beds. *Water* **2020**, *12*, 2249. [CrossRef]
2. Maleki, S.; Fiorotto, V. Hydraulic Jump Stilling Basin Design over Rough Beds. *J. Hydraul. Eng.* **2021**, *147*, 04020087. [CrossRef]
3. Dey, S.; Sarkar, A. Response of velocity and turbulence in submerged wall jets to abrupt changes from smooth to rough beds and its application to scour downstream of an apron. *J. Fluid Mech.* **2006**, *556*, 387–419. [CrossRef]
4. Bayon-Barrachina, A.; Lopez-Jimenez, P.A. Numerical analysis of hydraulic jumps using OpenFOAM. *J. Hydroinform.* **2015**, *17*, 662–678. [CrossRef]
5. Bai, R.; Ning, R.; Liu, S.; Wang, H. Hydraulic Jump on a Partially Vegetated Bed. *Water Resour. Res.* **2022**, *58*, e2022WR032013. [CrossRef]
6. Li, P.; Zhu, D.Z.; Xu, T.; Zhang, J. Air Demand of a Hydraulic Jump in a Closed Conduit. *J. Hydraul. Eng.* **2022**, *148*, 04021058. [CrossRef]
7. Bantacut, A.Y.; Azmeri, A.; Jemi, F.Z.; Ziana, Z.; Muslem, M. An experiment of energy dissipation on USBR IV stilling basin—Alternative in modification. *J. Water Land Dev.* **2022**, *53*, 68–72. [CrossRef]
8. Mazumder, S. Hydraulic jump control using stilling basin with Adverse slope and positive step. *ISH J. Hydraul. Eng.* **2020**, *28*, 18–20. [CrossRef]
9. Macián-Pérez, J.F.; Bayón, A.; García-Bartual, R.; López-Jiménez, P.A.; Vallés-Morán, F.J. Characterization of Structural Properties in High Reynolds Hydraulic Jump Based on CFD and Physical Modeling Approaches. *J. Hydraul. Eng.* **2020**, *146*, 04020079. [CrossRef]
10. Elsayed, H.; Helal, E.; El-Enany, M.; Sobeih, M. Impacts of multi-gate regulator operation schemes on local scour downstream. *ISH J. Hydraul. Eng.* **2018**, *27*, 51–64. [CrossRef]
11. Ead, S.A.; Rajaratnam, N. Hydraulic Jumps on Corrugated Beds. *J. Hydraul. Eng.* **2002**, *128*, 656–663. [CrossRef]
12. Ead, S.A.; Rajaratnam, N. Plane Turbulent Wall Jets on Rough Boundaries with Limited Tailwater. *J. Eng. Mech.* **2004**, *130*, 1245–1250. [CrossRef]
13. Ead, S.A.; Rajaratnam, N. Plane turbulent surface jets in shallow tailwater. *J. Fluids Eng.* **2000**, *123*, 121–127. [CrossRef]
14. Chen, J.-G.; Zhang, J.-M.; Xu, W.-L.; Wang, Y.-R. Numerical simulation of the energy dissipation characteristics in stilling basin of multi-horizontal submerged jets. *J. Hydrodyn.* **2010**, *22*, 732–741. [CrossRef]
15. Balachandar, R.; Kells, J.; Thiessen, R.; Gunal, M.; Guven, A.; Kells, J.A.; Hagel, K.P. The effect of tailwater depth on the dynamics of local scour. *Can. J. Civ. Eng.* **2000**, *27*, 138–150. [CrossRef]
16. Sarker, M.; Rhodes, D. Calculation of free-surface profile over a rectangular broad-crested weir. *Flow Meas. Instrum.* **2004**, *15*, 215–219. [CrossRef]
17. Dey, S.; Sarkar, A. Characteristics of Turbulent Flow in Submerged Jumps on Rough Beds. *J. Eng. Mech.* **2008**, *134*, 599. [CrossRef]
18. Liu, M.; Rajaratnam, N.; Zhu, D.Z. Turbulence Structure of Hydraulic Jumps of Low Froude Numbers. *J. Hydraul. Eng.* **2004**, *130*, 511–520. [CrossRef]
19. Habibzadeh, A.; Loewen, M.R.; Rajaratnam, N. Performance of Baffle Blocks in Submerged Hydraulic Jumps. *J. Hydraul. Eng.* **2012**, *138*, 902–908. [CrossRef]
20. Pagliara, S.; Lotti, I.; Palermo, M. Hydraulic jump on rough bed of stream rehabilitation structures. *J. Hydro-Environ. Res.* **2008**, *2*, 29–38. [CrossRef]

21. Zobeyer, A.H.; Jahan, N.; Islam, Z.; Singh, G.; Rajaratnam, N. Turbulence characteristics of the transition region from hydraulic jump to open channel flow. *J. Hydraul. Res.* **2010**, *48*, 395–399. [CrossRef]
22. Abbaspour, A.; Dalir, A.H.; Farsadizadeh, D.; Sadraddini, A.A. Effect of sinusoidal corrugated bed on hydraulic jump characteristics. *J. Hydro-Environ. Res.* **2009**, *3*, 109–117. [CrossRef]
23. Mignot, E.; Cienfuegos, R. Energy Dissipation and Turbulent Production in Weak Hydraulic Jumps. *J. Hydraul. Eng.* **2010**, *136*, 116–121. [CrossRef]
24. Babaali, H.; Shamsai, A.; Vosoughifar, H. Computational Modeling of the Hydraulic Jump in the Stilling Basin with Convergence Walls Using CFD Codes. *Arab. J. Sci. Eng.* **2014**, *40*, 381–395. [CrossRef]
25. Habibzadeh, A.; Wu, S.; Ade, F.; Rajaratnam, N.; Loewen, M.R. Exploratory Study of Submerged Hydraulic Jumps with Blocks. *J. Hydraul. Eng.* **2011**, *137*, 706–710. [CrossRef]
26. Ali, A.M.; Mohamed, Y.A. Effect of stilling basin shape on the hydraulic characteristics of the flow downstream radial gates. *Alex. Eng. J.* **2010**, *49*, 393–400. [CrossRef]
27. Mishra, K. 3DNumerical Modelling of Energy Dissipation in Flexible Apron of Barrages. *J. Inst. Eng. Ser. A* **2015**, *96*, 47–56. [CrossRef]
28. Alikhani, A.; Behrozi-Rad, R.; Fathi-Moghadam, M. Hydraulic jump in stilling basin with vertical end sill. *Int. J. Phys. Sci.* **2010**, *5*, 25–29.
29. Elsaeed, G.; Ali, A.; Abdelmageed, N.; Ibrahim, A. Effect of End Step Shape in the Performance of Stilling Basins Downstream Radial Gates. *J. Sci. Res. Rep.* **2016**, *9*, 1–9. [CrossRef]
30. Tiwari, H.L.; Panwar, A.; Gehlot, B.; Singh, J. Study of Shape of Intermediate Sill on the Design of Stilling Basin Model. *Int. J. Res. Eng. Technol.* **2014**, *3*, 133–138. [CrossRef]
31. Tiwari, H.L.; Gehlot, V.K.; Tiwari, S. Effect of Height of Triangular Siil on the Performance of Stilling Basin Model. *Int. J. Res. Eng. Technol.* **2014**, *3*, 868–873. [CrossRef]
32. Hager, W.H.; Li, D. Sill-controlled energy dissipator. *J. Hydraul. Res.* **1992**, *30*, 181. [CrossRef]
33. Herrera-Granados, O.; Kostecki, S.W. Numerical and physical modeling of water flow over the ogee weir of the new Niedów barrage. *J. Hydrol. Hydromechanics* **2016**, *64*, 67–74. [CrossRef]
34. Zulfiqar, C.; Kaleem, S.M. Launching/Disappearance of Stone Apron, Block Floor Downstream of the Taunsa Barrage and Unprecedent Drift of the River towards Kot Addu Town. *Sci. Technol. Dev.* **2015**, *34*, 60–65. [CrossRef]
35. Chaudary, Z.A.; Sarwar, M.K.; Barrage, R.T. Rehabilitated Taunsa Barrage: Prospects and Concerns. *Sci. Technol. Dev.* **2014**, *33*, 127–131.
36. Chaudhry, Z.A. Surface Flow Hydraulics of Taunsa Barrage: Before and after Rehabilitation. *Pak. J. Sci.* **2010**, *62*, 116–119.
37. Witt, A.; Gulliver, J.; Shen, L. Simulating air entrainment and vortex dynamics in a hydraulic jump. *Int. J. Multiph. Flow* **2015**, *72*, 165–180. [CrossRef]
38. Yoo, S.; Hong, M.; Wang, H. 3-Dimensional numerical study of a flow patterns around a multipurpose dam. In Proceedings of the International Conference on Hydro Informatics, Cardiff, UK, 1–5 July 2002; pp. 353–357.
39. Aydogdu, M.; Gul, E.; Dursun, O.F. Experimentally Verified Numerical Investigation of the Sill Hydraulics for Abruptly Expanding Stilling Basin. *Arab. J. Sci. Eng.* **2022**, *48*, 4563–4581. [CrossRef]
40. Mukha, T.; Almeland, S.K.; Bensow, R.E. Large-Eddy Simulation of a Classical Hydraulic Jump: Influence of Modelling Parameters on the Predictive Accuracy. *Fluids* **2022**, *7*, 101. [CrossRef]
41. El Azim, N.A.; Saleh, O.; Tohamy, E.; Mahgoub, S.; Ghany, S. Effect of Vertical Screen on Energy Dissipation and Water Surface Profile Using Flow 3D. *Egypt. Int. J. Eng. Sci. Technol.* **2022**, *38*, 20–25.
42. Kosaj, R.; Alboresha, R.S.; Sulaiman, S.O. Comparison between Numerical Flow3d Software and Laboratory Data, for Sediment Incipient Motion. In Proceedings of the IOP Conference Series: Earth and Environmental Science, Baghdad, Iraq, 3–4 November 2022; p. 961. [CrossRef]
43. Mirzaei, H.; Tootoonchi, H. Experimental and numerical modeling of the simultaneous effect of sluice gate and bump on hydraulic jump. *Model. Earth Syst. Environ.* **2020**, *6*, 1991–2002. [CrossRef]
44. Macián-Pérez, J.F.; García-Bartual, R.; Bayon, B.H.A.; Vallés-Morán, F.J. Analysis of the flow in a typified USBR II stilling basin through a numerical and physical modeling approach. *Water* **2020**, *12*, 227. [CrossRef]
45. Daneshfaraz, R.; Ghader, A. Numerical Investigation of Inverse Curvature Ogee Spillway. *Civ. Eng. J.* **2017**, *3*, 1146. [CrossRef]
46. Anjum, N.; Ghani, U.; Pasha, G.A.; Rashid, M.U.; Latif, A.; Rana, M.Z.Y. Reynolds stress modeling of flow characteristics in a vegetated rectangular open channel. *Arab. J. Sci. Eng.* **2018**, *43*, 5551–5558. [CrossRef]
47. Dargahi, B. Experimental Study 3DNumerical Simulations for a Free-Overflow Spillway. *J. Hydraul. Eng.* **2006**, *132*, 899–907. [CrossRef]
48. Karim, O.A.; Ali, K.H.M. Prediction of flow patterns in local scour holes caused by turbulent water jets. *J. Hydraul. Res.* **2000**, *38*, 279–287. [CrossRef]
49. Liu, X.; García, M.H. Three-Dimensional Numerical Model with Free Water Surface and Mesh Deformation for Local Sediment Scour. *J. Waterw. Port Coast. Ocean. Eng.* **2008**, *134*, 203–217. [CrossRef]
50. Bayon, A.; Valero, D.; García-Bartual, R.; Vallés-Morán, F.J.; López-Jiménez, P.A. Performance assessment of OpenFOAM and FLOW-3D in the numerical modeling of a low Reynolds number hydraulic jump. *Environ. Model. Softw.* **2016**, *80*, 322–335. [CrossRef]

51. Bayon-Barrachina, A.; Valles-Moran, F.J.; Lopes-Jiménez, P.A.; Bayn, A.; Valles-Morn, F.J.; Lopes-Jimenez, P.A. Numerical Analysis and Validation of South Valencia Sewage Collection System. In Proceedings of the 36th IAHR World Congress, The Hague, The Netherlands, 2015; Volume 17, pp. 1–11.
52. Nguyen, V.T.; Nestmann, F.; Scheuerlein, H. Three-dimensional computation of turbulent flow in meandering channels and rivers. *J. Hydraul. Res.* **2007**, *45*, 595–609. [CrossRef]
53. Riad, P.; Roelvink, D.J.A.; Ahmed, A.; Boeru, P. Use of Mathematical Models to Improve the Design of the Stilling Basin. Available online: https://doi.org/10.13140/RG.2.2.34562.94409 (accessed on 27 September 2022).
54. Chatila, J.; Tabbara, M. Computational modeling of flow over an ogee spillway. *Comput. Struct.* **2004**, *82*, 1805–1812. [CrossRef]
55. Cassan, L.; Belaud, G. Experimental and Numerical Investigation of Flow under Sluice Gates. *J. Hydraul. Eng.* **2012**, *138*, 367–373. [CrossRef]
56. Carvalho, R.F.; Lemos, C.M.; Ramos, C.M. Numerical computation of the flow in hydraulic jump stilling basins. *J. Hydraul. Res.* **2008**, *46*, 739–752. [CrossRef]
57. World Bank. Announcement of World Bank Provides United States (U.S.) One Hundred Twenty-Three Million Dollars Loan to Rehabilitate and Modernize Taunsa Barrage in Pakistan on 15 March 2005 (English). Available online: http://documents.worldbank.org/curated/en/099915203252274440/IDU01bcc88300a93504d270b1a20dbc553c584df (accessed on 25 March 2022).
58. Zaidi, S.; Khan, M.A.; Rehman, S.U. Planning and Design of Taunsa Barrage Rehabilitation Project. In Proceedings of the Pakistan Engineering Congress, 71st Annual Session Proceedings, Pakistan, Lahore, 2004; Paper No. 687. pp. 228–286.
59. Aloini, D.; Dulmin, R.; Mininno, V.; Ponticelli, S. Supply chain management: A review of implementation risks in the construction industry. *Bus. Process Manag. J.* **2012**, *18*, 735–761. [CrossRef]
60. Ullah, F.; Ayub, B.; Siddiqui, S.Q.; Thaheem, M.J. A review of public-private partnership: Critical factors of concession period. *J. Financ. Manag. Prop. Constr.* **2016**, *21*, 269–300. [CrossRef]
61. Ayub, B.; Thaheem, M.J.; Din, Z.U. Dynamic Management of Cost Contingency: Impact of KPIs and Risk Perception. *Procedia Eng.* **2016**, *145*, 82–87. [CrossRef]
62. Muneeswaran, G.; Manoharan, P.; Awoyera, P.O.; Adesina, A. A statistical approach to assess the schedule delays and risks in Indian construction industry. *Int. J. Constr. Manag.* **2020**, *20*, 450–461. [CrossRef]
63. Gündüz, M.; Nielsen, Y.; Özdemir, M. Quantification of Delay Factors Using the Relative Importance Index Method for Construction Projects in Turkey. *J. Manag. Eng.* **2013**, *29*, 133–139. [CrossRef]
64. Ahmad, Z.; Thaheem, M.J.; Maqsoom, A. Building information modeling as a risk transformer: An evolutionary insight into the project uncertainty. *Autom. Constr.* **2018**, *92*, 103–119. [CrossRef]
65. Siddiqui, S.Q.; Ullah, F.; Thaheem, M.J.; Gabriel, H.F. Six Sigma in construction: A review of critical success factors. *Int. J. Lean Six Sigma* **2016**, *7*, 171–186. [CrossRef]
66. Sciences, F. *FLOW 3D User Manual, Version 10.1*; Flow Sciences: Santa Fe, NM, USA, 2013.
67. Yakhot, V.; Thangam, S.; Gatski, T.B.; Orszag, S.A.; Speziale, C.G. Development of turbulence models for shear flows by a double expansion technique. *Phys. Fluids A* **1991**, *4*, 1510–1520. [CrossRef]
68. Training, E. Water & Environment Training Guide 1: CFD Project Workflow Guide, n.d. Available online: https://www.flow3d.com/wp-content/uploads/2017/03/cfd-project-workflow-guide-v112.pdf (accessed on 7 August 2021).
69. Kometa, S.T.; Olomolaiye, P.O.; Harris, F.C. Attributes of UK construction clients influencing project consultants' performance. *Constr. Manag. Econ.* **1994**, *12*, 433–443. [CrossRef]
70. Bennett, N.D.; Croke, B.F.W.; Guariso, G.; Guillaume, J.H.A.; Hamilton, S.H.; Jakeman, A.J.; Marsili-Libelli, S.; Newham, L.T.H.; Norton, J.P.; Perrin, C.; et al. Characterising performance of environmental models. *Environ. Model. Softw.* **2013**, *40*, 1–20. [CrossRef]
71. Nikmehr, S.; Aminpour, Y. Numerical simulation of hydraulic jump over rough beds, Period. *Polytech. Civ. Eng.* **2020**, *64*, 396–407. [CrossRef]
72. Savage, B.M.; Johnson, M.C. Flow over Ogee Spillway: Physical and Numerical Model Case Study. *J. Hydraul. Eng.* **2001**, *127*, 640–649. [CrossRef]
73. Johnson, M.C.; Savage, B.M. Physical and Numerical Comparison of Flow over Ogee Spillway in the Presence of Tailwater. *J. Hydraul. Eng.* **2006**, *132*, 1353–1357. [CrossRef]
74. Ebrahimiyan, S.; Hajikandi, H.; Bejestan, M.S.; Jamali, S.; Asadi, E. Numerical Study on the Effect of Sediment Concentration on Jump Characteristics in Trapezoidal Channels. *Iran. J. Sci. Technol.—Trans. Civ. Eng.* **2021**, *45*, 1059–1075. [CrossRef]
75. Bricker, J.D.; Nakayama, A. Contribution of Trapped Air Deck Superelevation, and Nearby Structures to Bridge Deck Failure during a Tsunami. *J. Hydraul. Eng.* **2014**, *140*, 05014002. [CrossRef]
76. Soori, S.; Babaali, H.; Soori, N. An Optimal Design of the Inlet and Outlet Obstacles at USBR II Stilling Basin. *Int. J. Sci. Eng. Appl.* **2017**, *6*, 134–142. [CrossRef]
77. Nguyen, V.T. 3D numerical simulation of free surface flows over hydraulic structures in natural channels and rivers. *Appl. Math. Model* **2015**, *39*, 6285–6306. [CrossRef]
78. Geun, K.D.; Jae Hyun, P. Analysis of Flow Structure over Ogee-Spillway in Consideration of Scale and Roughness Effects by Using CFD Model. *KSCE J. Civ. Eng.* **2005**, *9*, 161–169.
79. D'Agostino, V.; Ferro, V. Scour on Alluvial Bed Downstream of Grade-Control Structures. *J. Hydraul. Eng.* **2004**, *130*, 24–37. [CrossRef]

80. Chen, Q.; Dai, G.; Liu, H. Volume of Fluid Model for Turbulence Numerical Simulation of Stepped Spillway Overflow. *J. Hydraul. Eng.* **2002**, *128*, 683–688. [CrossRef]
81. Lin, P.; Xu, W. NEWFLUME: A numerical water flume for two-dimensional turbulent free surface flows NEWFLUME: Un canal numérique hydraulique pour écoulements bidimensional à surface libre turbulents. *J. Hydraul. Res.* **2006**, *44*, 79–93. [CrossRef]
82. Chaudhry, Z.A. Performance Assessment of Taunsa Barrage Subsidiary Weir for Long Term Rehabilitation Planning. *Pak. J. Engg. Appl. Sci.* **2010**, *7*, 65–70.
83. Adduce, C.; Sciortino, G. Scour due to a horizontal turbulent jet: Numerical and experimental investigation. *J. Hydraul. Res.* **2006**, *44*, 663–673. [CrossRef]
84. Guan, D.; Melville, B.W.; Friedrich, H. Flow Patterns and Turbulence Structures in a Scour Hole Downstream of a Submerged Weir. *J. Hydraul. Eng.* **2014**, *140*, 68–76. [CrossRef]
85. Neisi, M.; Sajjadi, S.M.; Shafai Bejestan, M. Experimental Investigation of Discharge and Flow Energy Dissipation of Elliptical Lopac Gate in Free Flow Condition. *Irrig. Sci. Eng.* **2020**, *46*, 91–101. [CrossRef]
86. Grimaldi, C.; Gaudio, R.; Cardoso, F.C.A.H. Control of Scour at Bridge Piers by a Downstream Bed Sill. *J. Hydraul. Eng.* **2009**, *135*, 13–21. [CrossRef]
87. Kabiri-Samani, A.; Ansari, A.; Borghei, S.M. Hydraulic behaviour of flow over an oblique weir. *J. Hydraul. Res.* **2010**, *48*, 669–673. [CrossRef]
88. Gaudio, R.; Marion, A. Time evolution of scouring downstream of bed sills. *J. Hydraul. Res.* **2003**, *41*, 271–284. [CrossRef]
89. Stojnic, I.; Pfister, M.; Matos, J.; Schleiss, A.J. Hydraulic jump downstream of a stepped chute: An experimental study. In Proceedings of the E-Proceedings 38th IAHR World Congr, Panama City, Panama, 1–6 September 2019; pp. 2056–2065. [CrossRef]
90. Wüthrich, D.; Chamoun, S.; De Cesare, G.; Schleiss, A.J. Behaviour of a scour protection overlay with randomly distributed concrete prisms in plunge pools downstream of mobile barrages for exceptional operation conditions. In Proceedings of the 7th IAHR International Symposium Hydraulic Structures, ISHS 2018, Aachen, Germany, 15–18 May 2018; pp. 150–158. [CrossRef]
91. Marion, A.; Lenzi, M.A.; Comiti, F. Effect of sill spacing and sediment size grading on scouring at grade-control structures. Earth Surf. Process. *Landforms* **2004**, *29*, 983–993. [CrossRef]
92. Goel, A.; Pal, M. Application of support vector machines in scour prediction on grade-control structures. *Eng. Appl. Artif. Intell.* **2009**, *22*, 216–223. [CrossRef]
93. Goel, A. Design of stilling basin for circular pipe outlets Can. *J. Civ. Eng.* **2008**, *35*, 1365–1374. [CrossRef]
94. Kucukali, S.; Chanson, H. Turbulence measurements in the bubbly flow region of hydraulic jumps. *Exp. Therm. Fluid Sci.* **2008**, *33*, 41–53. [CrossRef]
95. Zaidi, S.M.A.; Amin, M.; Ahmadani, M.A. Performance Evaluation of Taunsa Barrage Emergency Rehabilitation and Modernization Project. Paper No. 705. 2011, pp. 650–682. Available online: https://pecongress.org.pk/images/upload/books/PERFORMANCE%2520EVALUATION%2520OF%2520TAUNSA%2520BARRAGE%2520EMERGENCY%2520REHABILIT.pdf (accessed on 3 October 2023).

fluids

MDPI

Article

Understanding the Influence of the Buoyancy Sign on Buoyancy-Driven Particle Clouds

Ali O. Alnahit [1], Nigel Berkeley Kaye [2,*] and Abdul A. Khan [2,*]

1 Civil Engineering Department, King Saud University, P.O. Box 800, Riyadh 11421, Saudi Arabia; alialnaheet@ksu.edu.sa

2 Glenn Department of Civil Engineering, Clemson University, Clemson, SC 29631, USA

* Correspondence: nbkaye@clemson.edu (N.B.K.); abdkhan@clemson.edu (A.A.K.)

Abstract: A numerical model was developed to investigate the behavior of round buoyancy-driven particle clouds in a quiescent ambient. The model was validated by comparing model simulations with prior experimental and numerical results and then applied the model to examine the difference between releases of positively and negatively buoyant particles. The particle cloud model used the entrainment assumption while approximating the flow field induced by the cloud as a Hill's spherical vortex. The motion of individual particles was resolved using a particle tracking equation that considered the forces acting on them and the induced velocity field. The simulation results showed that clouds with the same initial buoyancy magnitude and particle Reynolds number behaved differently depending on whether the particles were more dense or less dense than the ambient fluid. This was found even for very low initial buoyancy releases, suggesting that the sign of the buoyancy is always important and that, therefore, the Boussinesq assumption is never fully appropriate for such flows.

Keywords: particle clouds; dredging; two-phase flows; integral models

Citation: Alnahit, A.O.; Kaye, N.B.; Khan, A.A. Understanding the Influence of the Buoyancy Sign on Buoyancy-Driven Particle Clouds. *Fluids* **2024**, 9, 101. https://doi.org/10.3390/fluids9050101

Academic Editors: Jaan H. Pu, Prashanth Reddy Hanmaiahgari, Manish Pandey, Mohammad Amir Khan and Sourabh V. Apte

Received: 28 February 2024
Revised: 10 April 2024
Accepted: 18 April 2024
Published: 23 April 2024

1. Introduction

Particle cloud dynamics play an important role in many natural and human-induced processes. For instance, particle cloud dynamics are relevant in activities such as dumping dredged sediment waste in assigned water areas or placing sand in water for land reclamation purposes [1]. Overall, large amounts of sediment are removed from inland waterways through dredging operations every year, particularly in estuarine areas [2]. Storing the dredged sediment on land is expensive; therefore, releasing it into seawater has become an attractive option [2]. However, even if the sediment is not contaminated, releasing it into seawater can have consequences such as increased turbidity and disruption of biological habitats [3–6]. Therefore, it is crucial to understand the fate of dredged sediment releases to manage estuarine and coastal zones effectively.

Previous research has examined the behavior of sediments that are released in stagnant homogeneous and stratified ambient conditions [7–11]. When particles are released into a fluid ambient (e.g., water), they behave as a source of negative buoyancy. In such cases, the cloud of particles is often considered to be a continuous, single phase of uniform density that is no different from a heavy fluid that has been released with the same average density. During the initial phase, the particle cloud accelerates and expands rapidly since the cloud behaves as a homogeneous dense fluid cloud. It has been shown in many studies that the duration of this phase depends on the initial buoyancy, and the cloud reaches the self-preserving phase after a depth equivalent to one to three times its source diameter [8]. In the self-preserving phase, the fluid and particle cloud undergoes deceleration due to the rapid entrainment of ambient fluid. As there is no representative length scale in this region, it is typically assumed that the buoyant cloud has reached a state of self-similarity where all lengths are proportionate [12]. Finally, as the descending particle cloud decelerates and its velocity approaches the settling velocity of individual particles, the circulation is insufficient to keep the particles suspended. Hence, the particles settle out of the cloud, descending as a particle "swarm". A schematic diagram of this flow is shown in Figure 1 for both positively and negatively buoyant particle clouds.

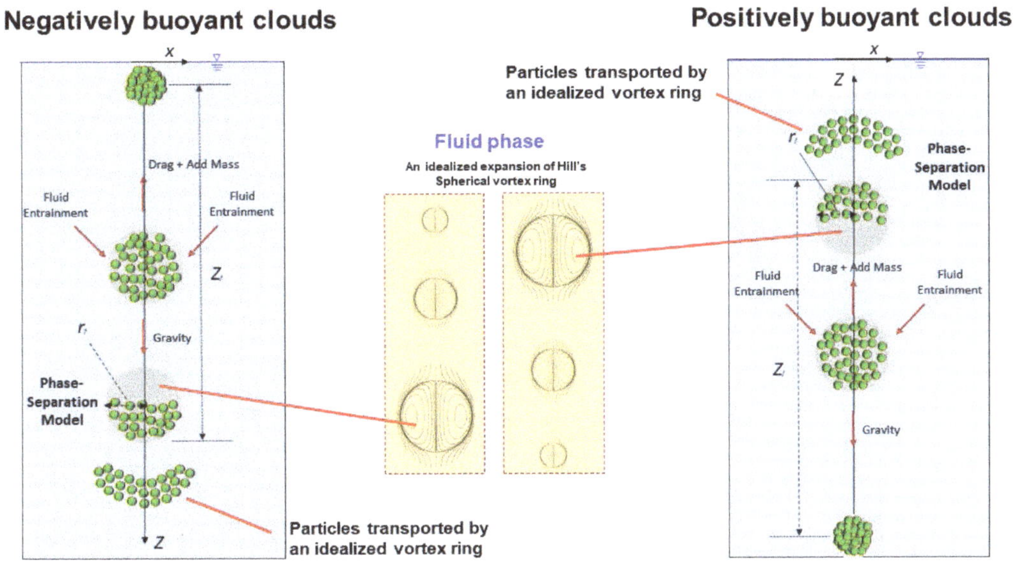

Figure 1. Schematic diagram showing the particle cloud falling and spreading out with the particles eventually falling out of the vortex.

Sediment clouds have been the subject of several past investigations. Studies have investigated the effects of different release conditions on the behavior of particle clouds in the ambient field. For example, the impact of water content in released sediments [11] and the effect of cloud momentum generated by releasing dry sediments at a height above the water surface [10]. The influence of various ambient conditions, such as ambient stratification, on particle clouds has also been researched [7,13]. The influence of ambient waves was studied in [14], ambient cross-flow in [15], and the two-phase characteristics of a particle cloud in [16].

Several studies using computational fluid dynamics (CFD) have enhanced the modeling of sediment clouds [17–20]. For instance, ref. [8] created a CFD model that treated discrete particles as a continuous density field and used a mixing length model for turbulence closure. Turbulence coefficients were estimated through calibration with Scorer's previous experimental data [21]. Li's findings showed that fine particles with low settling velocity exhibit vortex motion (a vortex ring). Gu and Li [22] developed a CFD model combining Eulerian–Lagrangian methods to study sediment clouds with multiple particle sizes. Fluid phase motion is computed using a two-equation turbulence model, while solid phase (particles) motion is computed by assuming the particle's velocity to be the sum of the fluid random velocity and the particle settling velocity. The two phases are coupled using the multiphase particle-in-cell method. In addition to CFD modeling, a number of researchers have developed numerical models to evaluate the environmental impact of sediment disposal in open-water activities. For example, ref. [16] developed a two-phase sediment cloud model that was based on Hill's spherical vortex, and they used particle tracking equations to follow the particles. They assumed that particle clouds had uniform particle sizes but later expanded the model to include poly-disperse releases [1].

In this study, we extend the cloud model proposed by Lai et al. [16] to investigate the characteristics of buoyant clouds with positive and negative buoyancy effects. Our objective is to test whether particle clouds with the same initial conditions, buoyancy magnitude, and Reynolds number show different behaviors depending on their density relative to the ambient fluid. The goal of this paper is to assess the appropriateness of the Boussinesq assumption for these types of flows.

The Boussinesq assumption hypothesizes that, when density differences are small, they can be ignored in the momentum terms and only need to be considered in the buoyancy term. One result of this assumption is that the flow behavior is independent of the sign of the buoyancy term. Therefore, two buoyant clouds that are identical other than the sign of the reduced gravity will behave identically other than the direction of flow. So, if two particle clouds are released with the same number of particles each with the same diameter and each with the same difference in density with respect to the ambient fluid but with one set of particles lighter than the ambient and the other set denser, then the cloud velocity and diameter will evolve in the identical way for each cloud. Herein, we test this assumption through a detailed parametric study of cloud development for a range of particle sizes and concentrations and show that, even for small density differences, clouds do not evolve in the same way when the buoyancy sign is reversed.

This paper is structured as follows: In Section 1, we present the introduction. In Section 2, we provide a brief description of the Lai et al. [16] model and discuss its enhancement and extension to include the release of both positively and negatively buoyant particle clouds. Section 3 presents the simulation results that we conducted to validate our model. Then, in Section 4, we apply our model to conduct a parametric study of the similarities and differences in positively and negatively buoyant particle clouds. Finally, in Section 5, we discuss our findings and draw conclusions.

2. Model Development

2.1. Flow Field Model

Several models for particle cloud behavior have been presented in the literature [1,7,16]. In this study, we consider a volume of spherical particles with a total mass (m_o) and density (ρ_p) released from the rest into a quiescent ambient fluid of density (ρ_w) with acceleration due to gravity (g). The ambient is homogeneous, where the ambient fluid density is constant through the depth of the fluid column. After an initial acceleration, the total buoyancy of the particles $B_0 = \frac{m_0}{\rho_p}(\rho_p - \rho_w)g$ induces a buoyant vortex ring structure. In this study, we refer to the fluid field (entrained into the buoyant vortex ring) as the "fluid phase", while particles are referred to as the "solid phase".

Experimental evidence suggests that the fluid field can be approximated as an expanding Hill's spherical vortex [12,23]; therefore, we estimated the flow field analytically assuming it behaves as an expanding Hill's vortex. The flow field is modeled in three dimensions using Cartesian coordinates (x, y, and z). The flow field is modeled within and outside the cloud. The flow is considered to be within the cloud if R is less than r_c (the radius of the particle cloud), where R is the radial distance from the cloud center and is calculated as $R^2 = x^2 + y^2 + (z - z_c)^2$ where z_c is the height of the cloud centroid.

For the flow within the cloud, the mean velocities (u_x, u_y, and u_z) are calculated using the following equations:

$$u_x = \frac{3u_c}{2r_c^2}x(z - z_c), \tag{1}$$

$$u_y = \frac{3u_c}{2r_c^2}y(z - z_c), \tag{2}$$

and

$$u_z = -\frac{3u_c}{4}\left[4\left(\frac{x^2 + y^2}{r_c^2}\right) + 2\left(\frac{z - z_c}{r_c}\right)^2 - \frac{10}{3}\right] \tag{3}$$

The mean velocity of the flow field outside the cloud in the x, y, and z directions is calculated as follows:

$$u_x = \frac{3u_c r_c^3 x(z - z_c)}{2[(z - z_c)^2 + x^2 + y^2]^{5/2}}, \tag{4}$$

$$u_y = \frac{3u_c r_c^3 y(z - z_c)}{2[(z - z_c)^2 + x^2 + y^2]^{5/2}}, \tag{5}$$

and

$$u_z = \frac{u_c r_c^3}{2} \frac{2(z - z_c)^2 - x^2 - y^2}{[(z - z_c)^2 + x^2 + y^2]^{5/2}} \tag{6}$$

refs. [12,16], where u_c is the velocity of the cloud center.

2.2. Cloud Characteristics Model

The particle cloud is described by three main variables: velocity (u_c), radius (r_c), and centroid depth (z_c), which are predicted using an integral model. This model is used to determine the three variables of the particle cloud at each time step. The cloud initially has a total excess mass of m_0, with a volume of $V_0 = \frac{m_0}{\rho_p}$ and total momentum of $M_0 = 0$. The volume of the cloud is modeled using the entrainment assumption [24], which can be expressed as

$$\frac{dV}{dt} = 4\pi\alpha r_c^2 u_c \tag{7}$$

where u_c is the cloud velocity, r_c is the cloud radius, V is the cloud volume, and α is the entrainment coefficient and can be related to r_c and u_c as follows

$$\frac{dr_c}{dz_c} = \alpha \tag{8}$$

Due to the continuous particles raining out of the cloud, the entrainment coefficient α can be a function of the cloud number N_c (see [1,9]). Here, N_c is defined as

$$N_c = u_c r_c \left(\frac{\rho_w}{B_o}\right)^{1/2} \tag{9}$$

The settling velocity of particles has a significant impact on their N_c value. Particles with a high settling velocity usually have an N_c value closer to one, while particles with a low settling velocity have a lower N_c value. The relationship between N_c and α needs to be determined using experimental data. We used the best-fit curve from the experimental work of [1,16]. It was found that when all particles fell out of the particle clouds, the value of α was 0.007 [16]. For particles within the cloud, the value of α can be calculated using the following equation:

$$\alpha = 0.27(1 - 0.28N_c^{1.64}) \tag{10}$$

The momentum M of the cloud depends on the buoyancy contributed by the particles and can be calculated as

$$\frac{dM}{dt} = \frac{d}{dt}(u_c V \rho_w) = B \tag{11}$$

where B is the total buoyancy of the cloud. The vertical position change is determined by the cloud's vertical velocity.

$$\frac{dz_c}{dt} = u_c \tag{12}$$

The total buoyancy inside the cloud is contributed by the particles. Over time, the buoyancy gradually decreases since the particles (the source of buoyancy) gradually drift out of the cloud. Therefore, B can be expressed as follows:

$$B = B_o\phi \tag{13}$$

where ϕ is the fraction of the initial number of particles that remain inside the cloud.

The radius and velocity of the cloud can be expressed using the following two equations.

$$r_c = \left(\frac{3V}{4\pi}\right)^{1/3} \quad (14)$$

and

$$u_c = \frac{M}{\rho_w V} \quad (15)$$

2.3. Particle Trucking Model

The motion of particles can be predicted using the particle tracking equation with the computed flow field and accounts for different forces acting on the particle to calculate the acceleration of the particle. Ignoring the added mass, inertial, and history forces on the particle, the resulting equations for the time variation of the location of a particle (x_P, y_P, z_P) in x, y, and z are given by

$$\frac{d^2x_p}{dt^2} = \left(\frac{f}{u_t}\right)(u_x - u_p), \quad (16)$$

$$\frac{d^2y_p}{dt^2} = \left(\frac{f}{u_t}\right)(u_y - v_p), \quad (17)$$

and

$$\frac{d^2z_p}{dt^2} = \left(\frac{f}{u_t}\right)(u_z - w_p) - \left(1 - \frac{\rho_w}{\rho_p}\right)g \quad (18)$$

where f is a function of the drag coefficient (CD) and the particle Reynolds number (Re_p), u_t is the settling velocity of the particles, and subscripts p and w represent the 'particle' and 'fluid'.

The particle Reynolds numbers based on the particle diameter, the relative velocity of the particle, and the flow field are given as

$$Re_p = \frac{u_d d}{v} \quad (19)$$

where d is the particle diameter, v is the kinematic viscosity, and u_d is the difference between the resultant velocities of the fluid and the individual particles. u_d can be expressed as

$$u_d = \sqrt{(u_x - u_p)^2 + (u_y - v_p)^2 + (u_z - w_p)^2} \quad (20)$$

The drag coefficient is calculated using the Swamee and Ojha formula [25] and can be expressed as

$$CD = 0.5[16(\Phi_1 + \Phi_2)^{2.5} + (\Phi_3 + 1)^{-0.25}]^{0.25} \quad (21)$$

where

$$\Phi_1 = \left(\frac{24}{Re_p}\right)^{1.6}, \quad \Phi_2 = \left(\frac{130}{Re_p}\right)^{0.72}, \quad \text{and} \quad \Phi_3 = \left(\frac{40,000}{Re_p}\right)^2 \quad (22)$$

Now, the f value can be estimated as

$$f = \frac{CDRe_p}{24} \quad (23)$$

The settling velocity u_t is calculated using Dietrich's (1982) equation [26] for spherical particles. The formula is applicable to both laminar and turbulent flow regimes.

$$log_{10}W = -3.76715 + 1.92944log_{10}D - 0.09815log_{10}D^2 - 0.00557log_{10}D^3 - 0.00056log_{10}D^4 \quad (24)$$

where

$$W = \frac{u_t^3}{g' v}, \quad D = \frac{g' d^3}{v^2}, \quad \text{and} \quad g' = g\left(\frac{\rho_p}{\rho_w} - 1\right) \tag{25}$$

where g' is the reduced gravity of the particles. The particle Reynolds numbers based on the individual particle settling speeds are given as

$$Re_{P_t} = \frac{u_t d}{v} \tag{26}$$

Note that the particle Reynolds numbers presented in the later results section are based on the particle falling at its terminal velocity in a quiescent environment and are used to characterize the particle properties. During each run, the model calculates the Reynolds number of each particle based on the velocity of the flow relative to the particle. This Reynolds number is used to calculate the drag coefficient for each particle at each time step.

2.4. Turbulent Dispersion of the Particles

Previous studies have shown that the particle cloud can cause a strong mixing of particles. To consider the impact of turbulent mixing within the cloud, we have incorporated a turbulent dispersion term into our particle tracking model using a random walk model [27].

$$x_p(t + \Delta t) = x_p(t) + u_p \Delta t + \varsigma\sqrt{2K\Delta t}, \tag{27}$$

$$y_p(t + \Delta t) = y_p(t) + v_p \Delta t + \varsigma\sqrt{2K\Delta t}, \tag{28}$$

$$z_p(t + \Delta t) = z_p(t) + w_p \Delta t + \varsigma\sqrt{2K\Delta t} \tag{29}$$

where ς is a normally distributed random variable with zero mean and unit variance and K is the dispersion coefficient, which is assumed to be constant in our model. The second part of the equation $u_p \Delta t$ is the "advection term" as a result of the mean flow field of the cloud. The last part of the equation $\varsigma\sqrt{2K\Delta t}$ is the "diffusion term", which results from the turbulence within the cloud. The value of K is calculated based on the results of Lai et al. [1], where

$$K = 0.02\sqrt{\frac{B_0}{\rho_w}}. \tag{30}$$

3. Model Implementation and Validation

The model was solved numerically using MATLAB (R2023b) built-in functions for solving systems of ordinary differential equations. In the model, the initial conditions were set up so the cloud had a finite volume. For a single-phase buoyant cloud [24], there is a similar solution to the differential equations for the volume and velocity of the cloud. The solution has the cloud radius growing linearly with distance from its source. The limiting case for the source is singular with zero volume (V_0) and infinite reduced gravity (g_0') but finite total buoyancy ($g_0' V_0$). To avoid the singularity at our initial conditions, the simulations were started with the cloud located 10 cm from its virtual (singular) source, with all the particles uniformly distributed in a grid within the spherical cloud. The cloud velocity and radius were calculated based on the similarity solution for a single-phase cloud with the same total buoyancy. The model tracks each particle separately as well as the cloud size, velocity, and location. While the numerical implementation calculated the location of each particle, there was no attempt to track particle collisions. That is, even if particles overlapped in space, they were assumed not to be influenced by the other particles, only by the flow in and around the cloud.

We have conducted a comparison between our model and Lai's experimental and numerical results [16]. Specifically, we have compared the two models in terms of tracking the particle cloud characteristics, including the depth of the cloud (z_c) and the half-width of the cloud (r_c) over time. The validation was carried out for two particle sizes—0.725 mm and 0.513 mm—with all particles having a density of 2.5 g/cm^3. Figure 2 displays the vertical distance traveled by the cloud and the cloud's radius as functions of time (Size A and Size B). Although there may be minor differences in these values, our model accurately models the cloud bulk parameters, which is consistent with prior models. The results suggest that our model captures the essential physical features of the flow.

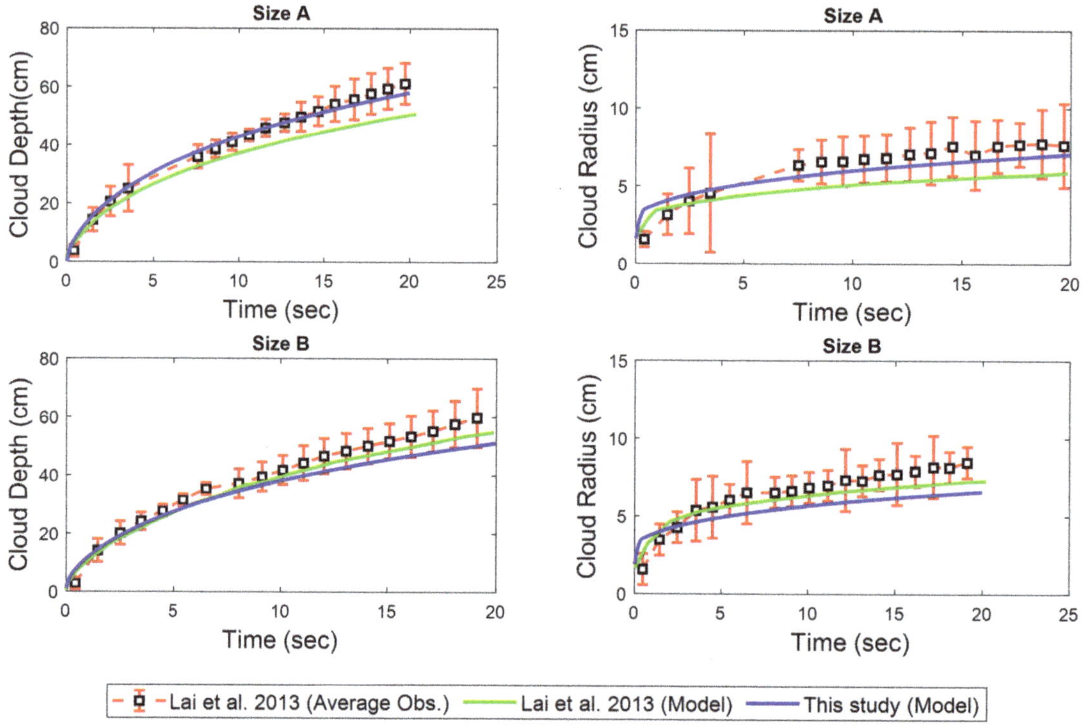

Figure 2. Transient depth and half-width of the particle cloud (in the entrained fluid phase). Lines prediction, symbols observation. Size A represents particles with a diameter of 0.725 mm, while size B represents particles with a diameter of 0.513 mm [16].

4. Parametric Study Results

We now present the results of a parametric study designed to understand the similarities and differences between particle clouds with nominally identical bulk characteristics but with buoyancy of different signs. That is, we compare clouds made of particles that are lighter than the ambient fluid and particles that are denser than the ambient fluid but are otherwise identical. By identical, we mean that the particles have the same diameter, density ratio with the ambient fluid, terminal velocity (though with opposite sign), and Reynolds number and that the clouds have the same initial number of particles in the cloud and the total buoyancy of the cloud has the same magnitude with the opposite sign.

4.1. Parametric Study Details

Simulations were run for 10 different density ratios, five with the particles denser than the ambient fluid and five with the particles less dense than the ambient fluid. For each density ratio, simulations were run for seven different particle Reynolds numbers, which were calculated based on the terminal velocity of an isolated particle. The parameters simulated are listed in Table 1. A total of 70 simulations were run.

Table 1. List of density ratios and Reynolds numbers simulated.

Buoyancy Sign	ρ_p/ρ_w	Re_p
Positive	0.9, 0.8, 0.7, 0.6, 0.5	0.48, 1.0, 1.56, 2.36, 12.76, 25.6, 43.5
Negative	1.1, 1.2, 1.3, 1.4, 1.5	0.48, 1.0, 1.56, 2.36, 12.76, 25.6, 43.5

4.2. Qualitative Results

Figure 3 shows examples of particle clouds that are released from the rest for both positive and negative particles. In this case, the particle Reynolds number is $Re_p = 1.56$, and the pictures show three distinct density ratios for negatively (1.5, 1.3, and 1.1) and positively (0.5, 0.7, and 0.1) buoyant particles. The simulations continue until all particles have left the cloud, which occurs when the total buoyancy of the cloud reaches zero. The initial total cloud buoyancy (B_0) is the same in all cases. The images show snapshots of both positive and negative clouds at the same time. It is worth noting that negatively buoyant particles exit the cloud more quickly than their positively buoyant counterparts.

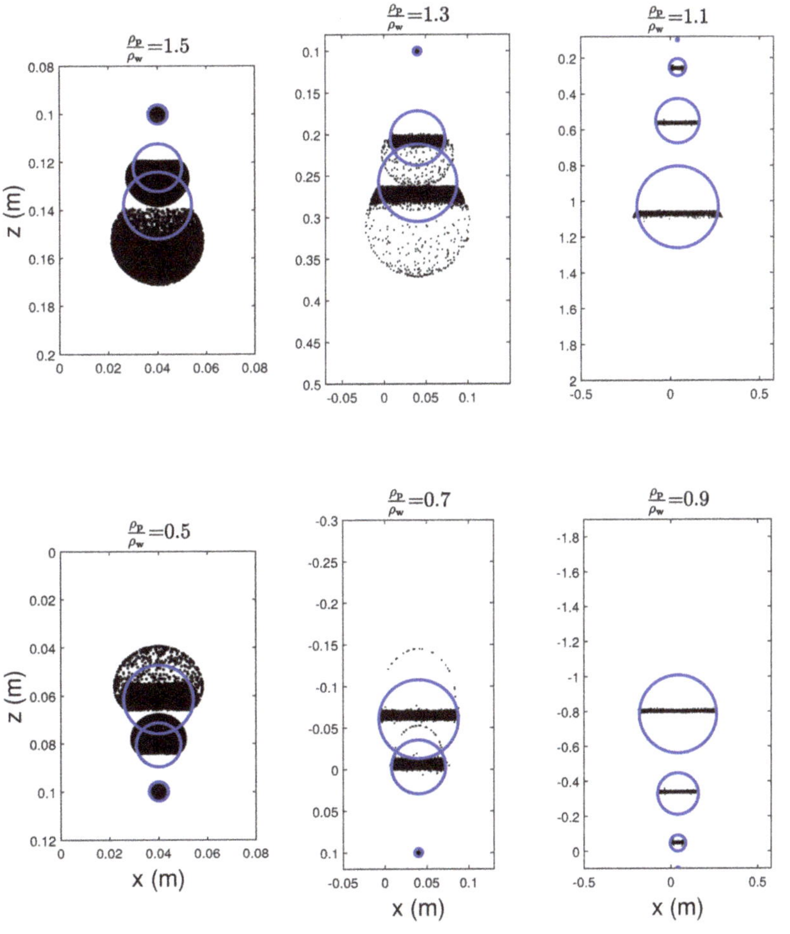

Figure 3. An example of model predictions for positively and negatively buoyant cloud particles with six different density ratios for $Re_p = 1.56$. The particles are represented in black, while the blue color shows the particle cloud growing over time due to entrainment. Note that the vertical scales differ in each column in order to show the full behavior over time for each cloud.

At high-density ratios, such as 1.5 and 0.5, the particles initially remain together in the cloud as a single entity. Later, they precipitate in a swarm vortex ring shaped like a bowl. However, the particles stay in the cloud longer for small-density ratios like 1.1 and 0.9. This results in a significant increase in the cloud size due to entrainment. Eventually, the particles rain out of the cloud but dissipate differently from the swarm-shaped formation observed in high-density ratios.

For the intermediate density difference case (center column in Figure 3), some particles are ejected from the cloud at an early time and travel ahead of the cloud. However, the number ejected is much higher for the denser particles (top row with $\rho_p/\rho_w = 1.3$) compared to the number initially ejected for the lighter particles (bottom row with $\rho_p/\rho_w = 0.7$). This behavior is also observed to a greater or lesser extent for the other cases shown in Figure 3, though it is less clear in the two-dimensional plots of the three-dimensional flow for these cases.

4.3. Quantitative Results

To better compare positively and negatively buoyant cloud cases, we introduce a time and length scale based on cloud buoyancy and particle terminal velocity. The scales are given by

$$T = \frac{\sqrt{\frac{B}{\rho_w}}}{u_t^2} \quad \text{and} \quad L = \frac{\sqrt{\frac{B}{\rho_w}}}{u_t}. \tag{31}$$

Based on these scales, several non-dimensional variables can be defined. These are as follows: the cloud buoyancy

$$\beta = \frac{B}{Bo}, \tag{32}$$

travel time

$$\tau = \frac{t}{T}, \tag{33}$$

cloud velocity

$$\nu = \frac{u_c}{u_t}, \tag{34}$$

cloud radius

$$\zeta = \frac{r_c}{L}, \tag{35}$$

and cloud height

$$\Lambda = \frac{z_c}{L}. \tag{36}$$

Figure 4 displays the dimensionless cloud buoyancy β, velocity ν, and radius ζ as a function of time for the cases shown in Figure 3. In all six cases, there is an initial ejection of a large number of particles (top row of Figure 4) followed by their re-entrainment into the cloud. For the lowest density difference (right-hand column), almost all the particles are re-entrained and remain within the cloud for a prolonged period before settling out. However, despite having the same terminal velocity, the denser (negatively buoyant) particles (red line) fall out much earlier than the lighter (positively buoyant) particles (blue line). For the larger density differences, the deviation between the red and blue lines occurs much sooner, as fewer of the denser particles are re-entrained back into the cloud following the initial ejection. Note that the ejection and re-entrainment cycle produces a significant oscillation in the cloud buoyancy (β) (Figure 4 middle column top row) but only a small amplitude fluctuation in the velocity (middle row), while the displacement plot is smooth (bottom row). This is because the buoyancy fluctuation is integrated over time to obtain the velocity. Also, the similarity solution to the single-phase buoyant cloud model indicates the velocity scales on the square root of the cloud buoyancy [12]. Therefore, the amplitude of the fluctuations in β is attenuated in the velocity signal.

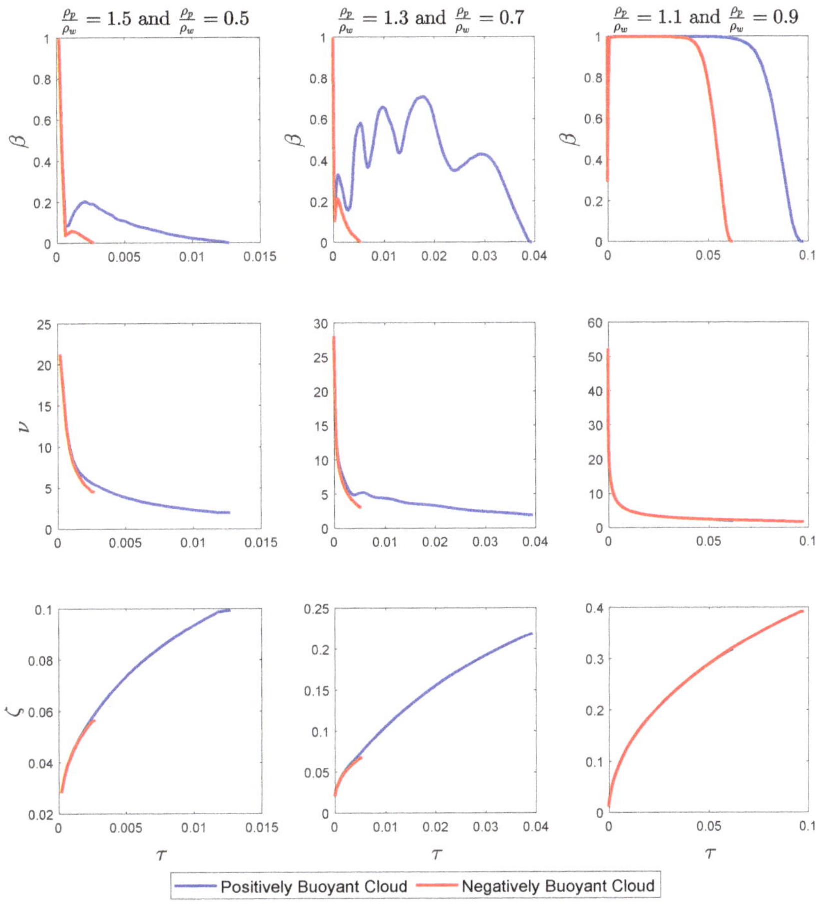

Figure 4. Dimensionless cloud buoyancy β, velocity ν, and radius ζ as a function of time for the cases shown in Figure 3.

For each simulation listed in Table 1, the fraction of particles that remain within the cloud (β) are tracked over time. These data are plotted in Figure 5 for all the cases simulated. Each row of plots represents a pair of density ratios that produce the same initial conditions in every way except the sign of the buoyancy. Therefore, the only difference in behavior between each pair of plots is the sign of the particle buoyancy.

For the largest density difference particles, shown in the top row of Figure 5, there is a rapid decrease in the number of particles in the cloud as they are ejected from the vortex ring. This is particularly true for the denser particles (right-hand plot), where the vast majority of the particles are removed from the cloud by $\Lambda = 1.2$. For the cloud formed by lighter particles, this is also true for most particle Reynolds numbers. However, for the smallest particles, $Re_P = 0.48$, the particles are constantly being ejected and re-entrained into the cloud as indicated by the oscillating line.

As the density difference between the particles and the ambient fluid decreases (lower rows), the time for particles to be fully removed increases. This is because the particle terminal velocity is smaller and it takes longer for the cloud to slow down to the terminal velocity of the particles. The same difference between heavy and light particles is still seen with the smaller lighter particles being re-entrained back into the cloud for longer periods of time compared to the equivalent denser particles. Further, as the density difference decreases, higher Reynolds number particles start to exhibit the ejection re-entrainment cycle. For $\rho_P/\rho_w = 1.3$, even some of the denser particles start to be re-entrained.

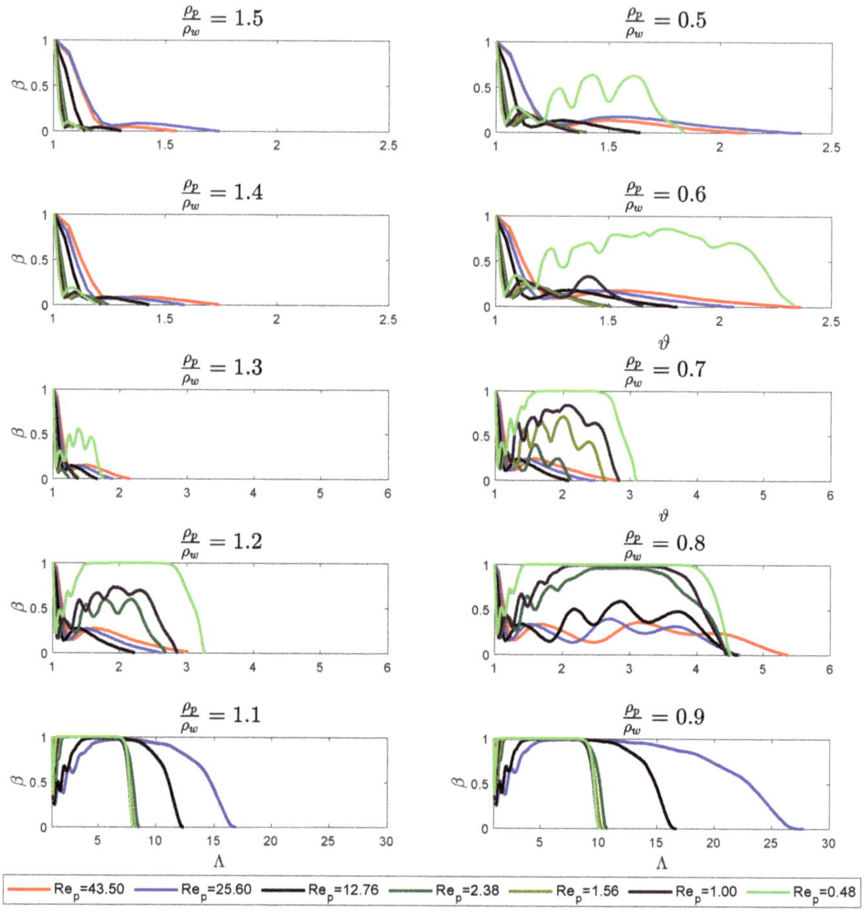

Figure 5. Dimensional cloud buoyancy β, velocity ν, and radius ζ for the cases shown in Figure 3.

For the lowest density difference cases, $\rho_P/\rho_w = 1.1$ and $\rho_P/\rho_w = 0.9$, a different behavior is observed. In these cases, some of the particles are initially ejected, but all particles are re-entrained and then remain circulating in the cloud for some time before eventually falling out of the cloud. However, as stated before, equivalent pairs of particles, that is the same Re_P, have the same terminal velocity, and each cloud has the same initial buoyancy. However, the denser particles leave the cloud sooner than the lighter particles. Therefore, this process is not due solely to the cloud velocity falling below the particle terminal velocity. If that were the case, the ejection height would be the same for heavy and light particles with the same Re and density difference.

One possible explanation for this difference in behavior is the response of the particles to being pushed along the curved streamlines of the spherical Hill's vortex. The particles that are denser than the ambient fluid will find it harder to follow the curved path and will tend to move away from the center of curvature of the streamlines and out of the particle cloud, whereas the particles that are less dense will be drawn further into the cloud and have a longer residence time.

The results in Figure 5 indicate that, even for particle clouds that have a bulk density ratio that is relatively small, their behavior may differ significantly depending on whether or not the particle cloud is positively or negatively buoyant. That is, a particle cloud that would typically be regarded as Boussinesq based on its bulk density ratio will behave differently depending on whether the constituent particles are positively or negatively buoyant. This is even true for small particle density ratios. In the bottom row of Figure 5, the density ratios are $\rho_p/\rho_w = 1.1$ and $\rho_p/\rho_w = 0.9$ and the time at which all the particles have left the cloud is up to 60% longer for the positively buoyant (lighter) particles compared to the negatively buoyant (heavier) particles. As such, the behavior of the cloud depends on the direction of the sign of the buoyancy of the clouds. This is a hallmark of non-Boussinesq flows indicating that two-phase density-driven clouds never fully behave in line with the Boussinesq assumption.

5. Discussion and Conclusions

A model for particle clouds has been developed and validated by the numerical and experimental work of Lai et al. [16]. The cloud was modeled as a buoyant vortex ring, with the velocity field computed by approximating the buoyant vortex ring as an expanding Hill's spherical vortex. The rate of growth of the cloud radius in the fluid phase is assumed to be a function of the entrainment coefficient, which is a function of the cloud number. This was obtained from the experimental work of Lai et al. [1]. The growth rate of the vortex ring in the subsequent dispersive regime was also obtained experimentally by Lai et al. [16]. The particle tracking equation is then used to track the motion of particles with different forces acting on them. The contribution of buoyancy of all particles inside the cloud is accounted for by summing the number of particles within the cloud at each time step. The model included a random walk turbulent dispersion term for the particles and was validated using previously published experimental data (see Figure 2).

The validated model was used to conduct a parametric study to examine the difference in behavior between clouds that are nominally identical other than the sign of the buoyancy term. Simulation results indicate that particles are ejected and/or fall out of the buoyant cloud more rapidly when they are denser than the ambient fluid compared to identical clouds wherein the particles are less dense than the ambient fluid (see Figure 5). This is even the case for the smallest density differences and lowest particle Reynolds numbers modeled. Therefore, even for cases where the particle settling velocity and density differences are small, the behavior of the cloud is quite different depending on the sign of the particle buoyancy.

This would suggest that the Boussinesq assumption may not be appropriate for such flows and that positively and negatively buoyant clouds will behave differently. This difference is due to the response of the particles to ejection and re-entrainment into the cloud. Denser particles are harder to re-entrain into the cloud because of their higher inertia. Another result of this study is that simplified modeling approaches that treat the sediment cloud as a continuum and assume that the particles are retained in the cloud until the cloud velocity falls below the settling velocity, may over-predict the time for particles to settle out because they do not model the ejection and re-entrainment process.

Therefore, it is important to understand the size distribution of dredged materials before making a decision about how to model the flow, as even fairly small particles with small density differences can result in flows that differ from continuum models even when the nominal conditions for the Boussinesq assumption have been satisfied.

Although this study modeled the formation of a particle cloud as a thermal, with a particle tracking model to quantify particle ejection and re-entrainment and a random walk model for both positive and negative particle clouds, this study assumed that the particles have the same diameters, that there are no collisions between particles, and that they can overlap on top of each other in the space domain. In future studies, the aim is to extend the model to the case of polydisperse particles and analyze elastic/inelastic collisions in three dimensions. There is also a clear need for more experimental studies over a broader range of parameters to identify the limitations of applying the Boussinesq assumption to sediment clouds.

Author Contributions: A.O.A. developed the numerical code that simulated the model, produced the results figures and collaborated with N.B.K. on the writing and problem conception and definition. N.B.K. suggested the initial topic for the study and collaborated with A.O.A. on the detailed problem conception, problem definition, model refinement, results presentation, and writing. A.A.K. worked on the initial problem development and writing. All authors have read and agreed to the published version of the manuscript.

Funding: This research received no external funding.

Data Availability Statement: The data present in this study are available on request from the first author.

Conflicts of Interest: The authors declare no conflicts of interest.

Nomenclature

Latin Symbols	
B	Cloud buoyancy
B_0	Total particle buoyancy
K	Dispersion coefficient
f	Particle force parameter
g	Gravitational acceleration
g'	Cloud reduced gravity
L	Length scale
m_0	Total particle mass
M_0	Initial cloud momentum
N_c	Cloud number
r_c	Particle cloud radius
Re_p	Particle Reynolds number
T	Time scale
u_c	Velocity of the cloud center
u_d	Magnitude of the relative velocity of the particle and fluid
u_t	Particle settling velocity
u_p, v_p, w_p	Particle velocity
u_x, u_y, u_z	Fluid velocity within the cloud
V_0	Initial cloud volume
V	Cloud volume
x, y	Horizontal cartesian coordinates
x_p, y_p, z_p	Particle location
z	Vertical coordinate
z_c	Vertical location of the cloud center
Greek Symbols	
α	Entrainment coefficient
β	Non-dimensional cloud buoyancy
Δt	Time step
Λ	Non-dimensional cloud height
ν	Non-dimensional cloud velocity
ϕ_1	Fraction of particles remaining in the cloud
Φ_1, Φ_2, and Φ_3	Drag coefficient parameters
ρ_p	Particle density
ρ_w	Fluid density
τ	Non-dimensional time
υ	Kinematic viscosity of the fluid
ς	Normally distributed random variable
ζ	Non-dimensional cloud radius

References

1. Lai, A.C.; Wang, R.Q.; Law, A.W.K.; Adams, E.E. Modeling and experiments of polydisperse particle clouds. *Environ. Fluid Mech.* **2016**, *16*, 875–898. [CrossRef]
2. Polrot, A.; Kirby, J.; Birkett, J.; Sharples, G. Combining sediment management and bioremediation in muddy ports and harbours: A review. *Environ. Pollut.* **2021**, *289*, 117853. [CrossRef] [PubMed]
3. Lohrer, A.M.; Wetz, J.J. Dredging-induced nutrient release from sediments to the water column in a southeastern saltmarsh tidal creek. *Mar. Pollut. Bull.* **2003**, *46*, 1156–1163. [CrossRef] [PubMed]
4. Nayar, S.; Miller, D.; Hunt, A.; Goh, B.; Chou, L. Environmental effects of dredging on sediment nutrients, carbon and granulometry in a tropical estuary. *Environ. Monit. Assess.* **2007**, *127*, 1–13. [CrossRef] [PubMed]
5. Manap, N.; Voulvoulis, N. Environmental management for dredging sediments–The requirement of developing nations. *J. Environ. Manag.* **2015**, *147*, 338–348. [CrossRef] [PubMed]
6. Todd, V.L.; Todd, I.B.; Gardiner, J.C.; Morrin, E.C.; MacPherson, N.A.; DiMarzio, N.A.; Thomsen, F. A review of impacts of marine dredging activities on marine mammals. *ICES J. Mar. Sci.* **2015**, *72*, 328–340. [CrossRef]
7. Bush, J.W.; Thurber, B.; Blanchette, F. Particle clouds in homogeneous and stratified environments. *J. Fluid Mech.* **2003**, *489*, 29–54. [CrossRef]
8. Li, C.W. Convection of particle thermals. *J. Hydraul. Res.* **1997**, *35*, 363–376. [CrossRef]
9. Rahimipour, H. Dynamic behavior of particle clouds. In Proceedings of the 11th Australian Fluid Mechanics Conference, University of Tasmania, Hobart, Australia, 14–18 December 1992.
10. Zhao, B.; Law, A.W.; Eric Adams, E.; Shao, D.; Huang, Z. Effect of air release height on the formation of sediment thermals in water. *J. Hydraul. Res.* **2012**, *50*, 532–540. [CrossRef]
11. Ruggaber, G.J. Dynamics of Particle Clouds Related to Open-Water Sediment Disposal. Ph.D. Thesis, Massachusetts Institute of Technology, Cambridge, MA, USA, 2000.
12. Turner, J. Buoyant plumes and thermals. *Annu. Rev. Fluid Mech.* **1969**, *1*, 29–44. [CrossRef]
13. Noh, Y. Sedimentation of a particle cloud across a density interface. *Fluid Dyn. Res.* **2000**, *27*, 129. [CrossRef]
14. Zhao, B.; Law, A.; Lai, A.; Adams, E. On the internal vorticity and density structures of miscible thermals. *J. Fluid Mech.* **2013**, *722*, R5. [CrossRef]
15. Gensheimer, R.J.; Adams, E.E.; Law, A.W. Dynamics of particle clouds in ambient currents with application to open-water sediment disposal. *J. Hydraul. Eng.* **2013**, *139*, 114–123. [CrossRef]
16. Lai, A.C.; Zhao, B.; Law, A.W.K.; Adams, E.E. Two-phase modeling of sediment clouds. *Environ. Fluid Mech.* **2013**, *13*, 435–463. [CrossRef]
17. Hu, J.; Xu, G.; Shi, Y.; Wu, L. A numerical simulation investigation of the influence of rotor wake on sediment particles by computational fluid dynamics coupling discrete element method. *Aerosp. Sci. Technol.* **2020**, *105*, 106046. [CrossRef]
18. Balakin, B.V.; Hoffmann, A.C.; Kosinski, P.; Rhyne, L.D. Eulerian-Eulerian CFD model for the sedimentation of spherical particles in suspension with high particle concentrations. *Eng. Appl. Comput. Fluid Mech.* **2010**, *4*, 116–126. [CrossRef]
19. Chan, S.N.; Lee, J.H. A particle tracking model for sedimentation from buoyant jets. *J. Hydraul. Eng.* **2016**, *142*, 04016001. [CrossRef]
20. Sun, R.; Xiao, H. SediFoam: A general-purpose, open-source CFD–DEM solver for particle-laden flow with emphasis on sediment transport. *Comput. Geosci.* **2016**, *89*, 207–219. [CrossRef]
21. Scorer, R.S. Experiments on convection of isolated masses of buoyant fluid. *J. Fluid Mech.* **1957**, *2*, 583–594. [CrossRef]
22. Gu, J.; Li, C.W. Modeling instantaneous discharge of unsorted particle cloud in ambient water by an Eulerian—Lagrangian method. *J. Hydraul. Res.* **2004**, *42*, 399–405.
23. Lai, A.C.; Zhao, B.; Law, A.W.K.; Adams, E.E. A numerical and analytical study of the effect of aspect ratio on the behavior of a round thermal. *Environ. Fluid Mech.* **2015**, *15*, 85–108. [CrossRef]
24. Turner, J.S. Turbulent entrainment: The development of the entrainment assumption, and its application to geophysical flows. *J. Fluid Mech.* **1986**, *173*, 431–471. [CrossRef]
25. Swamee, P.K.; Ojha, C.S.P. Drag coefficient and fall velocity of nonspherical particles. *J. Hydraul. Eng.* **1991**, *117*, 660–667. [CrossRef]
26. Dietrich, W.E. Settling velocity of natural particles. *Water Resour. Res.* **1982**, *18*, 1615–1626. [CrossRef]
27. Kitanidis, P.K. Particle-tracking equations for the solution of the advection-dispersion equation with variable coefficients. *Water Resour. Res.* **1994**, *30*, 3225–3227. [CrossRef]

fluids

Article

Inclusive Hyper- to Dilute-Concentrated Suspended Sediment Transport Study Using Modified Rouse Model: Parametrized Power-Linear Coupled Approach Using Machine Learning

Sanny Kumar [1], Harendra Prasad Singh [1], Srinivas Balaji [2], Prashanth Reddy Hanmaiahgari [2,*] and Jaan H. Pu [3,*]

[1] Department of Water Engineering and Management, Central University of Jharkhand, Ranchi 835205, India
[2] Department of Civil Engineering, IIT Kharagpur, Kharagpur 721302, India
[3] Faculty of Engineering and Informatics, University of Bradford, Bradford BD7 1DP, UK
* Correspondence: hpr@civil.iitkgp.ac.in (P.R.H.); j.h.pu1@bradford.ac.uk (J.H.P.)

Abstract: The transfer of suspended sediment can range widely from being diluted to being hyper-concentrated, depending on the local flow and ground conditions. Using the Rouse model and the Kundu and Ghoshal (2017) model, it is possible to look at the sediment distribution for a range of hyper-concentrated and diluted flows. According to the Kundu and Ghoshal model, the sediment flow follows a linear profile for the hyper-concentrated flow regime and a power law applies for the dilute concentrated flow regime. This paper describes these models and how the Kundu and Ghoshal parameters (linear-law coefficients and power-law coefficients) are dependent on sediment flow parameters using machine-learning techniques. The machine-learning models used are XGboost Classifier, Linear Regressor (Ridge), Linear Regressor (Bayesian), K Nearest Neighbours, Decision Tree Regressor, and Support Vector Machines (Regressor). The models were implemented on Google Colab and the models have been applied to determine the relationship between every Kundu and Ghoshal parameter with each sediment flow parameter (mean concentration, Rouse number, and size parameter) for both a linear profile and a power-law profile. The models correctly calculated the suspended sediment profile for a range of flow conditions ($0.268\ \text{mm} \leq d_{50} \leq 2.29\ \text{mm}$, $0.00105\frac{\text{g}}{\text{mm}^3} \leq$ particle density $\leq 2.65\frac{\text{g}}{\text{mm}^3}$, $0.197\frac{\text{mm}}{\text{s}} \leq v_s \leq 96\frac{\text{mm}}{\text{s}}$, $7.16\frac{\text{mm}}{\text{s}} \leq u_* \leq 63.3\frac{\text{mm}}{\text{s}}$, $0.00042 \leq \bar{c} \leq 0.54$), including a range of Rouse numbers ($0.0076 \leq P \leq 23.5$). The models showed particularly good accuracy for testing at low and extremely high concentrations for type I to III profiles.

Keywords: Rouse number; mean concentration; suspended sediment transport; sediment size parameter; parameterized power-linear model; machine learning; decision tree regressor; support vector machines

Citation: Kumar, S.; Singh, H.P.; Balaji, S.; Hanmaiahgari, P.R.; Pu, J.H. Inclusive Hyper- to Dilute-Concentrated Suspended Sediment Transport Study Using Modified Rouse Model: Parametrized Power-Linear Coupled Approach Using Machine Learning. *Fluids* **2022**, *7*, 261. https://doi.org/10.3390/fluids7080261

Academic Editor: Chandrashekhar S. Jog

Received: 30 May 2022
Accepted: 21 July 2022
Published: 30 July 2022

1. Introduction

Sediment transport commonly occurs in unlined water conveyance systems such as rivers, streams, canals, and drainage channels. There are three sediment transport modes: wash load, bedload, and suspended load. In wash load, particles do not exist on the bed, therefore, the characterization and prediction of wash load composition is highly difficult. The bedload transfer is almost always in contact with the bed. Bedload transport takes place if critical friction velocity (u_{*c}) is less than the friction velocity (u_*). The suspended load, which is part of the total load, moves without continuously being in contact with the bed. Turbulence is the main flow property that keeps the sediment in suspension [1–3]. The turbulence is characterized by the magnitude of root mean square velocity (u'). For significant suspension to occur u' near the bed must exceed the sediment fall velocity (v_s). Sediment particle concentration distribution along the depth is very important for predicting the sediment transport rate that is taking place in the river. However, the problem with suspended load transport is that it is not fully understood, because of limitations in modelling techniques such as diffusion theory [4]. Rouse [5] developed a suspended sediment concentration profile based on diffusion theory for steady uniform flows by using the eddy viscosity model to relate Reynolds stress and log-law for the velocity profile. Despite the drawbacks of diffusion theory, the Rouse model has become a standard for calculating suspended sediment concentration profiles and a basis for many models developed later.

Two-phase flows have complex interactions between the phases. Modelling fluid–particle interactions and particle–particle attractions is difficult therefore they are excluded. Huang et al. [6], Hsu et al. [7] and Rouse [5] resolved this issue based on the differential continuity equation of suspended sediment diffusion in the 2D steady turbulent flow. They considered that the sediment diffusion coefficient is related to both the mixing length and the turbulence intensity. These models could not account for turbulence or forces operating on the sediment particle surfaces, such as collisions between particles, and as a result, their accuracy was limited. To overcome this, several studies have addressed sediment distributions in stationary and uniform two-dimensional (2D) open channel flows by solving the momentum equation for sediment particles [4,8,9]. Most notably, the diffusion coefficient of sediment particles generated from the momentum conservation equation provides significantly more accurate, perhaps allowing for a better understanding of sediment particle dispersion. Generally, the sediment concentration profile attains the maximum at some distance above the bed surface, decreases with farther moving away from the bed, and finally attains the minimum at the free surface. To model suspended-sediment distribution along with the depth, parameters such as particle fall velocity, particle diameter, turbulence intensity, shear velocity, Rouse number and mean concentration are required [10,11]. Kundu and Ghosal [4] and Pu and Lim [12] found that two-dimensional incompressible flow modelling over a sediment bed with a uniform slope predicts reasonably accurate suspended sediment transport since full three-dimensional modelling involves a lot more complexity. In addition, the diffusion theory with appropriate modifications such as a two-layer theoretical model based on diffusion theory or the fractional advection-diffusion equation can predict suspended sediment concentration profiles with reasonable accuracy. Goree et al. [13] used equations of motion and considered drift flux, which is a fluid that consists of multiple phases or volume fractions. The mixture flow consists of different volume fractions and each volume fraction has a different transport velocity. This velocity depends on the particle size and the total volumetric concentration of solids. Goree et al. [13] used LES for turbulence modelling however the computed results were not accurate at the wall. Another type of modelling concept is kinetic theory. Models established based on kinetic theory for granular flows and two-fluid models i.e., the probability density function (PDF) approach, in which both fluid and solid phases are considered as continuum media, allow classic continuum mechanics to be naturally employed to formulate the two-phase flows with fundamental conservation laws [4]. Therefore, the observable macro mechanical states of flows and sediment transport are completely determined by conservation equations of solid and liquid systems. In addition, other theories, such as a numerical investigation based on the kinetic theory of granular flows, in combination with a RANS (Reynolds-Averaged Navier–Stokes) turbulence model were investigated by Ekambara et al. [14] for a pipe flow using Ansys CFX which gave very good results. Ni et al. [10] developed a combined model of both kinetic theory and diffusion theory, which has given reasonable results.

The limitations of existing models to simulate suspended sediment transport can be efficiently overcome with data-driven models such as machine learning. Barati et al. [15] estimated the drag coefficient of a smooth sphere using multi-gene genetic programming. Alizadeh et al. [16] used ANN and Bayesian network models to predict pollutant transport in natural rivers. Sadeghifar and Barati [17] used soft-computing techniques with very good accuracy to predict sediment transport in the southern shorelines of the Caspian Sea. Cao et al. [18] developed a nonparametric machine-learning (ML) model to predict the settling velocity of noncohesive sediment which demonstrated the capability of the ML model for accuracy and consistency. Rushd et al. [19] used AI-based machine learning algorithms to develop a generalized model for computing the settling velocity of spheres in both Newtonian and non-Newtonian fluids. The above-mentioned studies demonstrate that data-driven models (AI and ML) and soft-computing techniques are powerful tools for predicting complicated processes such as sediment transport.

A central perspective advocated in this paper is that by exploiting foundational knowledge of suspended sediment profiles and physical constraints, data-driven approaches can be used to yield useful predictive results. In this paper, current models are analysed and the crucial flow parameters are estimated by using machine learning algorithms.

2. Model Review

The diffusion theory was utilised by van Rijn [20], Wang and Ni [21], McLean [22], and Zhong et al. [23] to define the transport of suspended material in numerical modelling. According to diffusion theory, sediment transport takes place from higher concentration areas to lower concentration areas [24]. Rouse [5] derived suspended sediment concentration in a steady uniform current. As per Rouse [5], the sediment is kept in suspension mainly by turbulence [7]. Rouse [5] used Prandtl's mixing length theory to estimate the vertical profile of suspended sediment. Rouse's [5] methodology is given as follows.

As shown in Figure 1, consider two sand particles 1 and 2 with a settling velocity v_s. In a unit time through a unit area on horizontal plan p-p, the sediment volumes going up (q_u) and down (q_d) are:

$$q_u = (w' - v_s)\left(c - \frac{1}{2}l\frac{dc}{dz}\right) \quad (1)$$

$$q_d = (w' + v_s)\left(c + \frac{1}{2}l\frac{dc}{dz}\right) \quad (2)$$

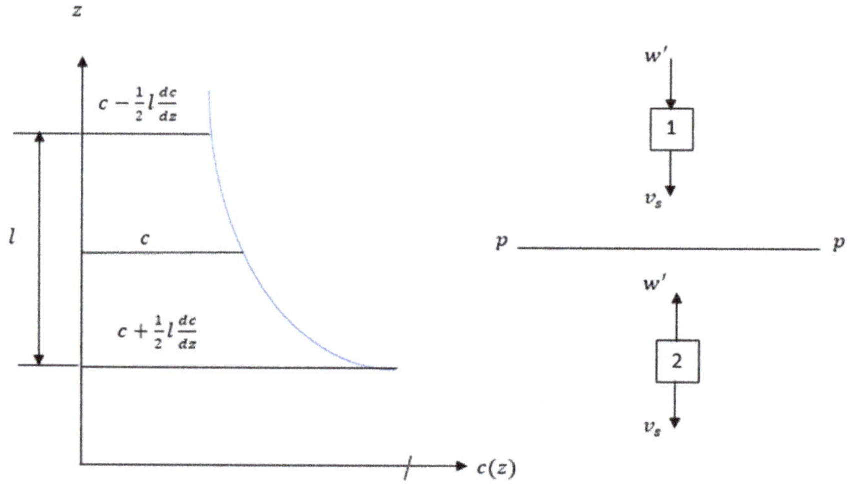

Figure 1. Suspended sediment concentration profile as governed by flow turbulence.

In an equilibrium state, q_u and q_d must be equal to each other, substracting Equation (2) from Equation (1) gives

$$cv_s + \frac{1}{2}w'l\frac{dc}{dz} = 0 \quad (3)$$

By assuming that

$$\frac{1}{2}w'l = u_*\left(1 - \frac{z}{H}\right)\kappa z \quad (4)$$

where mixing length $l = \kappa z$, $w' = u_*\left(1 - \frac{z}{H}\right)$, $\kappa = 0.4$ and u_* = friction velocity, H = flow depth.

Substituting Equation (3) into Equation (4) yields

$$cv_s + u_*\left(1 - \frac{z}{H}\right)\kappa z\frac{dc}{dz} = 0 \quad (5)$$

By integrating Equation (5), one obtains

$$c(z) = c_b\left(\frac{H-z}{z}\cdot\frac{b}{H-b}\right)^{\frac{v_s}{\kappa u_*}} \quad (6)$$

where c_b is the integration constant ($c|_{z=b} = c_b$) and $\frac{v_s}{\kappa u_*}$ = Rouse number (P). It is assumed that bedload transport takes place in the bedload layer from $z = 0$ to $z = b = k_s$. In the bedload layer, sediment concentration is given by c_b. Hsu et al. [7] suggested $\frac{b}{H}$ is approximately 0.005. The concentration distribution profile is thought to grow more uniform, according to Rouse [5]. Small settling velocity and minimal shear velocity can result in low Rouse numbers.

$$c(z) = c_b\left(\frac{b}{z}\cdot\frac{H-z}{H-b}\right)^{\frac{v_s}{\kappa u_*}} \quad (7)$$

Log-law can be written as

$$u(z) = \frac{u_*}{\kappa}\ln\left(\frac{z}{k_s}\right) + B \quad (8)$$

The flux per river cross-section per square meter can be written from Equations (7) and (8) as

$$\varnothing(z) = \left(\frac{u_*}{\kappa}\ln\left(\frac{z}{k_s}\right) + Bu_*\right)\left(\frac{b}{z}\cdot\frac{H-z}{H-b}\right)^{\frac{v_s}{\kappa u_*}} \quad (9)$$

For steady, uniform open channel flow, the sediment concentration varies with the distance z from the wall. The depth can be nondimensionalized as $\varepsilon = \frac{z}{H}$ where its range is $0 < \varepsilon \le 1$.

In addition, Kundu and Ghosal [11] recognized that sediment concentration profiles occur in three (type I, type II, and type III) types as shown in Figure 2. For the low-flow condition, the sediment carrying capacity is weak, therefore the maximum concentration occurs at the bed and decreases exponentially to the minimum concentration at the free surface. This type of profile is called type I. In medium-flow conditions, the sediment concentration is lower near the bed because of the gravity effect, sediment concentration attains a maximum concentration a little distance away from the bed and is the lowest near the free surface, and is classified as type II. The type III profile is attained during high-flow conditions subjected to hyper concentration. In the type III profile, sediment concentration is low near the bed as well as near the free surface, whereas the maximum concentration occurs in the middle region. According to Kundu and Ghoshal [11], the entire flow depth can be divided into two regions: the inner suspension region, where sediment concentration increases with a characteristic height from the sediment bed ($\varepsilon_b \le \varepsilon \le \varepsilon_{max}$). Here ε_{max} corresponds to c_{max} and ε_b corresponds to the nondimensional height at which the suspended sediment starts. The outer suspension region above the inner suspension region, where sediment concentration decreases with an increase in the characteristic height and $\varepsilon_{max} \le \varepsilon \le 1$. For each region, sediment concentration which is a function of characteristic height can be written as

$$\varnothing_1 = b_1\varepsilon^{a_1} + d_1 \text{ for inner suspension region } (\varepsilon_b \le \varepsilon \le \varepsilon_{max}) \quad (10)$$

$$\varnothing_2 = b_2\varepsilon^{a_2} + d_2 \text{ for outer suspension region } (\varepsilon_{max} \le \varepsilon \le 1) \quad (11)$$

where $a_1, b_1, d_1, a_2, b_2, d_2$ are experimental coefficients that can be found by the least squares method by analysing observed data.

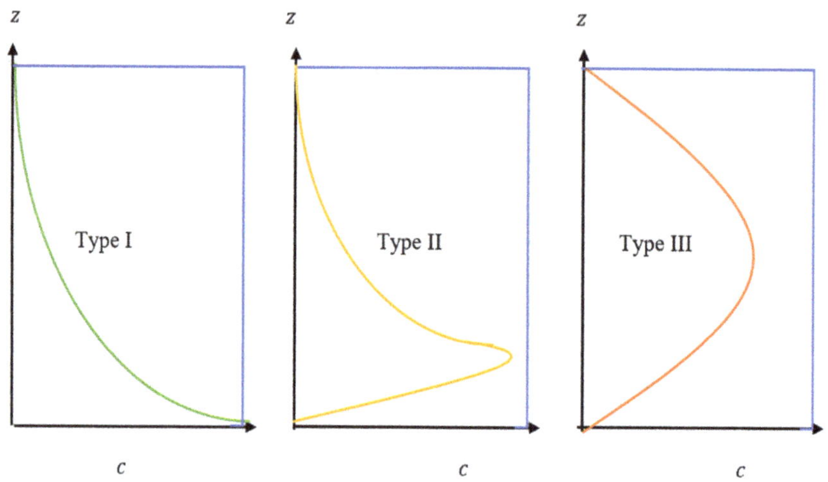

Figure 2. Type I, II, and III concentration distributions along with the depth.

In the asymptotic matching method (refer Figure 3), the final model of suspended sediment concentration for the entire flow region is given as follows [25].

$$\varnothing = \left(\frac{1}{\left(b_1 \varepsilon^{a_1} + d_1\right)^{-1} + \left(b_2 \varepsilon^{a_2} + d_2\right)^{-1}} \right) \tag{12}$$

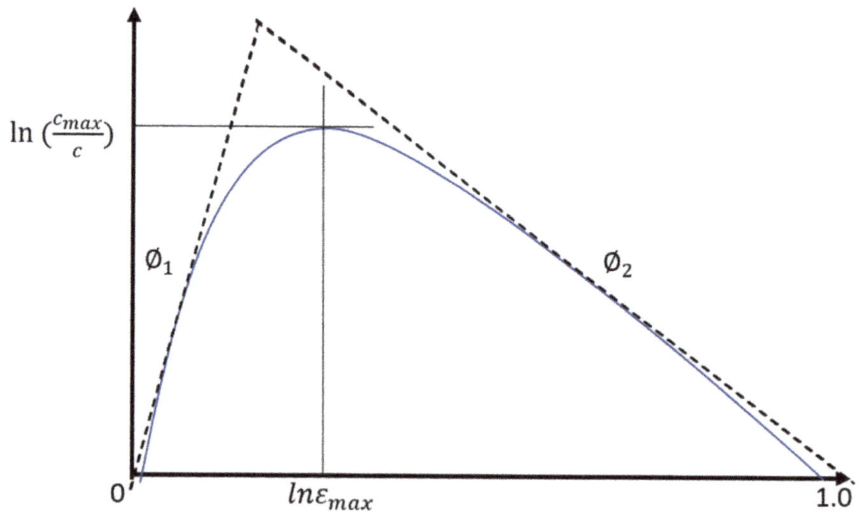

Figure 3. Schematic of the asymptotic matching method.

The aforementioned equation has also been used by Bouvard and Petkovic [26], Wang and Ni [21], and Einstein and Chein [27] to demonstrate that the sediment profile for a diluted flow should follow a power law (Equations (10) and (11)). In Equations (10) and (11), $d = 0$ produce a simplified form of power law as given below.

$$\varnothing = b\varepsilon^{a}. \tag{13}$$

Equation (13) is similar to Rouse Equation (Equation (7)). For type III, the sediment transport equation in power-law format is not suitable. Pu et al. [1] stated that the linear-type profile fits well. Therefore, the equation for the suspended sediment profile for type III is given as follows

$$\varnothing = b\varepsilon + d \tag{14}$$

This is similar to Equations (7) and (8) but a value is taken as unity. Rouse formulation has drawbacks at the riverbed and the free surface as shown by the experimental studies of Kironoto and Yulistiyanto [28], Goeree et al. [13], Greiimann and Holly [8], Jha and Bombardelli [9], and Sumer et al. [29]. Since the Rouse formula is derived from the diffusion theory, it is valid only for a single phase i.e., sediment phase. Therefore, the Rouse formula is applicable only for the type I profile with zero concentration at the free surface and infinite concentration at the bed [30]. According to Huang et al. [6], the Rouse formula produces an inaccurate estimation of sediment concentration near the bed for highly rough conditions. Kumbhakar et al., [30] showed that the Rouse formula can be improved by incorporating an additional factor to dampen the diffusion coefficient. Sumer et al. [29] tried to improve the reference height representation to better estimate suspended sediment modelling. Greiimann and Holly [8] stated that the Rouse formula excluded the particle–particle attractions in addition to the assumption in estimating diffusion coefficient resulting Rouse formula valid only for $\bar{c} < 0.1$. Rouse formula gives a good understanding as long as sediment particles have small inertia. Wang and Ni [21], Ni et al. [10], and Zhong et al. [23] models showed significant differences in the shape of the vertical profile. The concentration calculated by the Wang model changes slowly under 0.05 h but the Rouse model changes dramatically. The Zhong model, the power-law model, the Rouse model, and the two-phase flow model provide similar results near the water's surface.

The proposed study will look into the relationships between different flows and sediment factors to better describe them in Equations (10)–(12). This will result in a parameterized expression of the final characteristic model for suspended particles and enable accurate profile prediction. Rouse number P, size parameter S, and mean concentration $\bar{c}$ are the flow parameters that will be examined.

3. Proposed Model

Numerous studies have examined the factors that should be taken into account when determining a concentration profile, such as Rouse number, particle size, mean concentration, and flow depth [5,29,31,32]. By using a modified Equation (12), the variables that are thought to be connected to the power-linear law coefficients are as follows:

$$b_1,\ b_2,\ a_1,\ a_2,\ d_1, d_2 = f(P, S,\ \bar{c}) \tag{15}$$

where d_{50} is the diameter of the sediment particle, $\bar{c}$ is the mean concentration, P is the Rouse number, H is the flow depth, and S is the size parameter ($S = d_{50}/H$). In order to examine the distribution of each power-linear law coefficient toward the physical parameters of Rouse number, particle size, and mean concentration and to derive a modified Rouse model for validation tests, we collected the data from numerous reported experimental studies (as detailed in Table 1). Table 1 further shows that the data sources employed to cover a broad range of tested parameters. Particularly, the range of $\bar{c}$ in the referenced literature is from 0.00013 to 0.147, which provides an accurate picture of flow conditions from diluted to hyper-concentrated.

Table 1. Sources of data for parameterised modelling [1].

Data Sources	H	d_{50} (mm)	v_s (cm/s)	u_* (cm/s)	$\bar{c}$ ($\times 10^{-3}$)
Bouvard and Petkovic [26]	7.5	2.00–9.00	1.81–2.70	2.54–5.41	2.1–4.5
Cellino and Graf [33]	12.0	0.135	1.20	4.30–4.50	96–147
Coleman [32]	17.0–17.4	0.21–0.42	1.23–1.31	4.10	0.13–0.28
Muste et al. [34]	2.1	0.21–0.25	0.06	4.00–4.30	0.46–1.62

3.1. Machine-Learning Algorithms

In trying to investigate the relationship between power-linear law coefficients of the Kundu and Ghoshal model and sediment parameters such as the Rouse number, particle size, mean concentration, and flow depth, various machine-learning models have been used as follows:

3.1.1. XGboost Classifier

XGBoost is a decision-tree-based ensemble machine-learning technique that uses gradient boosting. It may be used to solve a variety of regression and classification predictive modelling challenges. It is a fast implementation of the stochastic gradient boosting algorithm with several hyperparameters that allows the fine tuning of the model's training process.

3.1.2. Linear Regressor (Ridge)

Linear regression uses a best-fit straight line to establish a link between a dependent variable (y) and one or more independent variables (x) (also known as a regression line). Ridge regression is a multicollinear data analysis technique (independent variables are highly correlated).

3.1.3. Linear Regressor (Bayesian)

While Ridge regression utilises L2 norm regularisation, Bayesian regression is a probabilistic linear regression model with explicit priors on the parameters. Priors can have a regularising impact, for example, the application of Laplace priors for coefficients is comparable to L1 regularisation.

3.1.4. K Nearest Neighbours

The supervised machine learning algorithm K-nearest neighbours (kNN) can be used to address both classification and regression problems. Because it delivers highly precise predictions, this algorithm can compete with the most accurate models. As a result, the kNN algorithm can be utilised in applications where great accuracy is required but a human-readable model is not required. The accuracy of the predictions is influenced by the distance measured.

3.1.5. Decision Tree Regressor

In the corporate world, the Decision Tree algorithm has become one of the most widely utilised machine learning algorithms. Both classification and regression issues can benefit from the use of the Decision Tree. In the shape of a tree structure, the Decision Tree constructs regression or classification models. It incrementally cuts down a dataset into smaller and smaller sections while also developing an associated decision tree. A tree with decision nodes and leaf nodes is the result.

3.1.6. Support Vector Machines (Regressor)

SVM stands for Support Vector Machine and is a supervised machine learning method capable of classification, regression, and even outlier detection. SVM classifiers have a high level of accuracy and can anticipate events quickly. They also utilise less memory in the decision phase because they only use a subset of training points. With a clear separation margin and a large dimensional space, SVM performs well.

The complete flow chart describing the workflow of using machine-learning algorithms for estimating coefficients for linear and power laws is shown in Figure 4.

3.2. Theoretical Description of Proposed Models

The three broad categories of machine-learning approaches are supervised, unsupervised, and semi-supervised learning. Classified data is needed in supervised learning systems so that the algorithm may be trained. In contrast, the unsupervised learning approach discovers features in the data sets without requiring training data to be classified beforehand. These methods have been practised and are well established. While supervised and unsupervised learning techniques are combined in semi-supervised learning systems. The theoretical foundation of the suggested ML models is provided in the paragraph below in this subsection.

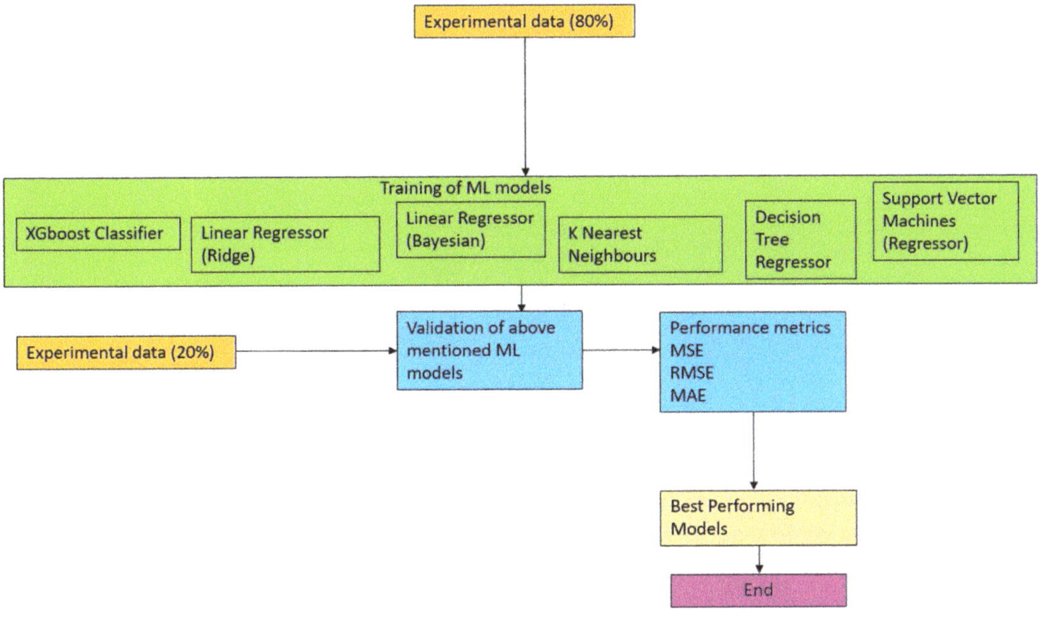

Figure 4. Flowchart for prediction of linear-law and power-law coefficients using the proposed models.

Adaptive Boosting (AdaBoost), a very popular boosting technique, focuses on combining several weak classifiers into a single strong classifier. It is the first binary classification boosting ensemble model. It applies to problems involving classification and regression. It can handle both textual and numerical data and is versatile. It is more susceptible to data noise and outliers than XGBoost is. It is more adaptable since it can be used to improve the weak classifiers' accuracy. In contrast to XGBoost, it minimises exponential loss function rather than differentiable loss function.

CatBoost, an ML algorithm is based on gradient-boosted decision trees. During the model's training, a collection of decision trees are built one after another. Compared to the prior trees, the loss decreases with each additional tree. The gradient boosting method is the foundation for the CatBoost Regressor implementation, which uses a decision tree as the primary predictor. Because there is not a sparse dataframe, the data, which contains many categorical features, is processed much more quickly than it would be if another technique, such XGBoost or Random Forest was being used. In CatBoost, parameters self-tune to save time modifying. With CatBoost, we can work more effectively with ML engineers and software engineers because we can essentially maintain the feature or column in its original state.

The K nearest neighbours (KNN), is a straightforward machine learning algorithm that examines all of the inputs and predicts the target based on features that are similar to them. As a nonparametric model, it has been used in statistical estimation and pattern recognition for more than 50 years. According to this method, the analyst must specify the neighbourhoods' size. To set the size that lowers the MSE to a minimal amount, cross-validation may also be used. Utilizing an inverse distance weighted average of the KNN is a straightforward use of KNN regression. It makes use of the same KNN classification distance functions. Equations (16)–(18) are used to represent the distance functions.

Euclidean

$$\sqrt{\sum_{i=1}^{k}(x_i - y_i)^2} \tag{16}$$

Manhattan

$$\sum_{i=1}^{k}|x_i - y_i| \tag{17}$$

Minkowski

$$\left(\sum_{i=1}^{k}(|x_i - y_i|)^q\right)^{\frac{1}{q}} \tag{18}$$

These three distance functions are applicable only for continuous variables while in the case of discrete or categorical data, the Hamming distance function is used. The Hamming distance function is given below by Equation (19) as:

$$D_H = \sum_{i=1}^{k}|x_i - y_i| \tag{19}$$

The Hamming distance function is zero if $x = y$, otherwise if $(x \neq y)$, it is equal to unity.

LightGBM, a decision-tree-based gradient boosting framework helps GPU learning while also maximising model efficiency by growing trees vertically (leaf-wise split), as opposed to other boosting approaches that grow trees horizontally, which increases accuracy (level-wise split). Since it transforms continuous values into discrete bins, the primary benefit of the LightGBM regressor is that it lowers memory usage without sacrificing forecasting speed or prediction accuracy.

3.3. Rouse Number Modelling

Shear velocity and settling velocity are the two key variables that influence drag on a sediment particle. The Rouse number as stated in Equation (6) is a dimensionless form of these parameters along with the von Karman constant. We may create the figures shown in Figures 5–8 below by testing each value in Equation (12) against P.

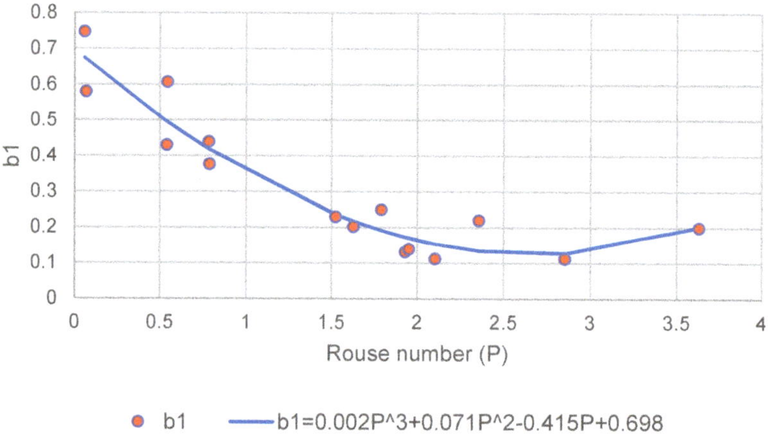

Figure 5. Rouse number vs. coefficient b_1 using SVM ($R^2 = 0.9148$).

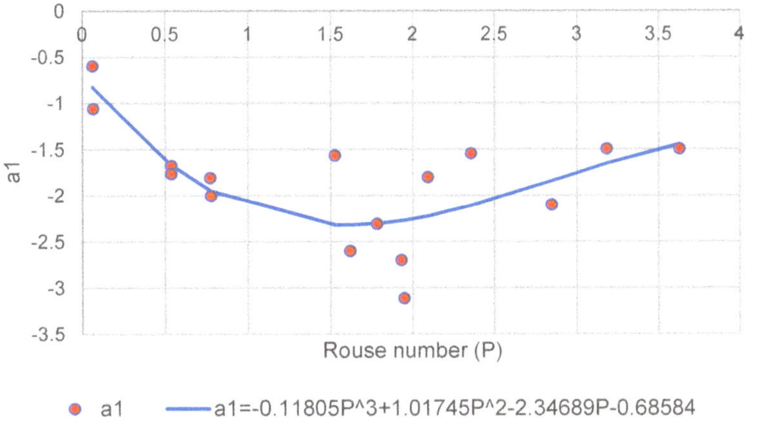

Figure 6. Rouse number vs. coefficient a_1 using KNN ($R^2 = 0.61$).

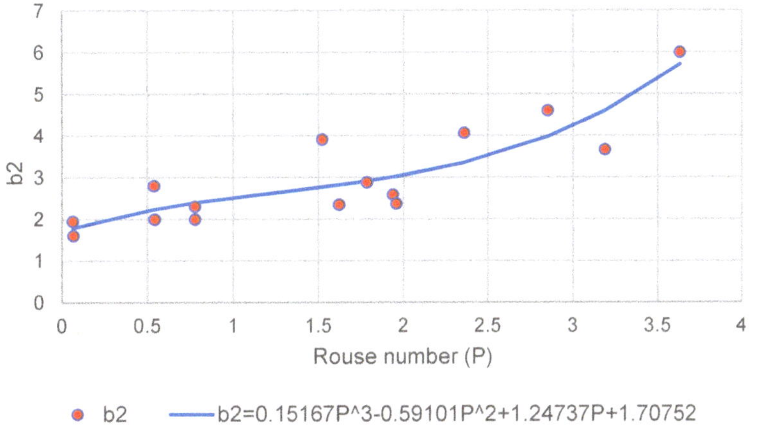

Figure 7. Rouse number regression vs. coefficient b_2 using KNN ($R^2 = 0.777$).

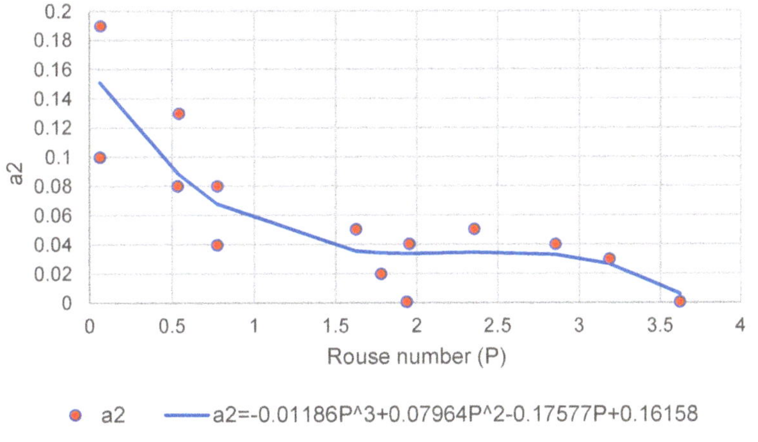

Figure 8. Rouse number regression vs. coefficient a_2 using decision trees ($R^2 = 0.7477$).

Figures 5–8 show the relationship between b_1, a_1, b_2 and a_2 against P. This result demonstrates that for a wide range of measured data, P offers a reasonable match which is represented by the power law and its coefficients.

3.3.1. Size Parameter Modelling

Another element that may impact sediment drag and lift is particle size. The effectiveness of interactive interactions acting on the particle can be influenced by the surface area of the particle, which is dictated by the diameter. The settling velocity of a particle is also influenced by its diameter [35]. The dimensionless S is utilised to model against the power-law coefficients in Figures 9–12 in this study. The effect of particle size has not been taken into account in the modified Rouse model [11] or the original Rouse technique [5] despite the fact that it is an important aspect that affects the behaviour of suspended sediment.

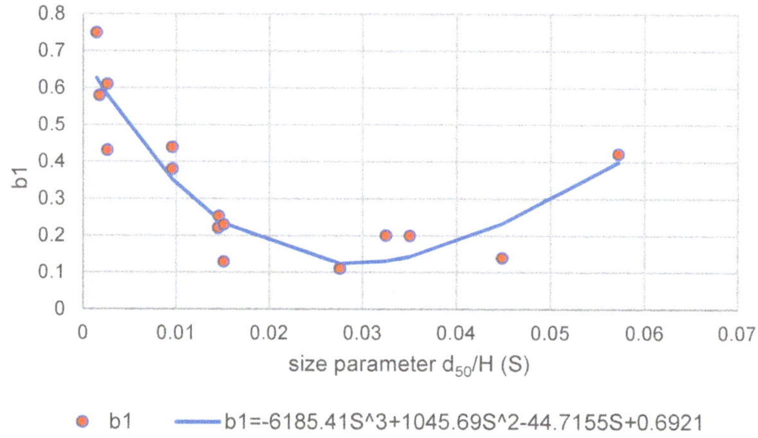

Figure 9. Size parameter vs. coefficient b_1 using decision trees ($R^2 = 0.856$).

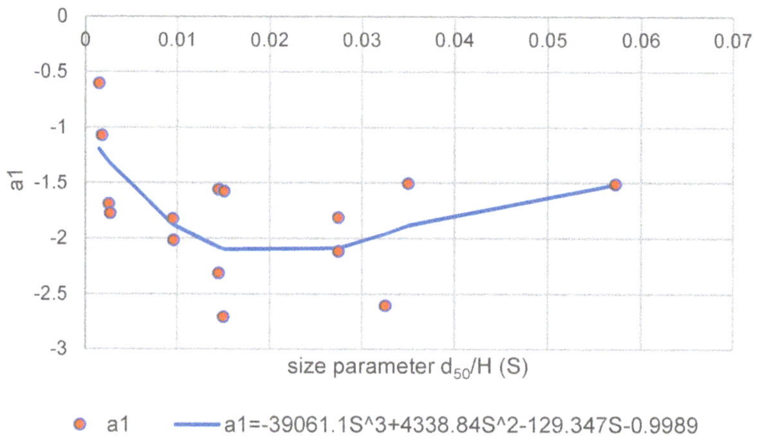

Figure 10. Size parameter vs. coefficient a_1 using decision trees ($R^2 = 0.4313$).

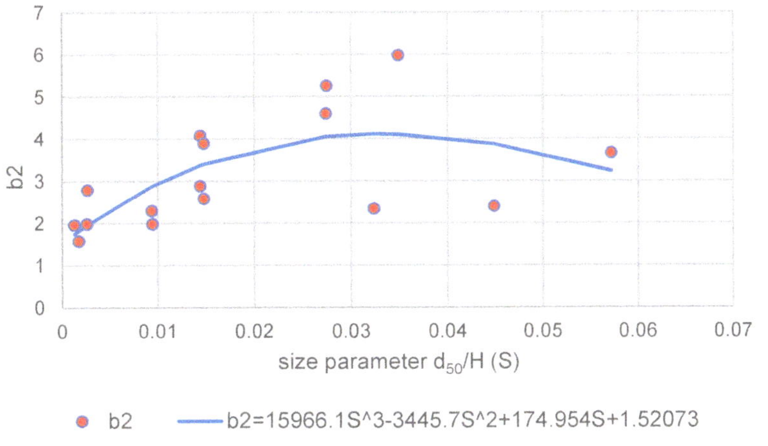

Figure 11. Size parameter vs. coefficient b_2 using KNN ($R^2 = 0.434$).

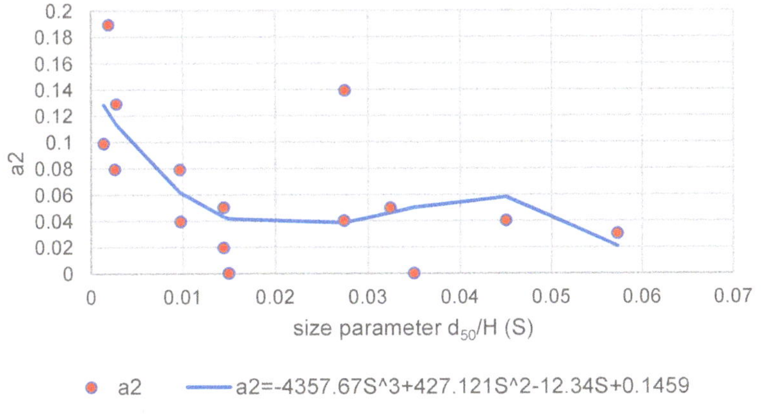

Figure 12. Size parameter vs. coefficient a_2 using KNN ($R^2 = 0.433$).

The relationship between b_1, b_2, a_1 and a_2 against S is depicted in Figures 9–12. Numerous analytical modelling studies, such as Wang and Ni [21] and Ni et al. [10], used a coefficient from the concentration equation that was deduced from the collected data to fit the particle diameter. Their experiments demonstrated that when different sediment diameters are investigated, it is challenging to capture the character of the concentration profile. This serves as more evidence of how challenging it is to identify a representative function of the particle size parameter studied here.

3.3.2. Mean Concentration Modelling

For hyper-concentrated flows, the recorded sediment concentration profiles were examined by Cellino and Graf [33], and Michalik [36]. Their findings demonstrated that, in contrast to the general power-law shown in earlier studies of diluted flows, the sediment profile for such flows follows a much more uniform and linear distribution. This study will develop the analytical approach by mean concentration to use in linear-law modelling in response to another empirical finding by Michalik [36] that the mean concentration has a key dominant impact changing the character of concentration distribution when compared to Rouse number and particle size. The fitting between the linear law coefficients and mean concentration was determined using the procedure below, with the findings displayed in Figures 13–16.

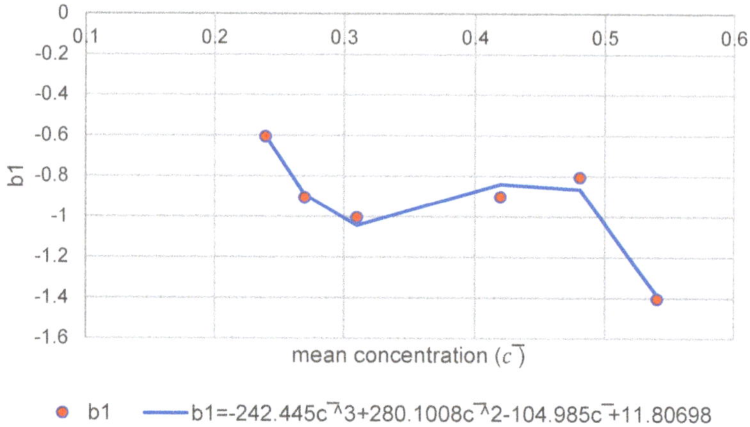

Figure 13. Mean concentration regression vs. coefficient b_1 using decision trees ($R^2 = 0.973$).

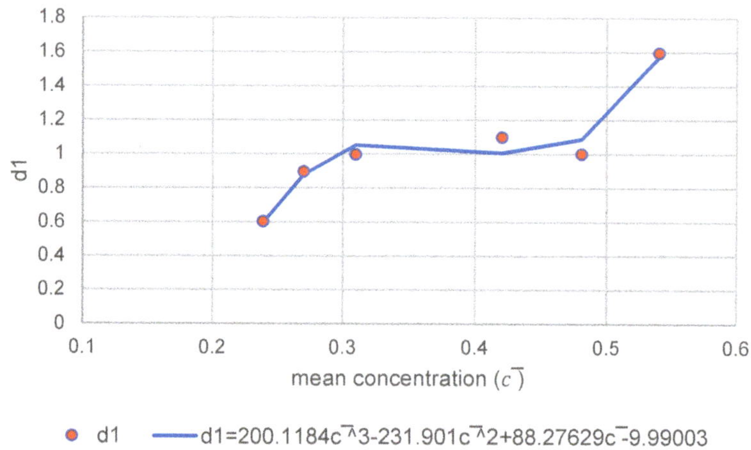

Figure 14. Mean concentration vs. coefficient d_1 using decision trees ($R^2 = 0.9617$).

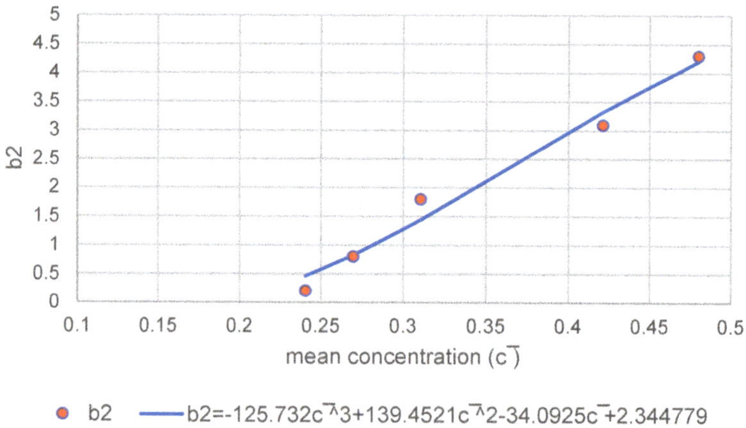

Figure 15. Mean concentration vs. coefficient b_2 using KNN ($R^2 = 0.982$).

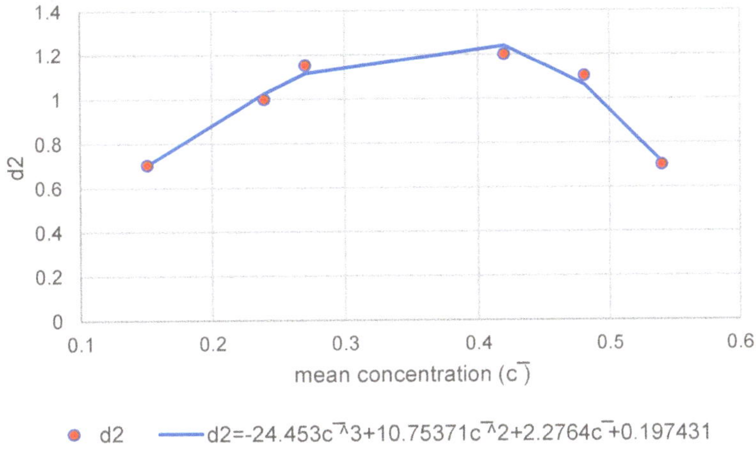

Figure 16. Mean concentration vs. coefficient d_2 using KNN ($R^2 = 0.979$).

The above Figures 13–16 show the relationship between b_1, b_2, d_1, and d_2. Here, the association between $\bar{c}$ and the hyper-concentrated profile is evident.

3.3.3. Modified Rouse Approach

Each coefficient is defined in the aforementioned sections before being adapted into Equation (12) to create a parameterized expression for the distribution of silt concentration over the flow depth. A linked approach controls the suggested model for sediment concentration prediction. When $0 < \bar{c} < 0.1$, a power law describes the concentration of diluted silt, whereas a linear law describes the concentration of dense hyper-concentration. This hyper-to-dilute boundary was established by comparing it to the research on diluted flow definition by Greiimann and Holly [8] and the Rouse model limit. The suggested strategy can be written as in Equation (20).

$$\frac{c}{\bar{c}} = \frac{1}{(b_1\varepsilon^{a_1} + d_1)^{-1} + (b_2\varepsilon^{a_2} + d_2)^{-1}} \tag{20}$$

where $0 < \bar{c} < 0.1$, and $d_1 = d_2 = 0$

$$\begin{aligned}
2b_1 &= 0.002P^3 + 0.07P^2 - 0.415P - 6185.41S^3 + 1045.69S^2 - 44.71S + 1.39,\\
2a_1 &= -0.11805P^3 + 1.0174P^2 - 2.3468P - 39061S^3 + 4338.84S^2 - 129.347S - 1.6847,\\
2b_2 &= 0.1516P^3 - 0.59101P^2 + 1.2473P + 15966.1S^3 - 3445.7S^2 + 174.95S + 3.228,\\
2a_2 &= -0.01186P^3 + 0.0796P^2 - 0.17577P - 4357.67S^3 + 427.12S^2 - 12.34S + 0.3075,
\end{aligned}$$

and when $\bar{c} \geq 0.1$, and $a_1 = a_2 = 1.0$

$$b_1 = -242.445\bar{c}^3 + 280.1008\bar{c}^2 - 104.985\bar{c} + 11.80698,$$

$$\begin{aligned}
d_1 &= 200.1184\bar{c}^3 - 231.901\bar{c}^2 + 88.2763\bar{c} - 9.99,\\
b_2 &= -125.73\bar{c}^3 + 139.45\bar{c}^2 - 34.0925\bar{c} + 2.3447,\\
d_2 &= -24.453\bar{c}^3 + 10.7537\bar{c}^2 + 2.276\bar{c} + 0.1974.
\end{aligned}$$

4. Model Validations

The link between parameters of the power-linear coupled model and sediment flow characteristics were discovered using a variety of machine-learning approaches. The distribution of suspended sediment over the characteristic height inside the flow was inclusively estimated from diluted to hyper-concentrated. The model can precisely calculate the suspended sediment profile for a range of flow conditions, and sediment sizes, including a range of Rouse numbers. The models show good accuracy for testing at low and extremely high concentrations for type I to III profiles. The best result-giving model's RMSE, MAE, and MSE values are displayed in Table 2.

Table 2. The following RMSE, MAE, and MSE scores of the best result-giving models are shown.

Relationship Modelled	Model Giving Best Result	MSE (Mean Square Error)	RMSE (Root Mean Square Error)	MAE (Mean Absolute Error)
a_1 and Rouse Number (P)	K-Nearest Neighbours	0.3876	0.6225	0.4522
b_1 and Rouse Number (P)	Support Vector Machines (Regressor)	0.0067	0.0820	0.0718
a_2 and Rouse Number (P)	Decision Trees	0.0014	0.0374	0.0225
b_2 and Rouse Number (P)	K-Nearest Neighbours	1.0384	1.0190	0.8427
a_1 and $(S = d_{50}/H)$	Decision Trees	0.2080	0.4561	0.4062
b_1 and $(S = d_{50}/H)$	Decision Trees	0.0064	0.0802	0.0506
a_2 and $(S = d_{50}/H)$:	K-Nearest Neighbours	0.0017	0.0416	0.0302
b_2 and $(S = d_{50}/H)$:	K-Nearest Neighbours	0.9923	0.9961	0.8027
b_1 and mean concentration $(\bar{c})$	Decision Trees	16.7333	4.0906	3.4095
d_1 and mean concentration $(\bar{c})$	Support Vector Machines (Regressor)	6.5332	2.5560	2.1114
d_2 and mean concentration $(\bar{c})$	K-Nearest Neighbours	3.3596	1.8329	1.4628
b_2 and mean concentration $(\bar{c})$	K-Nearest Neighbours	9.9020	3.1467	2.6971

Initially, the necessary python libraries were imported into the Google Colab environment for modelling. The dataset was then imported using the wget function. The dataset was preprocessed in a suitable format for modelling by separating the X (input feature) and Y (output feature). Using the train-test-split, the dataset was split into train and test sets. Cumulatively we obtained the X_train, Y_train, X_test and Y_test for training and validating the dataset. The models including XGBoost Classifier, Linear Regressor (Ridge), Linear Regressor (BayesianRidge), K-Nearest Neighbours, Decision Tree Regressor and Support Vector Machines (Regressor) were used.

The experimental data from Wang and Qian [37], Wang and Ni [31], and Michalik [36] were used to validate the model proposed in this study (Equation (20)). It was also contrasted with the models that Wang and Ni [31], Ni et al. [10], Zhong et al. [23], and Pu et al. [1] had previously presented. A theoretical distribution model derived from kinetic theory was presented by Wang and Ni [31]. Their model can only forecast type I and a subset of type II profiles since it is confined to diluted flow. They incorrectly assumed that particle interaction did not exist, and as a result, they blamed fluid-induced lift forces for the categorization of the distribution profile. Further, their analysis states that the distribution tends to follow the type I profile when the particle size is small.

The Ni et al. [10] model incorporated both continuum and kinetic theories. Kinetic theory employing the Boltzmann equation for the solid phase, and continuum theory for the fluid phase was used. The two-phase interactions in their derivation are represented by the forces acting on the sediment those are empirically weighted. It is claimed that the model can handle both diluted and dense flows. In comparison to the other two models, the model provided by Zhong et al. [23] is more sophisticated. It is founded on a three-part strategy in which the model can be made simpler for a given different empirically supported hypotheses. As a result, in their research, all type I, II, and III profiles are shown in the experimental testing. Their type III profile used varying empirical dampening function values to meet various flow conditions due to its complexity. Pu et al. [1] proposed a parameterized method using sediment size, mean concentration, Rouse number, and flow depth, which was effective as shown in the comparisons.

4.1. *Wang and Ni*

Wang and Ni [31] evaluated the diluted flow in a conduit using measured data. All of the investigated conditions are shown in Table 3; the quantities examined ranged from $0.00042 < \bar{c} < 0.0033$, and they were severely diluted. The ability of the suggested model to capture severely diluted flow is tested during this validation study. The sediments studied were granules and coarse sand with particle diameters ranging from 0.58 mm to 2.29 mm. The outcomes are displayed in Figure 17a–k. The measurements of Wang and Ni [31] show that for diluted flow the sediment concentration tends to follow the type I or II profile where the maximum concentration occurs at the near-bed region.

Table 3. Data by Wang and Ni [31].

Test No.	d_{50} (mm)	v_s (cm/s)	u_* (cm/s)	$\bar{c}$ ($\times 10^{-3}$)
1	1.80	2.56	3.28	3.30
A2	1.3	2.17	3.35	4.4
A3	1.40	6.90	4.76	3.10
A4	1.10	5.15	4.52	2.40
A5	0.58	4.51	4.79	0.97
A6	0.60	3.79	4.90	0.57
A7	2.29	6.15	4.83	0.44
B3	1.40	6.90	6.11	1.98
B4	1.10	5.15	6.15	2.00
B5	0.58	4.51	6.33	1.94
B6	0.60	3.79	6.23	0.42

The results shown in Figure 17a–k demonstrate that overall, there is a better fit between the proposed models and other models away from the near-bed regions. The proposed model corresponds to the experimental data well in the middle and free surface regions. In cases A1, A2, A6, A7, B3, and B4, the proposed model predicts experimental data better than the model proposed by Pu et al. [1] and other models. This is consistent with the literature [1,4,8] which states that possibility of particle–particle collisions increases near-bed to produce conditions that are more challenging to represent with mathematical modelling.

The biggest particles among all the measured data from Wang and Ni [31] were used in the experimental data presented in Figure 17g. Due to the higher surface area of the sediment particles, there would be stronger forces coming from the solid–fluid phase interaction. One can see that the concentration distribution for A7 shown by the proposed model (Figure 17g) is consistent with the observations. The suggested model exhibits the promising computation of big particles in observed data when compared to a model such as those by Zhong et al. [23] that does not account for particle size. In addition, the tests A5 and B5 are the smallest particle diameters, and the proposed model is doing better than Ni et al. [10] which relies on empirical particle interactions.

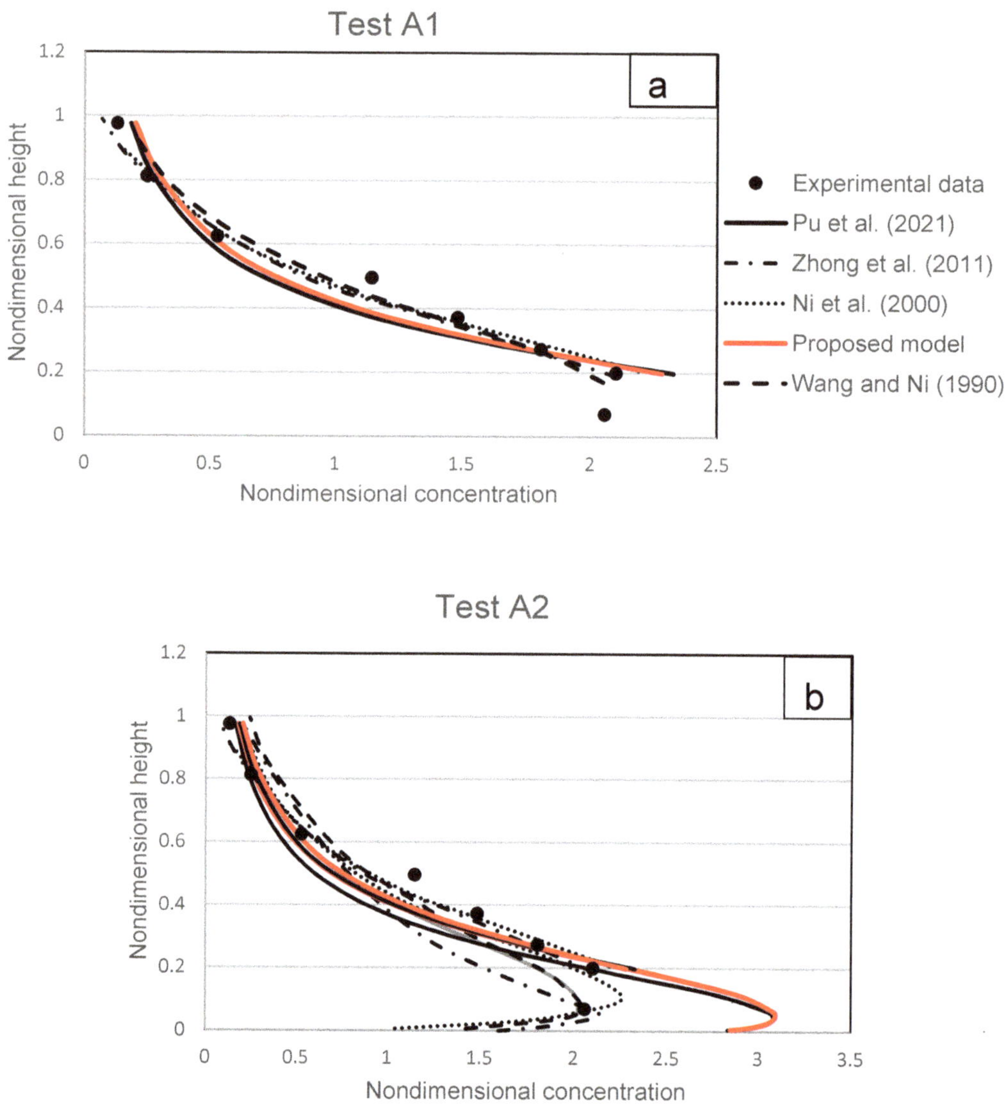

Figure 17. *Cont.*

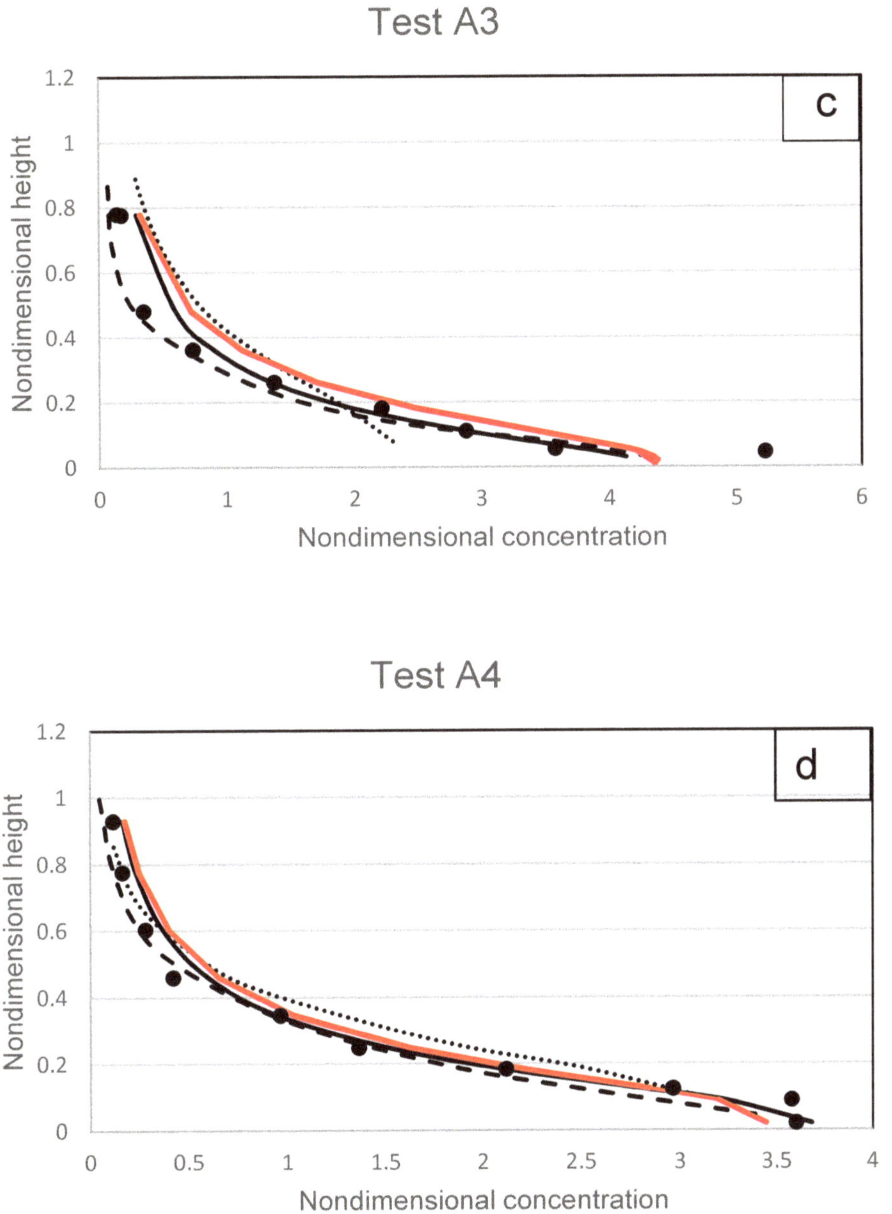

Figure 17. *Cont.*

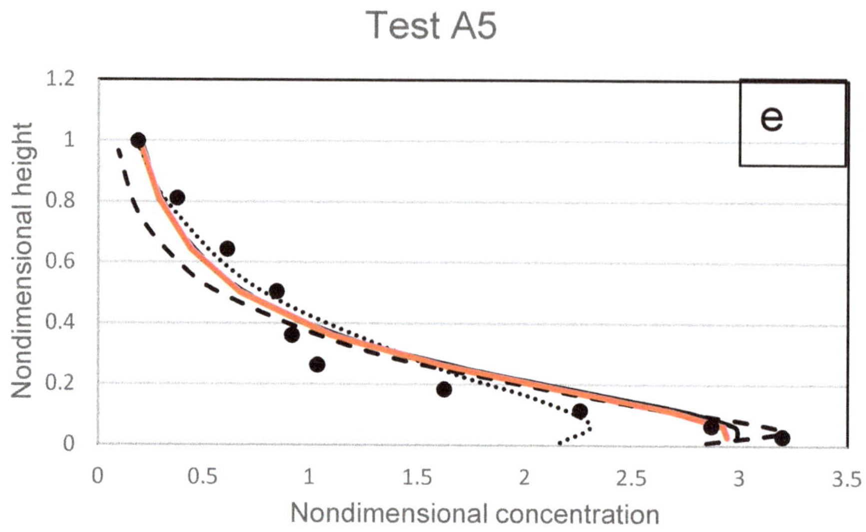

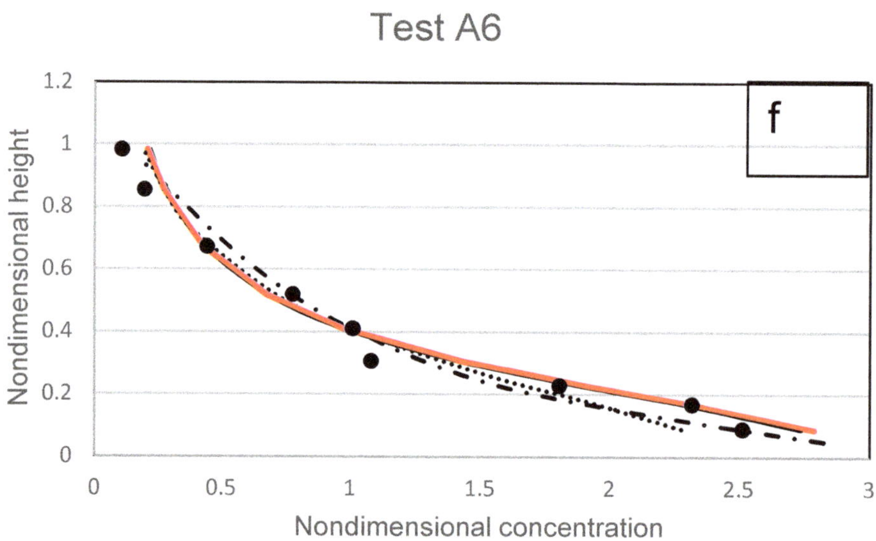

Figure 17. *Cont.*

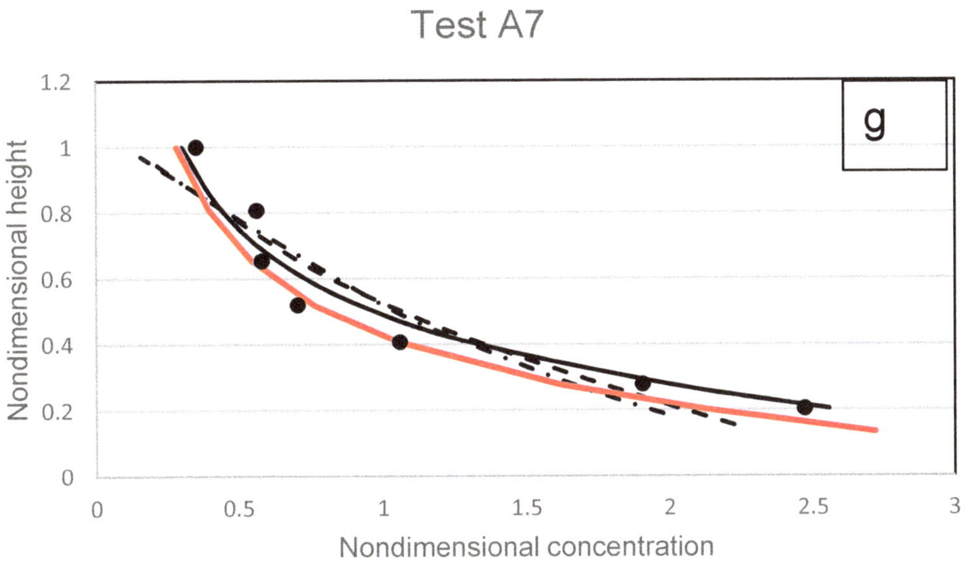

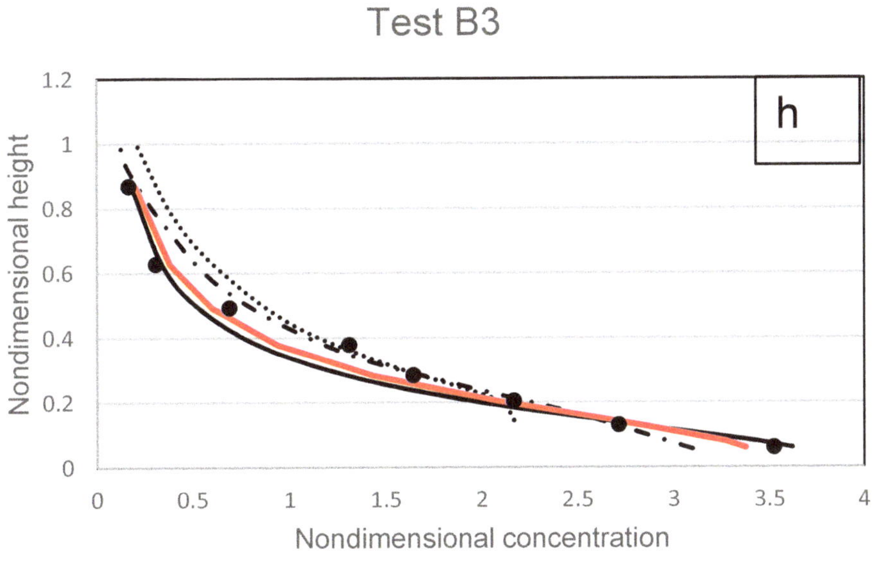

Figure 17. *Cont.*

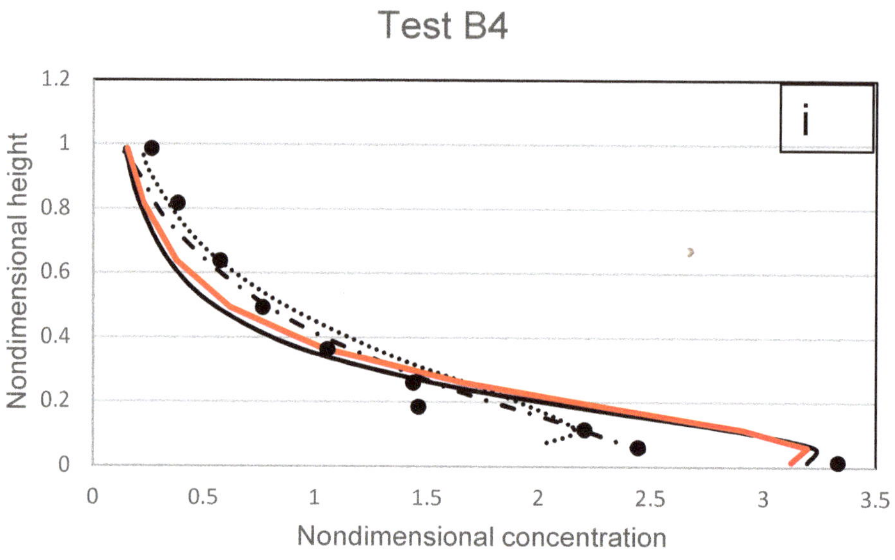

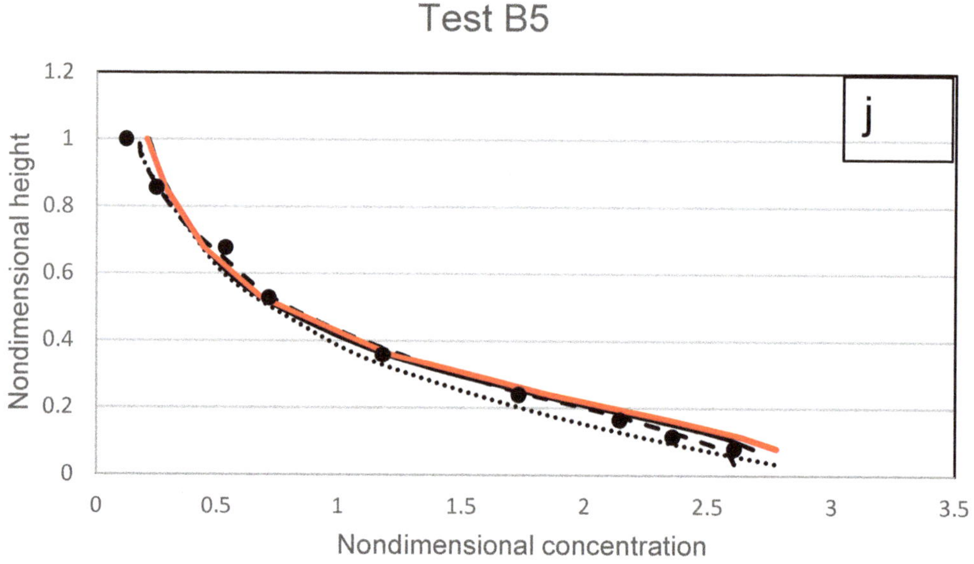

Figure 17. *Cont.*

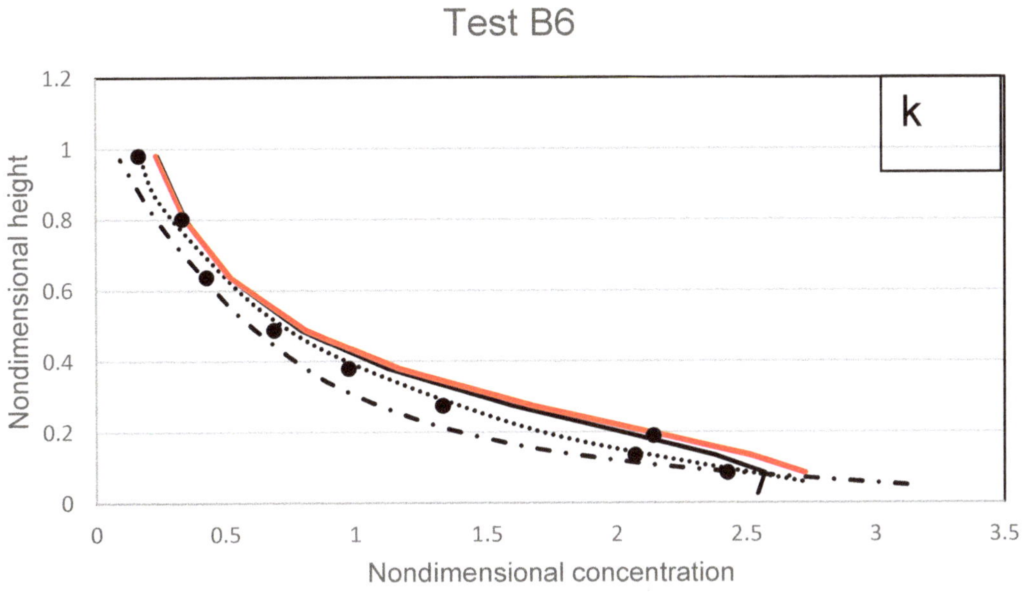

Figure 17. (**a–k**) The modelled results and comparisons against experimental data of Wang and Ni [31] and models of Wang and Ni [21], Pu et al. [1], Ni et al. [10] and Zhong et al. [23].

4.2. Wang and Qian

Using an experimental recirculating-tilting flume, Wang and Qian [37] investigated the impact of diluted to dense concentrations in an open channel flow. Their tests included silt with diameters ranging from 0.15 mm to 0.96 mm and concentrations ranging from 0.0102 to 0.0906. All the conditions tested are shown in Table 4. As a result, their tests have been used in this comparison with the suggested and other earlier models as the findings shown in Figure 18a–f. In general, the models' accuracy declines as the mean concentration rises across the flow depth. In particular, in the upper-flow zone, the experimental data are reasonably fit by the suggested model. The lift and drag caused by the fluid-induced forces as well as the inertia of the sediment particles are the principal forces acting on the particles in the upper layer. This demonstrates that the proposed model's incorporation of the Rouse and size parameters allows it to accurately calculate the solid-fluid interactions for individual particles.

Table 4. Data by Wang and Qian [37].

Test No.	d_{50} (mm)	v_s (cm/s)	u_* (cm/s)	$\bar{c}$ ($\times 10^{-3}$)
SF2	0.268	0.197	7.74	10.2
SF5	0.268	0.197	7.16	90.6
SM6	0.266	0.129	7.37	7.54
SM7	0.960	1.590	7.37	75.4
SC5	1.420	2.290	7.37	65.1
SC7	0.266	1.59	7.37	13.72

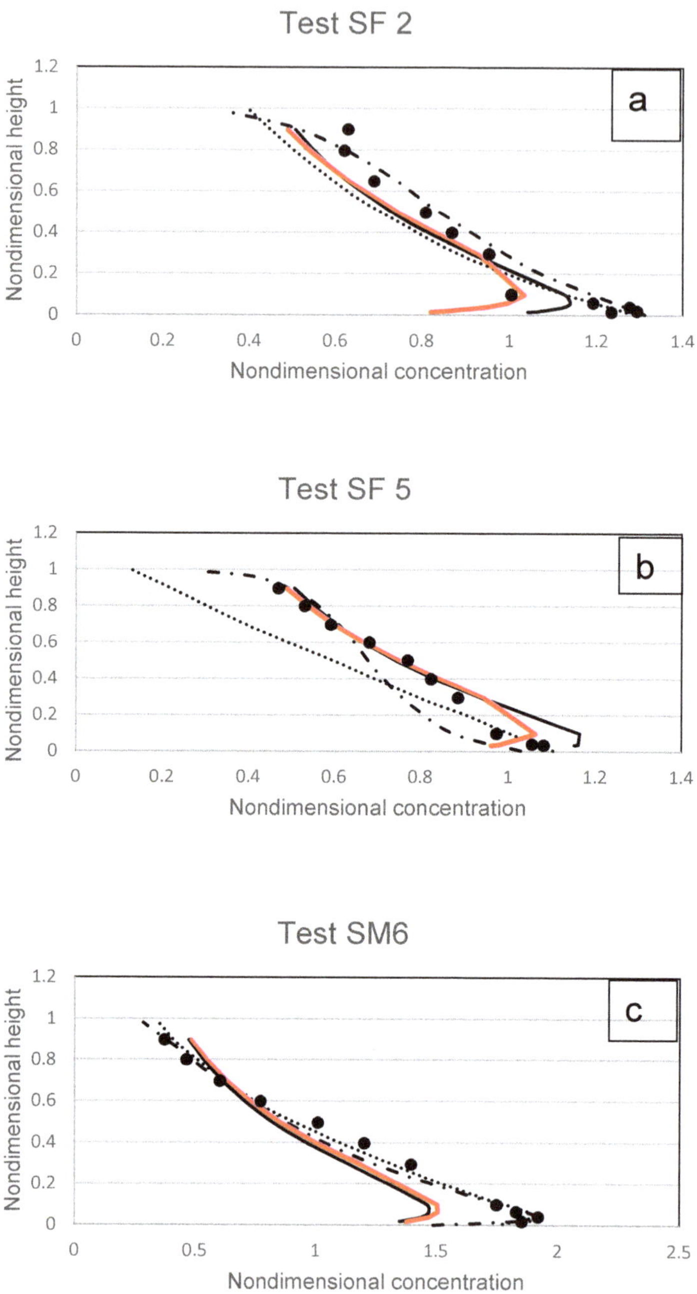

Figure 18. *Cont.*

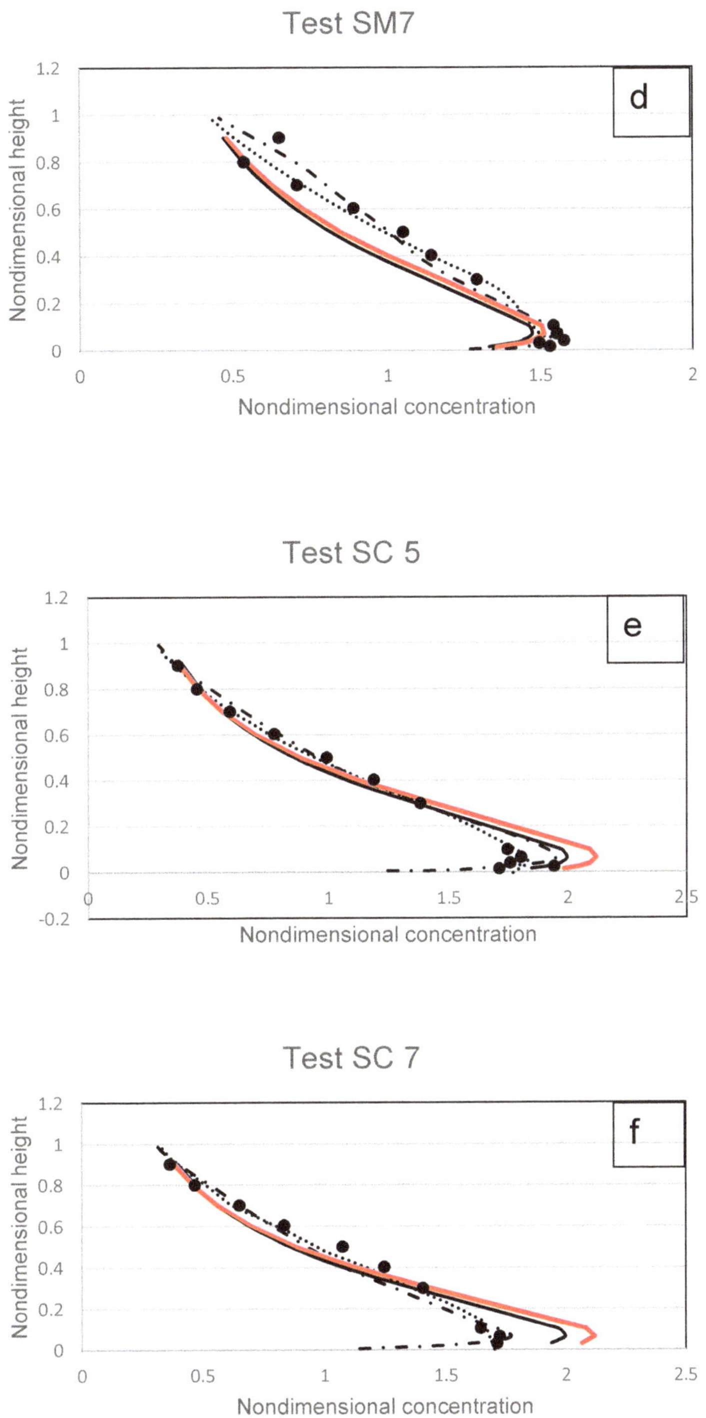

Figure 18. (**a**–**f**) Modelled results and comparisons against experimental data of Wang and Qian [37].

The accuracy of the various models, including the suggested model, decreases at the lower-suspension region, according to an overview of all the results shown in Figure 18. Compared to type III profiles, type II profiles have higher concentrations and a smaller turning point near the bed. Rather than suspended load, bedload behaviour governs the distribution of sediment along the wall boundary. In Wang and Qian's [37] investigation, plastic particles were also utilised in the flow, and theoretically, under the influence of particle–particle interacting forces, it produced less notable movement than the normal and natural sediment. This means that, in contrast to models by Ni et al. [10] and Zhong et al. [23], which incorporated empirical functions discovered from the relevant experiments into the modelling, the proposed model, which deals with the forces by coupling expressions of Rouse and size parameters, is unable to simulate its near-bed concentration reasonably for experiments SF2, SC5, and SC6 as shown in Figure 18a, Figure 18e, and Figure 18f respectively. Results of the proposed model are better than those from Pu et al. [1] for experiments SF5, SM6, and SM7 as shown in Figure 18b, Figure 18c, and Figure 18e.

4.3. Michalik

Michalik [36] researched the sediment profile of hyper-concentrated flows using experimental data. Sand with a diameter of 0.45 mm and concentrations ranging from 0.15 to 0.54 was employed as the sediment material (all the tested conditions are shown in Table 5). The linear law found in Equation (20) is used in this study to model Michalik's tests. The test findings are displayed in Figure 19a–h, where measurements and the proposed model results are contrasted.

Table 5. Data by Michalik [36].

Test No.	d_{50}	v_s (cm/s)	u_* (cm/s)	$\bar{c}$ ($\times 10^{-3}$)
1	0.45	6.15	15.56	150
2	0.45	6.15	15.56	240
3	0.45	6.15	15.56	270
4	0.45	6.15	15.56	310
5	0.45	6.15	15.56	420
6	0.45	6.15	15.56	450
7	0.45	6.15	15.56	480
8	0.45	6.15	15.56	540

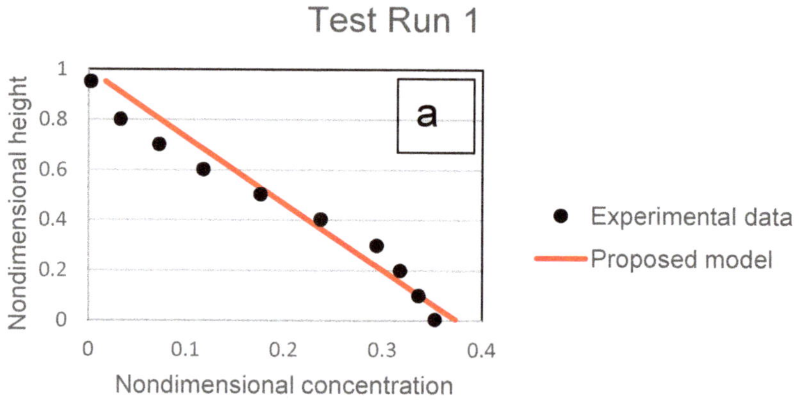

Figure 19. *Cont.*

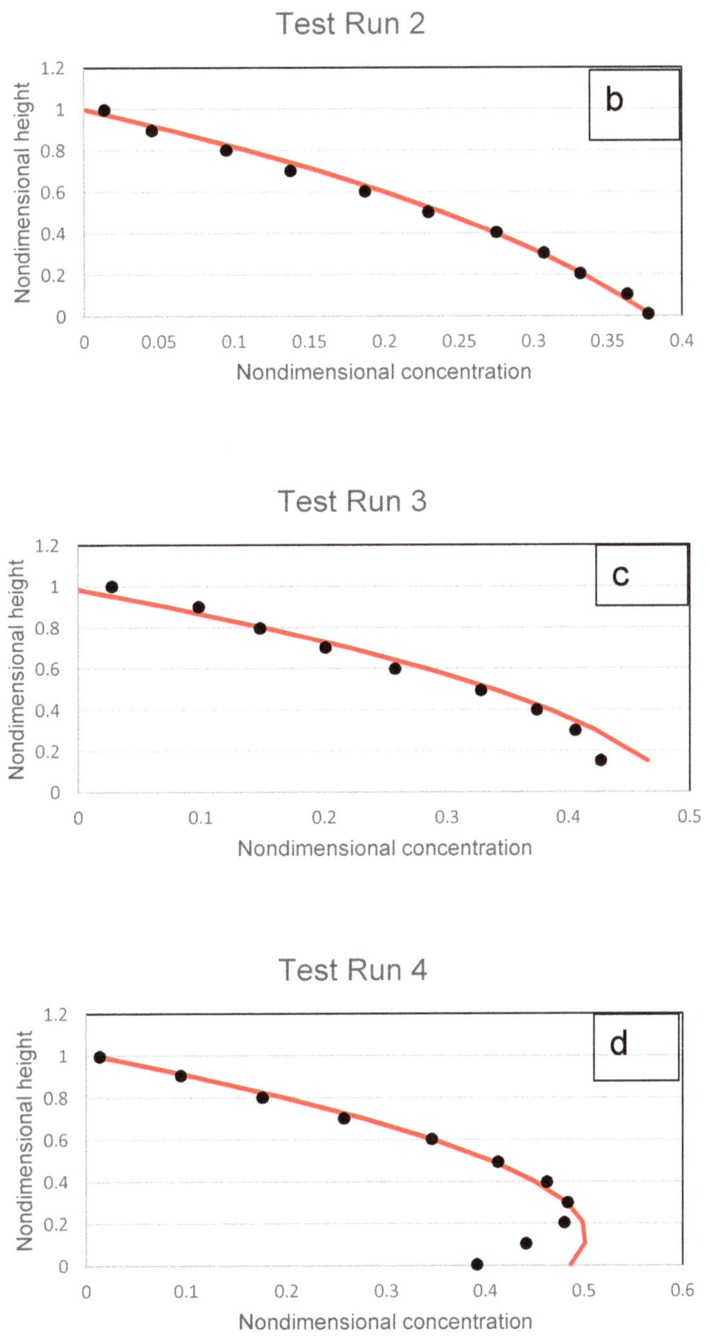

Figure 19. *Cont.*

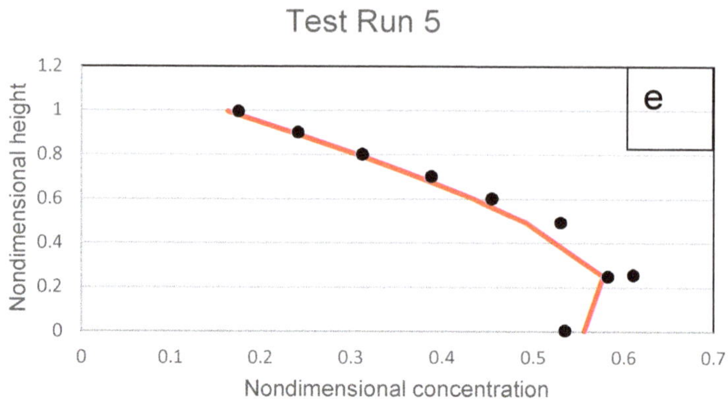

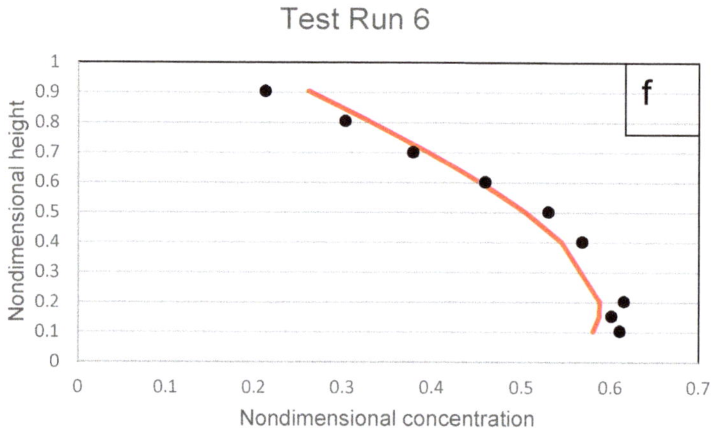

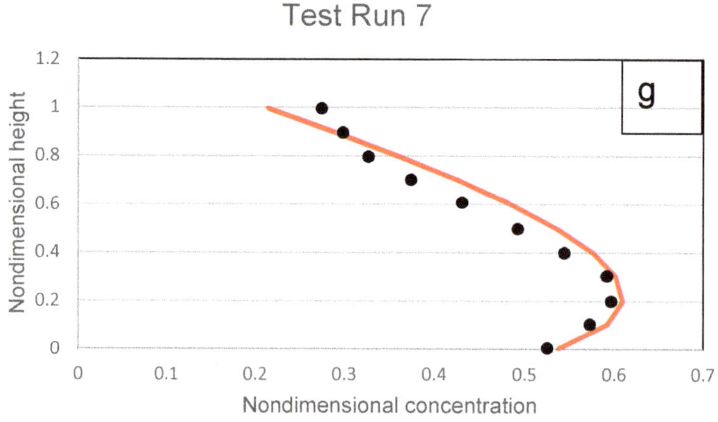

Figure 19. *Cont.*

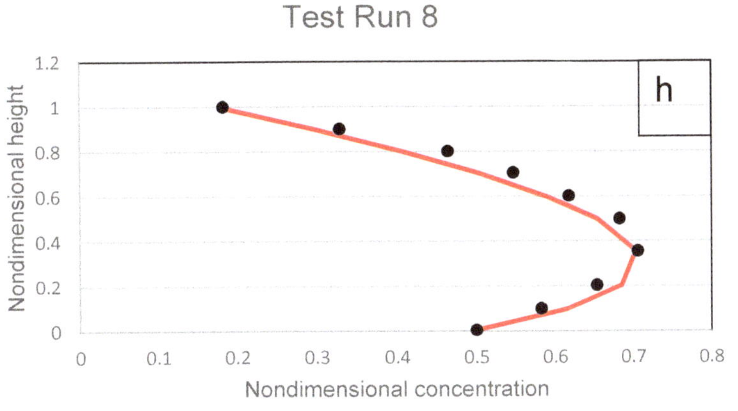

Figure 19. (**a–h**) The modelled results and comparisons for experimental data of Micahlik [36].

The sediment-concentration distribution from the investigated hyper-concentrated flows displays a type III characteristic. It is challenging to precisely calculate the concentration distribution's turning point in a hyper-concentrated flow. This is due to the fact that the boundary between the sediment's bed and suspended states is not clearly defined, but rather, the bedload will diffuse into the suspended state through a transitional zone. Because of this, the suspended load has occasionally been calculated as a portion of the bedload, which widens the gap between measurements and suspended sediment modelling. A correct assumption of ε_a and c_a is necessary in order to precisely describe the transition zone and, consequently, the position of the turning point [29]. Given that both the variables are constant in this study's proposed model and are not changing in an experiment, it is reasonable to draw the conclusion of ε_a and c_a [38].

The suggested model shows a decent fit of the type III profile to the hyper-concentrated distribution and offers a reasonable match to the experimental data. As demonstrated in Michalik [36], studies of hyper-concentrated flow demonstrate that as $\overline{c}$ increases, the sediment concentration distribution becomes more homogeneous. Power distributions may not be appropriate to model hyper-concentrated flow as a result. Rather, a linear model is used here, which has been shown to generate superior accuracy.

The height of the maximum concentration also increases as the mean concentration rises, possibly as a result of the growing possibility of a bedload layer close to the bed. The near-bed region becomes increasingly saturated as the mean concentration rises. The experiments in Figure 19 demonstrate that the settling velocity has no effect when $\overline{c} > 0.31$ and that the dominant interaction forces must result from particle–particle reactions. The suggested model accurately depicts the concentration distribution in these highly concentrated tests of $\overline{c} > 0.31$. In particular, for the majority of cases, the suggested model correctly predicts the height at which the greatest concentration (turning point) occurs.

5. Conclusions

For accurately estimating the diluted and hyper-concentrated distributions over typical heights within the flow, a parameterized power-linear coupled model with machine learning has been validated. The machine learning models used were XGboost Classifier, Linear Regressor (Ridge), Linear Regressor (Bayesian), K Nearest Neighbours, Decision Tree Regressor and Support Vector Machines (Regressor). The models were implemented using Google Colab. The models have been applied to investigate the relationship between every Kundu and Ghoshal [4] parameter with each sediment flow parameter namely, mean concentration, Rouse number, and size parameter. The distribution of suspended sediment over the characteristic height inside the flow was inclusively estimated from diluted to hyper-concentrated. Comparisons with experimental data from Wang and Ni [31] on highly diluted flows, Wang and Qian [37] on mixed diluted to dense flows, and Michalik [36] on hyper-concentrated flows for low and extremely high concentration testing that approximates type I to III profiles, the proposed power-law and linear-law models are more precise than Pu et al. [1] in some cases, and in other cases, just as precise as those of Pu et al. [1]. Finally, it is shown that the models produced from machine learning are capable of calculating the suspended sediment profile accurately under a variety of flow circumstances ($0.268\text{ mm} \leq d_{50} \leq 2.29\text{ mm}$, $0.00105\frac{\text{g}}{\text{mm}^3} \leq \text{particle density} \leq 2.65\frac{\text{g}}{\text{mm}^3}$, $0.197\frac{\text{mm}}{\text{s}} \leq v_s \leq 96\frac{\text{mm}}{\text{s}}$, $7.16\frac{\text{mm}}{\text{s}} \leq u_* \leq 63.3\frac{\text{mm}}{\text{s}}$, $0.00042 \leq \bar{c} \leq 0.54$), including different Rouse numbers ($0.0076 \leq P \leq 23.5$).

Author Contributions: S.K.: writing—original draft preparation, writing—review and editing, and data curation; funding acquisition; S.B.: writing—review and editing, and data curation; P.R.H.: writing—review and editing, project administration, data curation, and supervision; H.P.S.: writing—review and editing, data curation; J.H.P.: writing—review and editing, and data curation. All authors contributed to the work. All authors have read and agreed to the published version of the manuscript.

Funding: This research received no external funding.

Institutional Review Board Statement: Not applicable.

Informed Consent Statement: Not applicable.

Data Availability Statement: The data presented in this study are available on reasonable request from the corresponding author.

Conflicts of Interest: The authors declare no conflict of interest.

Notations

b	*Empirical coefficient*
c	*Concentration*
$\bar{c}$	*Mean concentration*
c_b	*Reference concentration*
d_{50}	*Sediment diameter*
D	*Coefficient of diffusivity*
H	*Flow depth*
P	*Rouse number*, $P = v_s/(\kappa u_*)$
d	*Empirical coefficient*
S	*Size parameter*, $S = d_{50}/H$
z	*Vertical distance from bed*
a	*Empirical coefficient*
ε	*Characteristic height*, $\varepsilon = z/H$
ε_b	*Reference characteristic height*
κ	*Von Karman constant*
u_*	*Shear velocity*
φ	*Function of characteristic height*
v_s	*Particle fall velocity*

References

1. Pu, J.H.; Wallwork, J.T.; Khan, M.A.; Pandey, M.; Pourshahbaz, H.; Satyanaga, A.; Hanmaiahgari, P.R.; Gough, T. Flood Suspended Sediment Transport: Combined Modelling from Dilute to Hyper-Concentrated Flow. *Water* **2021**, *13*, 379. [CrossRef]
2. Pu, J.H.; Hussain, K.; Shao, S.; Huang, Y. Shallow sediment transport flow computation using time-varying sediment adaptation length. *Int. J. Sediment Res.* **2014**, *29*, 171–183. [CrossRef]
3. Pu, J.H.; Wei, J.; Huang, Y. Velocity distribution and 3D turbulence characteristic analysis for flow over water-worked rough bed. *Water* **2017**, *9*, 668. [CrossRef]
4. Kundu, S.; Ghoshal, K. A mathematical model for type II profile of concentration distribution in turbulent flows. *Environ. Fluid Mech.* **2017**, *17*, 449–472. [CrossRef]
5. Rouse, H. Modern conceptions of the mechanics of fluid turbulence. *Trans. Am. Soc. Civ. Eng.* **1937**, *102*, 463–505. [CrossRef]
6. Huang, S.H.; Sun, Z.L.; Xu, D.; Xia, S.S. Vertical distribution of sediment concentration. *J. Zhejiang Univ. Sci. A* **2008**, *9*, 1560–1566. [CrossRef]

7. Hsu, T.J.; Jenkins, J.T.; Liu, P.L.F. On two-phase sediment transport: Dilute flow. *J. Geophys. Res.* **2003**, *108*, 2–14. [CrossRef]
8. Greiimann, B.P.; Holly, F.M. Two-phase Flow Analysis of Concentration Profiles. *J. Hydraul. Eng.* **2001**, *127*, 753–762. [CrossRef]
9. Jha, S.K.; Bombardelli, F.A. Two-phase modeling of turbulence in dilute sediment-laden, open-channel flow. *Environ. Fluid Mech.* **2009**, *9*, 237–266. [CrossRef]
10. Ni, J.R.; Wang, G.Q.; Borthwick, A.G.L. Kinetic theory for particles in dilute and dense solid-liquid flows. *J. Hydraul. Eng.* **2000**, *126*, 893–903. [CrossRef]
11. Kundu, S.; Ghoshal, K. Explicit formulation for suspended concentration distribution with. *Powder Technol.* **2014**, *253*, 429–437. [CrossRef]
12. Pu, J.H.; Lim, S.Y. Efficient numerical computation and experimental study of temporally long equilibrium scour development around abutment. *Environ. Fluid Mech.* **2014**, *14*, 69–86. [CrossRef]
13. Goree, J.C.; Keetels, G.H.; Munts, E.A.; Bugdayci, H.H.; Rhee, C.V. Concentration and velocity profiles of sediment-water mixtures using the drift flux model. *Can. J. Chem. Eng.* **2016**, *94*, 1048–1058. [CrossRef]
14. Ekambara, K.; Sanders, R.S.; Nandakumar, K.; Masliyah, J.H. Hydrodynamic Simulation of Horizontal Slurry Pipeline Flow Using ANSYS-CFX. *Ind. Eng. Chem. Res.* **2009**, *48*, 8159–8171. [CrossRef]
15. Barati, R.; Salehi Neyshabouri, S.A.A.; Ahmadi, G. Development of Empirical Models with High Accuracy for Estimation of Drag Coeeficient of Flow Around a Smooth Sphere: An Evolutionary Approach. *Powder Technol.* **2014**, *257*, 11–19. [CrossRef]
16. Alizadeh, M.J.; Shahheydari, H.; Kavianpour, M.R.; Shamloo, H.; Barati, R. Prediction of longitudinal dispersion coefficient in natural rivers using a cluster-based Bayesian network. *Environ. Earth Sci.* **2017**, *76*, 86. [CrossRef]
17. Sadeghifar, T.; Barati, R. Application of adaptive neuro-fuzzy inference system to estimate alongshore sediment transport rate (A real case study: Southern shorelines of caspian sea). *J. Soft Comput. Civ. Eng.* **2018**, *2*, 72–85. [CrossRef]
18. Cao, Z.; Wofram, P.J.; Rowland, J.; Zhang, Y.; Pasqualini, D. Estimating Sediment Settling Velocity from a Theoretically Guided Data-Driven Approach. *J. Hydraul. Eng.* **2020**, *146*, 04020067. [CrossRef]
19. Rushd, S.; Parvez, M.T.; Al-Faiad, M.A.; Islam, M.M. Towards optimal machine learning model for terminal settling velocity. *Powder Technol.* **2021**, *387*, 95–107. [CrossRef]
20. van Rijn, L.C. Sediment transport, part II: Suspended load transport. *J. Hydraul. Eng.* **1987**, *110*, 1190–1192. [CrossRef]
21. Wang, G.Q.; Ni, J.R. The kinetic theory for dilute solid/liquid two-phase flows. *Int. J. Multiph. Flow* **1991**, *17*, 273–281. [CrossRef]
22. McLean, S.R. On the calculation of suspended load for non-cohesive sediments. *J. Geophys. Res.* **1992**, *97*, 5759–5770. [CrossRef]
23. Zhong, D.Y.; Wang, G.Q.; Sun, Q.C. Transport equation for suspended sediment based on two-fluid model of solid/liquid two-phase flows. *J. Hydraul. Eng.* **2011**, *137*, 530–542. [CrossRef]
24. Fick, A. On liquid diffusion. *J. Membr. Sci.* **1995**, *100*, 33–38. [CrossRef]
25. Almedeij, J. Asymptotic Matching with a Case Study from Hydraulic Engineering. Ph.D. Thesis, Kuwait University, Kuwait, Kuwait, 2009.
26. Bouvard, M.; Petkovic, S. Vertical dispersion of spherical, heavy particles in turbulent open channel flow. *J. Hydraul. Res.* **1985**, *23*, 5–20. [CrossRef]
27. Einstein, H.A.; Qian, N. *Effects of Heavy Sediment Concentration Near the Bed on the Velocity and Sediment Distribution*; Army Engineer Division, Missouri River: Omaha, NA, USA, 1955; p. 45.
28. Kironoto, B.A.; Yulistiyanto, B. The validity of Rouse equation for predicting suspended sediment. In Proceedings of the International Conference on Sustainable Development for Water and Waste Water Treatment, Yogyakarta, Indonesia, 14–15 December 2009.
29. Sumer, B.M.; Kozakiewicz, A.; Fredsoe, J.; Deigaard, R. Velocity and concentration profiles in sheet-flow layer of movable bed. *J. Hydraul. Eng.* **1996**, *122*, 549–558. [CrossRef]
30. Kumbhakar, M.; Ghoshal, K.; Singh, V.P. Derivation of Rouse equation for sediment concentration using Shannon entropy. *Phys. A Stat. Mech. Its Appl.* **2017**, *465*, 494–499. [CrossRef]
31. Wang, G.Q.; Ni, J.R. Kinetic theory for particle concentration distribution in two-phase flows. *J. Eng. Mech.* **1990**, *116*, 2738–2748. [CrossRef]
32. Coleman, N.L. Effects of suspended sediment on the open-channel velocity distribution. *Water Resour. Res.* **1986**, *22*, 1377–1384. [CrossRef]
33. Cellino, M.; Graf, W.H. Sediment-laden flow in open-channels under noncapacity and capacity conditions. *J. Hydraul. Eng.* **1999**, *125*, 455–462. [CrossRef]
34. Muste, M.; Yu, K.; Fujita, I.; Ettema, R. Two-phase versus mixed-flow perspective on suspended sediment transport in turbulent channel flows. *Water Resour. Res.* **2005**, *41*, W10402. [CrossRef]
35. Cheng, N.S. Comparison of formulas for drag coefficient and settling velocity of spherical particles. *Powder Technol.* **2009**, *189*, 395–398. [CrossRef]
36. Michalik, A. Density patterns of the inhomogenous liquids in the industrial pipeline measured by means of radiometric scanning. *La Houille Blanche J.* **1973**, *1*, 53–57. [CrossRef]
37. Wang, X.K.; Qian, N. Turbulence characteristic of sediment-laden flows. *J. Hydraul. Eng.* **1989**, *115*, 781–799.
38. Ali, S.Z.; Dey, S. Mechanics of advection of suspended particles in turbulent flow. *Proc. R. Soc. A* **2016**, *472*, 20160749. [CrossRef]

fluids

MDPI

Article

Local Scour Patterns around a Bridge Pier with Cable-Wrapping

Valentine Muhawenimana [1,*], Nadine Foad [1], Pablo Ouro [2] and Catherine A. M. E. Wilson [1]

[1] Hydro-Environmental Research Centre, School of Engineering, Cardiff University, Cardiff CF24 3AA, UK
[2] Department of Mechanical, Aerospace and Civil Engineering, School of Engineering, University of Manchester, Manchester M1 7DN, UK
* Correspondence: muhawenimanav@cardiff.ac.uk

Abstract: The performance of cable flow-altering bed scour countermeasures was experimentally evaluated based on the scour reduction, bed morphology, and the effects on the flow field. An unprotected 40 mm diameter pier was compared to piers protected with spiral cables (2, 4, 6, 8 and 10 mm diameters) wrapped at a 15-degree angle for two-bed sediment sizes with median grain sizes of 0.86 and 1.83 mm, for a cylinder Reynolds number of 7120. The scour depth was reduced by the cables by up to 52 percent compared to the unprotected pier case, a reduction that increased with increasing cable diameter for both sediment beds. Scour depth and sediment deposition varied by sediment size, where the scour hole was up to 45 percent deeper for the finer sediment bed than that of the coarser bed. Velocity and turbulence statistics showed that cables attenuated the flow within the scour hole by diminishing the downflow and horseshoe vortex, whereas in the case of finer sediment, spatially averaged turbulent kinetic energy and Reynolds shear stresses were respectively up to 1.4 and 1.8 times higher for the unprotected pier than the protected pier, resulting in scour depth reduction. The presence of the cable also reduced the vortex shedding frequency in the pier wake as indicated by a Strouhal number of around 0.175. The results demonstrate the potential of cable threading as a flow-altering scour countermeasure to reduce bridge pier scour.

Keywords: bridge pier; deposition; horseshoe vortex; sediment; scour; scour countermeasures

Citation: Muhawenimana, V.; Foad, N.; Ouro, P.; Wilson, C.A.M.E. Local Scour Patterns around a Bridge Pier with Cable-Wrapping. *Fluids* **2023**, *8*, 3. https://doi.org/10.3390/fluids8010003

Academic Editors: Jaan H. Pu and Mehrdad Massoudi

Received: 13 October 2022
Revised: 14 December 2022
Accepted: 18 December 2022
Published: 21 December 2022

1. Introduction

Bridge pier scour remains problematic for the integrity and failure of bridge pier foundations and abutments, and several methods have been devised to attempt to mitigate it using a variety of scour countermeasures. As in classic cylinder flows, the flow around bridge piers creates a downflow that contributes to the formation of a horseshoe vortex at the bed boundary, the primary governing hydrodynamic mechanism in scour [1–3]. The downward flow results in a horseshoe vortex that spirals around the base of a bridge pier, and both mechanisms are generally accepted as responsible for the formation of the scour hole immediately upstream of the bridge pier due to enhanced shear stresses [1,4–6]. The sediment eroded in the formation of the scour hole is deposited in an exit dune in the pier wake. Several experimental investigations have examined the effectiveness of distinct scour countermeasure devices around bridge piers, including submerged vanes, bed sills, sacrificial piles, collars, pier slots and threaded cables, which are typically classified as flow altering or bed armouring measures [2,7–9]. However, studies that have examined the impact of a threaded cable wrapped around the pier, a form of the flow altering countermeasure to reduce downflow, show conflicting evidence on the effectiveness of their use in reducing local scour [10–13], while only one study has offered a physical understanding of how cables modify the flow field [10].

Under both live bed and clear water conditions, threaded piers have been shown to reduce scour depth by up to 48.1% for a cable diameter (b_c) to pier diameter (b) ratio (b_c/b) of 0.52 under wave conditions and 46.3% for a cable-pier ratio of 0.1 under steady current conditions, with pier diameter-based Reynolds number (Re_b) of 52,400 [10]. A scour reduction of up to 15% was observed with octagonal piers for b_c/b of 0.15 for Re_b = 74,568, and 75,462 by Farooq and Ghumman [13]. However, notably smaller scour reduction of 8.7% for Re_b = 10,480 [11] and 12.85% for Re_b = 12,000 [12] were observed in investigations conducted with smaller physical models than those of Dey et al. [10].

These variations have been attributed to the different physical scales of the studies, e.g., Reynolds number. Even though the cable diameter to pier diameter ratio (b_c/b) and the number of threads and threading angle were consistent between studies, it was postulated that the thinner cable used in the smaller flume study [11,12] was unable to dampen the downflow and weaken the induced horseshoe vortex to the same degree as in the larger flume study [11]. However, the sediment grain size, and more specifically the scaling of the pier width (b) to the median grain diameter (b/d_{50}) is well known to be an influential parameter governing the scour depth [14–16] and this is highly likely to be a reason for variations among the study findings. The larger scale study which showed greater scour depth reduction [10], was conducted with a considerably smaller median grain diameter than in the other two studies, thus having lower submerged weight, settling velocity, and for the particle Reynolds number examined, a lower Shields' number and threshold of motion. In addition, Dey et al. [10] used a pier width considerably greater, and hence the ratio of the pier width to the median grain diameter (b/d_{50}) was at least an order of magnitude greater than in the smaller physical model studies.

With the exception of the study by Dey et al. [10], the physical mechanisms by which these cable wrapping countermeasures alter the flow field have not been investigated, and there is a general inference that the downflow and horseshoe vortex is reduced by the presence of the pier-cable. The former study showed that the downflow was reduced by the presence of a pier-cable which not only resulted in scour depth reduction but also a change in the scour hole bathymetry. The presence of a cable resulted in the formation of a flattened base scour hole rather than a deepened groove immediately upstream of the pier [10,17]. Increasing the cable diameter to pier diameter ratio (b_c/b) has also been shown to consistently reduce the scour depth [10,12]. Furthermore, the use of a smaller cable threading angle to the horizontal and the employment of multiple cables wrapped around the pier, can both independently reduce the scour depth (10).

In this paper, clear-water scour experiments were conducted for two-bed particle sizes, two pier width (b) to the median grain diameter (b/d_{50}), and for a wider range of cable-pier diameter ratios than in previous studies. The tests were carried out at a cylinder Reynolds number (Re_b) of 7120, a comparable physical scale to previous studies [11,12] and using similar sediment particle sizes. Furthermore, the development of the bed morphological features over time was examined, including both the scour hole immediately upstream of the pier and the exit dune formed in the pier wake. The velocity fields upstream of the pier and in its wake were also examined to elucidate the cable impact on downflow, horseshoe vortex, von-Karman vortex shedding, and wake recovery, which drive the physical mechanisms governing the scour hole formation in the presence of a cable-wrapped pier and resulting downstream bed morphology.

2. Materials and Methods

2.1. Experimental Set-Up

Experiments were performed in a recirculating open channel flume of dimensions 4.1 m length, 0.6 m width and 0.2 m depth. The bed slope of the flume was set at a gradient of 1 in 1000. The flow was driven by a steady electric pump with a flow meter directly connected to the pump's water output displaying the flow rate. At the upstream end of the flume, the pump discharges water to the upstream reservoir, which flows along the length of the flume to the downstream end where a tailgate weir controls the surface water profile along the flume, before the water enters the downstream reservoir to be recirculated by the pump. Experiments were conducted using two sediment particle sizes with median grain diameters (d_{50}) of 1.83 mm and 0.86 mm (Test series A and B, respectively, Table 1), both corresponding to coarse sand. The grain size was graded uniformly corresponding to a geometric standard deviation of the grain size ($\sigma_g = (d_{84}/d_{16})^{0.5}$) of 1.22 and 1.21, respectively. A circular bridge pier of 40 mm diameter (b) was located 45 × b (1.8 m) downstream of the approach section. The center point of the bridge pier was positioned along the centerline of the flume, at a lateral distance of 300 mm from each sidewall. The flow depth was maintained at 70 mm and the discharge (Q) was fixed at 7.5 Ls^{-1} throughout the tests. The mean approach flow velocity U_0 was 0.178 ms^{-1}, and these flow conditions correspond to clear-water scour conditions and a pier diameter-based Reynolds number Re_b of 7120. The boundary shear stress ($\tau_0 = rgRS$) was 0.56 Nm^{-2}, and the corresponding shear velocity ($u_* = (\tau_0/r)^{0.5}$) was 0.024 ms^{-1}. The particle Reynolds numbers ($R_* = u_* d_{50}/\nu$) were 43.18 and 20.29 for sediment d_{50} of 1.83 mm and 0.86 mm with corresponding Shields numbers ($\tau_0/rg(S_s - 1)d_{50}$) of 0.02 and 0.04, respectively. The critical velocities according to the Hjulström curve were 0.30 and 0.14 ms^{-1}, resulting in flow intensity U_0/U_c of 0.4 and 0.58 for sediment d_{50} of 1.83 mm and 0.86 mm, respectively, which is below the peak threshold ratio for maximum scour of U_0/U_c = 1 [14]. The flow conditions were maintained below the level of significant suspended transport and significant bed erosion, to avoid excessive bed scour. Tests were performed for a range of cable diameters (b_c) from 2 mm to 10 mm increasing in intervals of 2 mm. The piers without cables were A1 and B1, and the piers with cables A2 to A6 and B2 to B6 (Table 1, Figure 1a). A single thread was wrapped around the pier at 15 degrees to the horizontal which is consistent with previous studies [10,12].

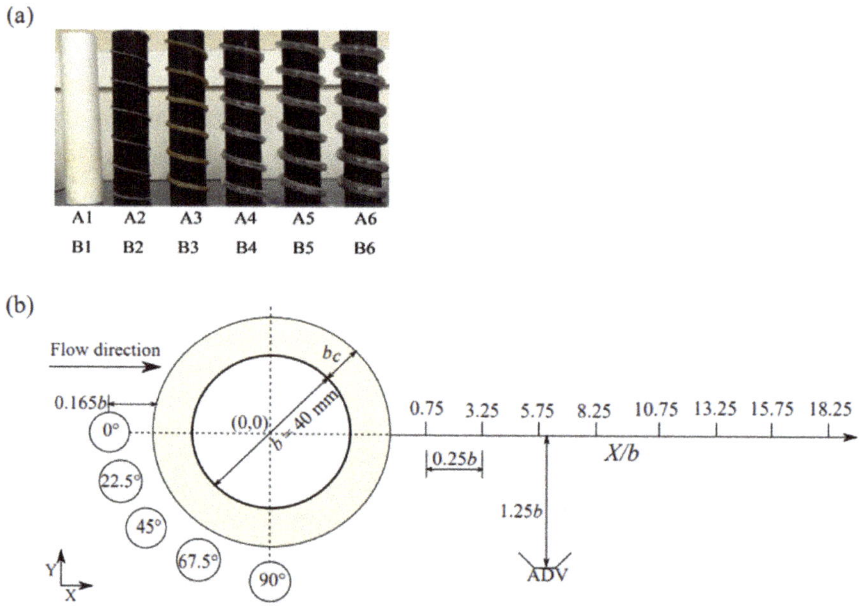

Figure 1. (**a**) Model bridge piers with diameter b = 40 mm of varying cable diameter to cable ratio (left to right): no cable (A1 and B1); 2 mm (A2 and B2); 4 mm (A3 and B3); 6 mm (A4 and B4); 8 mm (A5 and B5); and 10 mm (A6 and B6). (**b**) Plan view of the side-looking ADV probe sampling volume locations around the pier.

Table 1. Summary of experimental conditions where b_c/b is the cable diameter/pier diameter ratio, where the cylinder diameter b = 40 mm and cable diameters were 2, 4, 6, 8, and 10 mm. Tests were conducted with flow depth y = 70 mm, discharge Q = 7.5 Ls^{-1}, and pier-based Reynolds number ($Re = u.b/\nu$) of 7120. The type of test indicates the series for which bed levels and ADV velocity measurements were collected.

Series	Sediment d_{50} (mm)	Cable Diameter b (mm)	b_c/b (-)	Type of Test
A1	1.83	Without cable	-	Bed levels and ADV
A2	1.83	2	0.05	Bed levels
A3	1.83	4	0.10	Bed levels
A4	1.83	6	0.15	Bed levels
A5	1.83	8	0.20	Bed levels
A6	1.83	10	0.25	Bed levels and ADV
B1	0.86	Without cable	-	Bed levels and ADV
B2	0.86	2	0.05	Bed levels
B3	0.86	4	0.10	Bed levels
B4	0.86	6	0.15	Bed levels
B5	0.86	8	0.20	Bed levels
B6	0.86	10	0.25	Bed levels and ADV

2.2. Bed Level and Velocity Measurements

At the start of each experiment, the sediment bed was levelled using a board to form a uniform sediment bed thickness of 60 mm. The bed levels were measured with a Vernier pointer gauge to an accuracy of $\pm$0.1 mm and the velocity statistics were measured using a side-ways looking, four-beam Acoustic Doppler Velocimeter (ADV) Nortek Vectrino Plus with a sampling frequency of 200 Hz. Both instruments were mounted on an instrument rack which was set perpendicular to the flume's longitudinal axis.

Figure 1b illustrates the ADV sampling locations in the front and wake of the pier. The protected pier is shown with the cable area shaded, showing that the sampling locations remained the same irrespective of whether the cable was present and the cable diameter size. The ADV sampling volume, located 50 mm i.e., 1.25b from the probe, had a 6 mm diameter (0.165b) and 7 mm height. Upstream sampling points were located at 0°, 22.5°, 45°, 67.5° and 90° angles from the pier centerline. The wake sampling locations were spaced by 10 mm, and ranged from X/b = 0.75 to 18.25.

The first bed scour test for each setup had a total running time of 24 h. These tests were used to ascertain that at least 95% of the maximum scour depth was obtained within the first 10 and 12 h of the experiment for sediment d_{50} of 0.86 mm and 1.83 mm, respectively (Figure 2). Therefore, the remaining tests were performed over a total running time for each experiment of at least 12 h. The bed elevation was measured at the start of the test then every 15 min for the first 2 h, 30 min for the following 2 h, and then at 60 min intervals for the remainder of the test along the channel centerline and at a lateral distance of 40 mm offset from the centerline. To allow comparison with the results from former studies [10–12], percentage scour reduction efficiency is defined as the difference in scour depth at the deepest bed point between the protected cable-wrapped pier (D_{sp}) and the unprotected pier (D_s) divided by the bed level of the unprotected pier: $100 \cdot (D_s - D_{sp})/D_s$.

The sampling volume of the ADV velocimeter was located 50 mm away from the probe's center and comprised a cylinder of 7 mm height and 6 mm diameter (Nortek AS, 2013). Measurements were collected at five locations around the front of the pier circumference, and at five elevations within the water column corresponding to the sampling volume centroid being located at 10 mm, 15 mm, 20 mm, 25 mm, and 30 mm from the bed (see Figure 1b). Velocity measurements were taken after 12 h into the test. A sampling period of 300 s was used, which was sufficient to gain an accurate representation of the time-averaged velocity and turbulence statistics based on convergence tests. The sound-to-noise ratio (SNR) and correlation (COR) values were maintained above 20 dB and 70% respectively, which was achieved by seeding the water with silicate powder of 10 μm mean diameter (density of 1.1 g/cm^3). These SNR and COR thresholds were used for filtering the velocity time series, in addition to despiking using the Phase-Space Thresholding (PST) method of Goring and Nikora [18]. Power spectra analysis used MATLAB R2021a, where peak vortex shedding frequency was obtained by Fourier transform of wake data points sampled at the end of the 12-h experiments, at 40 mm (one cylinder diameter) above the initial bed level. The Fourier analysis was performed with the pwelch (Welch's power spectral density estimate) function in MATLAB which enables to sub-divide the signal into smaller time series and average of the resulting Fourier transform.

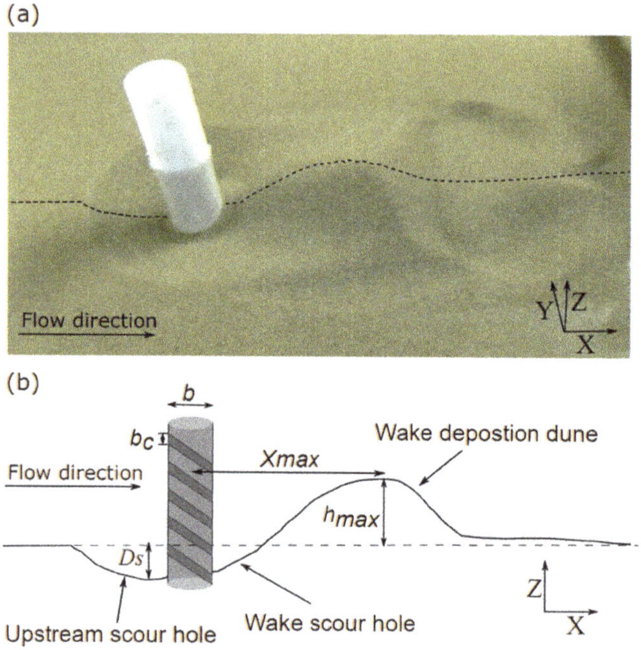

Figure 2. (**a**) Photograph of the scour pattern for the control pier without cable with sediment size d_{50} = 0.86 mm. (**b**) outline of the centerline of the bedforms at 12 h with a protected pier of diameter b and cable diameter b_c illustrated, and dimensions of the scour hole depth (D_s), the wake deposition dune peak height (h_{max}) and its longitudinal distance downstream of the pier (X_{max}).

3. Results and Discussion

3.1. Time Evolution of Bed Morphology and Scour Depth after 12 h

Figure 2 illustrates the observed bed formations including: the main scour hole immediately upstream of the pier, a smaller degree of scouring in the pier's wake, an exit dune formed immediately downstream of the wake scour hole, and two localized regions of scouring located downstream of the dune which straddled each side of the dune (Figure 2a).

Figure 3 shows the temporal evolution of the bed levels along the flume centreline (a and b) and at 1 pier diameter offset from the centreline (c and d) for the unprotected pier (test series B1) and b_c/b = 0.25 (test series B6), respectively for the small grain size sediment (d_{50} = 0.86 mm). Maximum scour depth D_s normalized by cylinder diameter b for both sediment sizes for series A1 to A5 and B1 to B6 are shown in Figure 4.

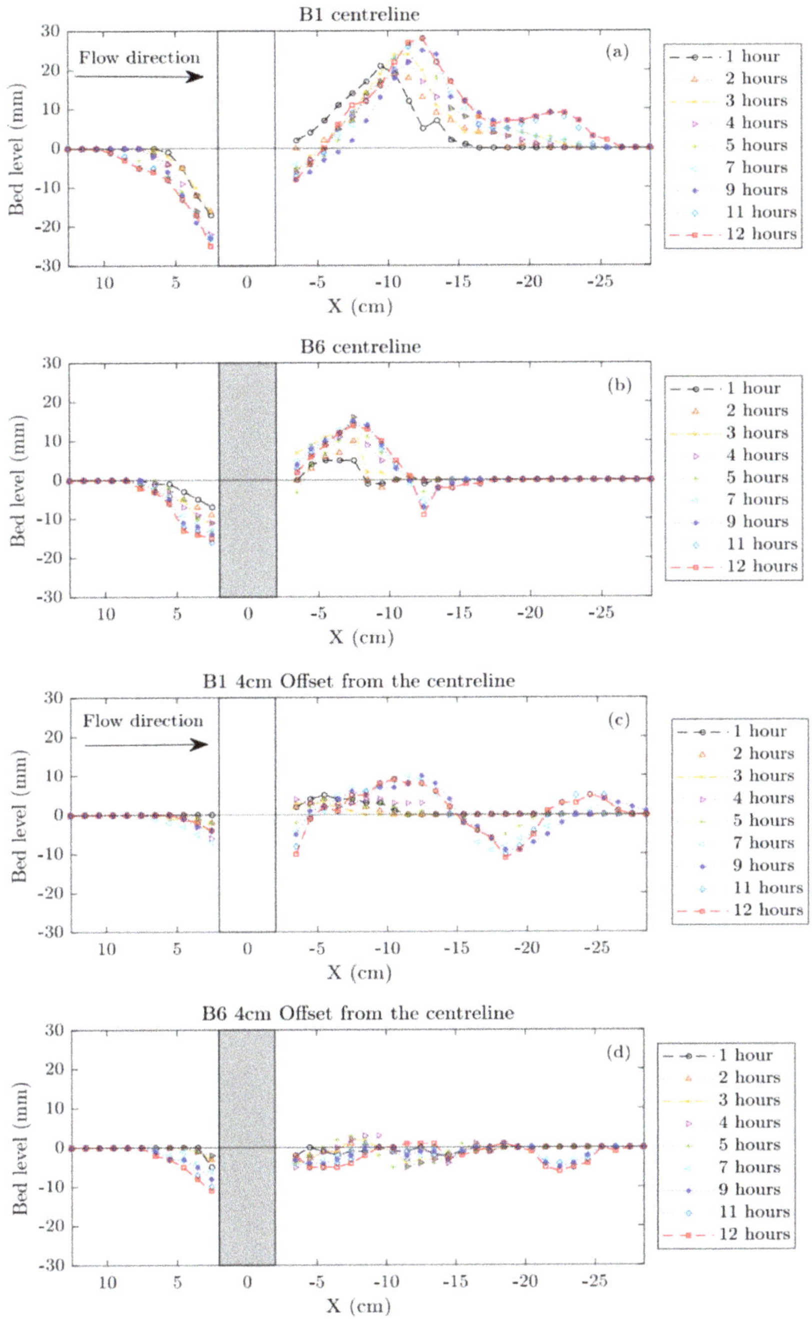

Figure 3. Longitudinal section of the bed development over time from 1 h to 12 h where (**a**,**b**) is along the channel and cylinder centerline (Y = 0) and (**c**,**d**) are one-cylinder diameter lateral offset from the centerline (Y = 40 mm) for the experiment with sediment d_{50} = 0.86 mm, where B1 is the unprotected pier and B6 is the protected pier with cable-pier diameter ratio b_c/b = 0.25. Flow is from left to right.

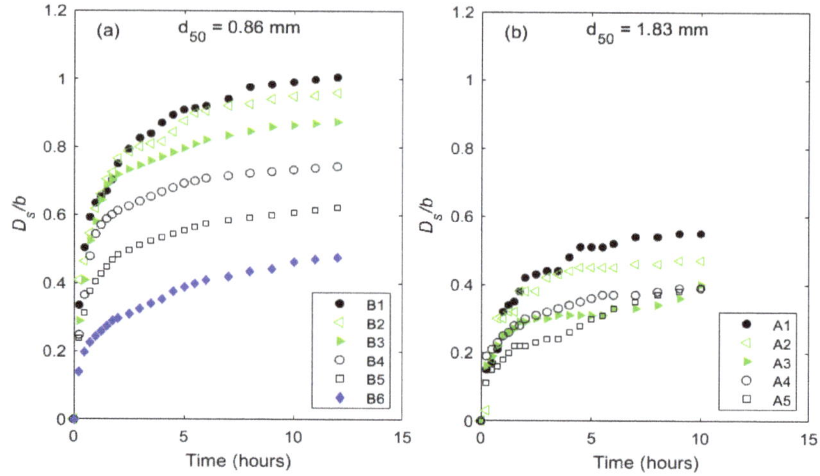

Figure 4. Time evolution of maximum scour for (**a**) the fine sediment with d_{50} = 0.86 mm and (**b**) coarse sediment with d_{50} =1.83 mm, where A1 and B1 are the unprotected pier and A2 to A5, and B1 to B6 are the cable protected piers.

The time evolution of bed levels showed similarities in prominent bed features, which were present for both the protected and unprotected pier cases, however, the features' extent and location varied between the two cases (Figure 3). The scour depth at 12 h depended on the ratio of d_{50} and cylinder diameter (d_{50}/b) as the unprotected pier showed D_s = 40.2 and 21.9 mm for the small grain size sediment (d_{50} = 0.86 mm) and the larger grain size sediment (d_{50} = 1.83 mm) respectively (Figure 4). This difference was reflected in the scour depth and wake depositions for cable-protected piers, whose exit dune maximum height and peak location downstream of the pier are shown in Figure 5, where the unprotected pier resulted in higher dune peak (h_{max}/h_o) (Figure 5a) that was located (X_{max}/h_o) further form the pier compared to the cable-protected piers (Figure 5b). Similar to the scour hole being longer, deeper and narrower for the unprotected pier than for the protected pier, the wake dune was taller, longer and wider than the dune formed for the protected pier and was located a further distance downstream from the pier, also presented in Figure 3, as sediment entrained from the upstream and in the near-wake is deposited downstream.

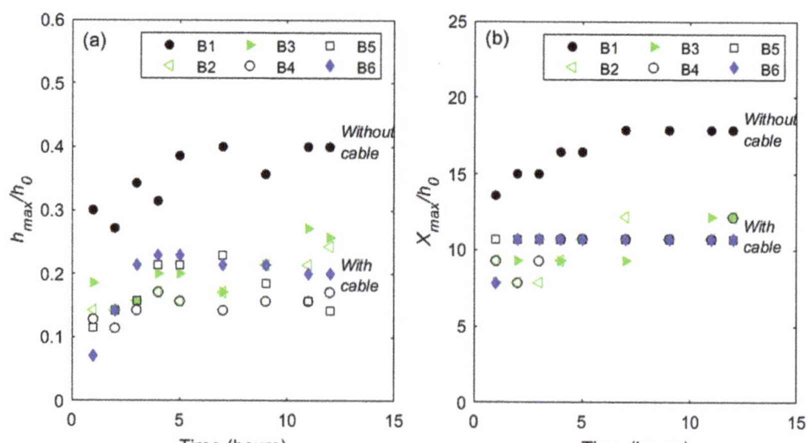

Figure 5. Wake dune height h_{max} (**a**) and downstream position x_{max}, (**b**) normalized with the approach flow depth h_0 for the B1 to B6 test series with sediment d_{50} = 0.86 mm.

For a bed composed of the small grain size sediment (d_{50} = 0.86 mm), the percentage scour depth reduction ($100 \cdot (D_s - D_{sp})/D_s$) increased with increasing cable-pier ratio for all cable-pier diameters examined. While for the coarser sediment bed (d_{50} = 1.83 mm) the scour depth reduction increased with increasing cable diameter up to a cable-pier diameter of 0.1 and remained nearly constant with further increases in cable-diameter ratio. For an intermediate cable-pier diameter ratio (b_c/b = 0.15), the presence of the cable had a similar impact on scour reduction independent of the bed sediment size (Figure 6). However, as the cable-pier diameter ratio increases further, the impact of the cable is greater for the finer bed sediment compared to the larger bed sediment, whereby the scour hole depth is halved for the highest cable-diameter ratio compared to the condition without cable pier (scour depth reduction of 52%). This compares to a scour hole depth reduction of around a third (scour depth reduction of 32%) for the coarser sediment particle size.

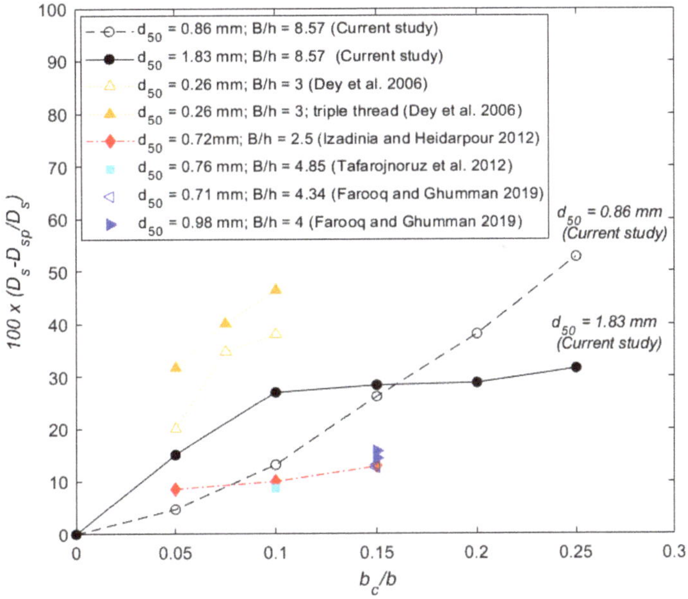

Figure 6. Variation in percentage scour depth reduction of the protected pier compared to the unprotected pier for different cable diameter to pier diameter ratios (b_c/b) and for different median sediment diameters (d_{50}). Data from the current study are compared with previous studies. The channel width to flow depth ratio (B/h) is given in the legend [10–13].

The upstream scour hole for the unprotected pier was deeper, longer, narrower, and more localized compared to the protected pier, which was shallower and wider with the scour depth maintained over a greater lateral distance away from the pier centerline, as displayed in the centreline and 1 pier diameter offset bed profiles (Figure 3), 12-h scour depth (Figure 4) and in final bathymetry contours at 12 h for test series B1 and B6 (Figure 7). For the protected pier, negligible scouring was present along the channel centerline immediately downstream of the pier (≤50 mm), whilst scouring occurred along the shear layer region separating the pier wake and freestream region (Figure 3). This contrasts with the unprotected pier where a scour hole formed in the recirculation zone immediately downstream of the pier. The development of the exit dune properties for d_{50} = 0.86 mm, B1 through B6 test series is shown in Figures 3 and 5, where for the unprotected pier, the downstream exit dune was approximately twice the height (Figure 5a), length and width of that from the protected pier and was formed further downstream from the pier than in the protected pier case (Figure 3).

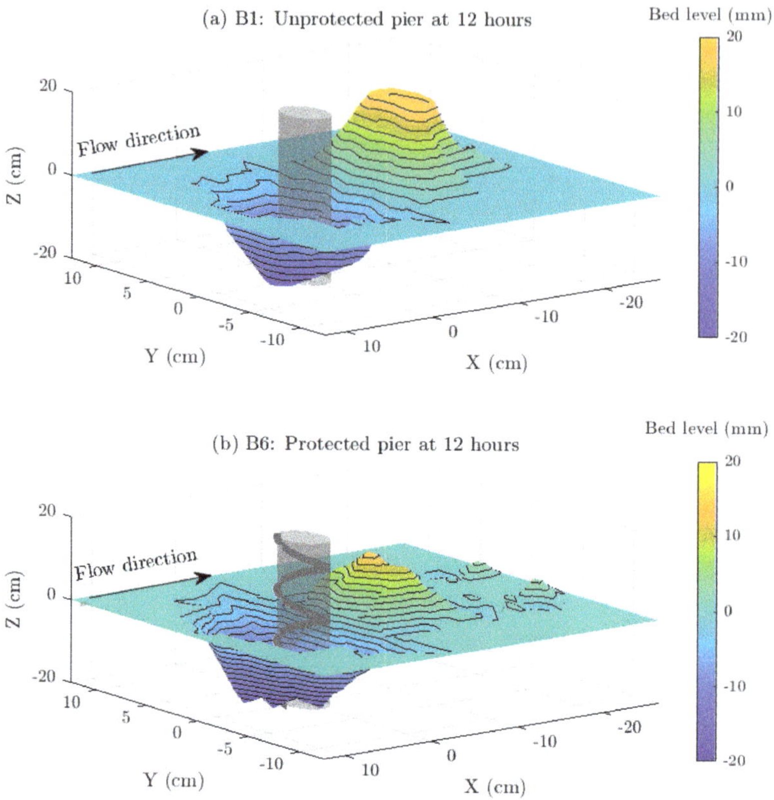

Figure 7. Local scour and deposition patterns for the (**a**) pier without cable and (**b**) cable-wrapped pier for the fine sediment (d_{50} = 0.86 mm) at 12 h. Bed levels are shown as contour levels in millimetres, where the initial bed level is zero, negative values indicate scour, and positive values indicate deposition. The flow direction is from left to right.

For the unprotected pier, the scour hole moved further downstream with time and closer to the pier's upstream face. Similarly, the exit dune formed in the pier wake moved progressively downstream with time (Figures 3 and 5). At around five hours, the height of the dune peak above the initial bed nearly reached its final height and downstream location and this bed feature induced the formation of two shallower to scour holes on either side of the dune (Figure 3b) which persisted throughout the remainder of the test (Figure 7b). The wake scour holes are deeper and located closer to the dune for the unprotected pier (Figure 7a) compared to the protected pier (Figure 7b); while for the protected pier, more prominent additional secondary scour holes are present downstream of the dune (Figures 3c and 7b).

3.2. Effect of Cable Wrapping on the Flow Field

The presence of the cable reduced the downflow on the upstream face of the pier, and this decrease was more pronounced at the sampling points closest to the bed, shown in Figure 8 for d_{50} = 0.86 mm, B1 unprotected and B6 protected piers. Indeed, the downflow is reduced on the upstream face of the pier, with the effective downflow vertical velocity being reduced in the presence of the wrapped cable. Furthermore, the turbulent kinetic energy ($k= 0.5(u'^2+ v'^2+ w'^2)$) and the turbulent shear stress ($(u'w'^2+u'v'^2)^{1/2}$) magnitudes were both reduced at the upstream face and over the pier's circumference (Figure 8c,d) and (e,f), respectively). Offset from the centerline from the point at 0° angle to that at 90°

angle, (see Figure 1b), the variation in both turbulence quantities over the lower half of the water column is much diminished and uniform for the condition with cable. The spatially averaged downflow velocity component for upstream locations above the scour hole after 12 h for the protected pier was 70% of the magnitude for the unprotected pier for the fine sediment tests (d_{50} = 0.86 mm).

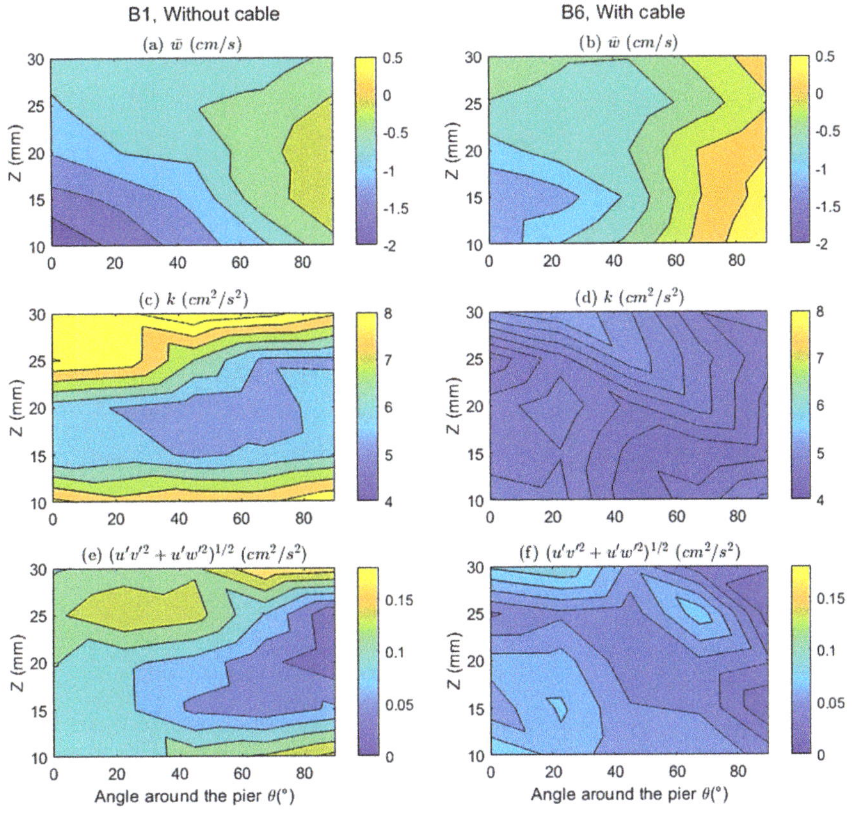

Figure 8. Time-averaged (**a**,**b**) vertical velocity, (**c**,**d**) turbulent kinetic energy, and (**e**,**f**) modulus of vertical and horizontal Reynolds stress components at points 1 (0°), 2 (22.5°), 3 (45°), 4 (67.5°) and 5 (90°) around the pier (see Figure 1b) and through the water column (z is measured from the bed) for the fine sediment (d_{50} = 0.86 mm) for the unprotected pier (B1) (**left**) and the protected pier with a 10 mm diameter wrapped cable (B6) (**right**).

In the pier wake, flow reversal in the recirculation zone downstream of the pier was present over a longer spatial extent for the unprotected configuration than the protected one, as negative streamwise velocities were higher for the unprotected pier for both bed sediments, as shown in Figure 9a versus (c) and (b) versus (d) for sediment test series A1 compared to A6 and B1 versus B6, respectively. The wake flow patterns also varied depending on sediment grain size due to the different final bed morphologies. As such, the deeper scour holes formed in the smaller grain sediment size (B1 and B6) increased the available cross-sectional area in front of the pier, resulting in less flow blockage and lower values of mean longitudinal (u) velocities than for the larger sediment (A1 and A6). The distribution of the vertical velocity component (w) in the pier wake (Figure 9e–h) showed larger values for coarser sediment (A cases) than the finer sediment (B cases) due to the variation of flow blockage effect of the various scour depths and hole geometries (see Figure 3).

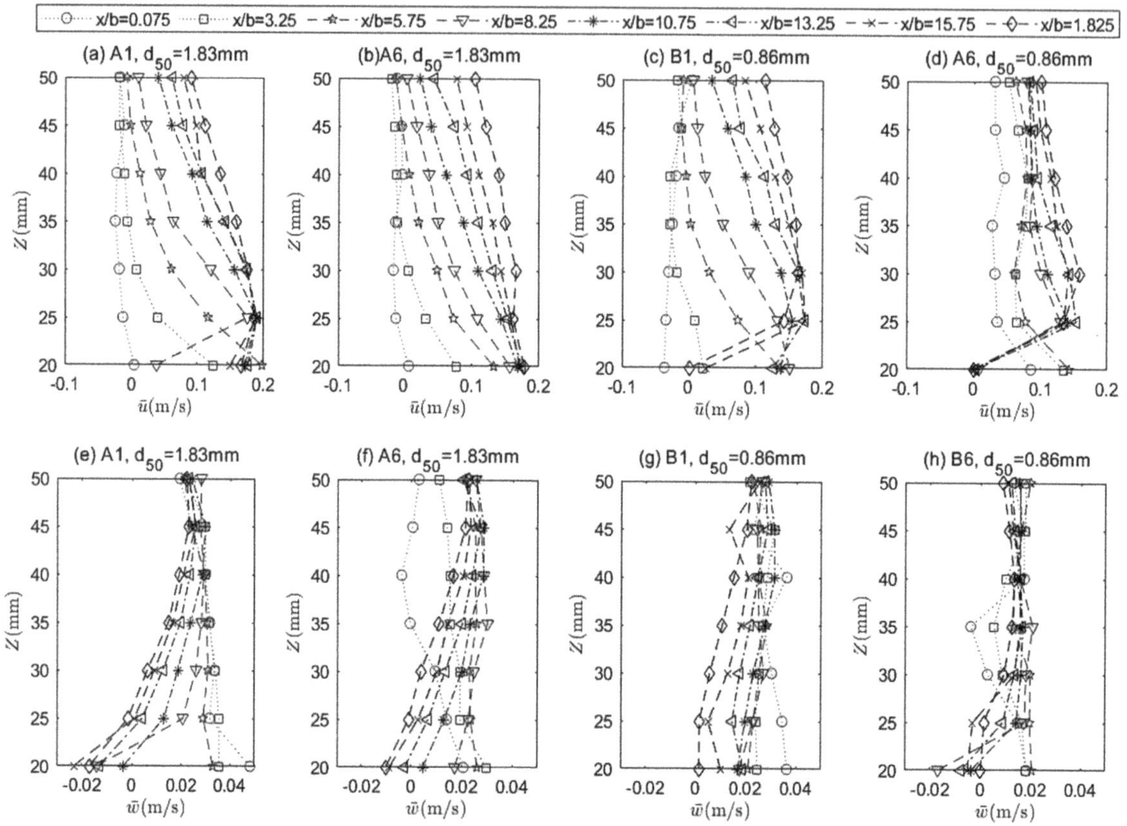

Figure 9. Profiles of the longitudinal u (**a–d**) and vertical w (**e–h**) velocity components along various longitudinal *X/b* distances in the pier wake for both sediment d_{50} = 1.83 mm (A1 and A6) (**left**) and d_{50} = 0.86 mm (B1 and B6) (**right**) for the unprotected pier (A1 and B1) and the protected pier with b_c/b = 0.25 (A6 and B6).

Strouhal numbers ($St = f{\cdot}b/U_0$, where f is the vortex shedding frequency) obtained from the time-series of three velocity components in the pier's wake at an elevation of one pier diameter above the bed are shown in Figure 10. The nominal pier diameter b = 40 mm was used to calculate St for the unprotected pier (A1 and B1), while an effective diameter of 45 mm was used for the cable-protected piers with 10 mm diameter cable in tests A6 and B6, to account for the presence of the cable around the pier. For the unprotected cases with coarse sediment, St was around 0.2 which coincides with that observed for unbounded cylinder flows. For the protected piers A6 and B6, St was nearly constant at around 0.175, which indicates a reduction in the shedding frequency of the wake vortices associated with the cable presence. The velocity components with the peak vortex shedding frequency in the power spectral density varied between cases, with more prominent vertical component w in the unprotected pier cases A1 and B1, while peak frequencies alternated between u, v, and w components for the protected pier cases A6 and B6.

The exit dune's peak height and its location from the pier (Figure 5) corresponded to the longer recirculation zone before the dune, generated by the unprotected pier for the fine sediment with d_{50} = 0.86 mm, and thus the von-Kármán vortices likely remain coherent over a longer downstream distance for the unprotected pier than for the protected pier. The cable interrupts the flow path of the upstream downflow and breaks up its energy, which diminishes the horseshoe vortex and its scouring potential, leading to reduced scour upstream. The reduced scour depth also results in a diminished exit dune in the pier wake, as the flow paths rejoin the free steam flow in the pier wake.

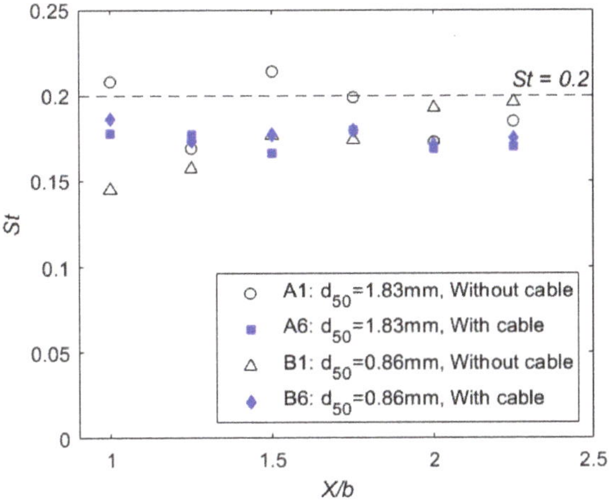

Figure 10. Peak vortex shedding frequencies f for points in the wake, located at one cylinder diameter, 40 mm, above the initial bed with increasing downstream distance (X/b) of the protected (A1 and B1) and unprotected piers (A6 and B6) for both d_{50} = 0.86 mm (B) and 1.83 mm (A) sediment sizes.

The current study corroborated the scour reductions due to cable countermeasures, in addition to evaluating the hydrodynamic impact of the cables on the scour processes. As previous studies have shown, cables utilized in combination with other flow-altering or bed-armoring countermeasures are highly efficient at reducing scour depth [12,13].

Much of the upstream scour hole depth was formed within the first five hours of flow action (Figures 3 and 4), which is typical for clear-water scour conditions [14]. The downstream dune progression was also similar to observations by Oliveto and Hager [19]. The scour reduction in terms of the scour efficiency $100 \cdot (D_s - D_{sp})/D_s$, demonstrated an overall increase with increasing b_c/b, although with varying extents with over 30% differences in scour reduction for $b_c/b \leq 0.1$ (Figure 6). b/d_{50} ratios were 46.5 and 21.85 in the current study for the finer and coarser sediments, respectively; while it was 769.23 [10]; 55.55 [12]; 63.15, 52.63 [11]; 70.4, 51.02, 102.04 and 140.84 [13] in previous studies shown in Figure 6. Therefore, differences in scour efficiencies are most likely due to differences in sediment grain size, particularly for the finer sediment used and the significantly greater pier to sediment grain ratio used in the study by Dey et al. [10] (Figure 6). Further differences in scour reduction might be attributed to the flow intensity being below the peak threshold condition of $U_0/U_c = 1$ [14].

Scour depth reductions resulting from the presence of the cable occurred due to the reduced effective downflow vertical velocity, turbulent kinetic energy and turbulent shear stress upstream of the pier. In addition, the reduced Strouhal number in the protected pier case suggests that the combined cable and pier diameter governed the wake vortex shedding and the wake bed formations.

Evaluating instantaneous flow fields, particularly the horseshoe vortex would further elucidate the impact of cables on the flow field and scour reduction, and evaluating the feasibility of the cable wrapping method, particularly the pier-cable diameter ratio at various scales, would further refine this scour countermeasure method.

4. Conclusions

Experiments were conducted to evaluate the effects of cable countermeasures on scour depth, flow field, vortex shedding frequency and bathymetry over time for two sediment sizes. Increasing the cable-pier diameter ratio resulted in a reduction in the scour depth, with reductions of 32% and 52% for the coarse and fine sediment sizes, respectively. Scour reduction reached a threshold at the cable-to-pier diameter ratio of 0.1 for the coarse sediment, while it continually increased up to a ratio of 0.25 for the fine sediment, indicating that benefits of scour reduction from increased cable-to-pier diameter ratio improve for finer sediments. The cables reduced scour depth upstream of the pier and in its wake over time, as well as reducing the downflow scour action in the sediment bed around the pier. In the wake, the exit dune's peak height increased with time and its maximum height was reached by 12 h of runtime. The exit dune had a smaller magnitude and was considerably closer to the downstream face of the pier for the protected pier than for the unprotected case. The vertical velocities, turbulent kinetic energy and Reynolds stress around and downstream of the pier reduced in magnitude for larger wrapping cables, indicating that the cables dampen the downflow and horseshoe vortex strength that govern pier scour formation, resulting in smaller changes in bed scour. The effect of the cable was also observed to reduce the vortex shedding frequency in the pier wake, where the Strouhal number decreased from 0.2 in the classic cylinder case to 0.175, indicating a reduction in wake vortices. Research into the instantaneous flow field is ongoing to further investigate the turbulent wake structure and the generated bed morphology.

Author Contributions: Conceptualization, V.M. and C.A.M.E.W.; methodology, V.M. and C.A.M.E.W.; Experimentation, V.M. and N.F.; validation, V.M., C.A.M.E.W., N.F. and P.O.; formal analysis, V.M., C.A.M.E.W., N.F. and P.O.; writing—original draft preparation, V.M. and C.A.M.E.W.; writing—review and editing, V.M., C.A.M.E.W., N.F. and P.O.; visualization, V.M.; supervision, C.A.M.E.W. and P.O.; funding acquisition, C.A.M.E.W. All authors have read and agreed to the published version of the manuscript.

Funding: The work was partly supported by Cardiff Undergraduate Research Opportunities Programme (CUROP) for N.F.

Data Availability Statement: All data are presented in the main text.

Acknowledgments: We thank Guglielmo Sonnino Sorisio, Ana Pinto Oliveira and Muhammad Hasanov for assistance with data collection during experiments.

Conflicts of Interest: The authors declare no conflict of interest.

References

1. Unger, J.; Hager, W.H. Down-flow and horseshoe vortex characteristics of sediment embedded bridge piers. *Exp. Fluids* **2006**, *42*, 1–19. [CrossRef]
2. Wang, C.; Yu, X.; Liang, F. A review of bridge scour: Mechanism, estimation, monitoring and countermeasures. *Nat. Hazards* **2017**, *87*, 1881–1906. [CrossRef]
3. Guan, D.; Chiew, Y.-M.; Wei, M.; Hsieh, S.-C. Characterization of horseshoe vortex in a developing scour hole at a cylindrical bridge pier. *Int. J. Sediment Res.* **2019**, *34*, 118–124. [CrossRef]
4. Hjorth, P. *Studies on the Nature of Local Scour: Institutionen för Teknisk Vattenresurslära*; Lunds Tekniska Högskola, Lunds Universitet: Lund, Sweden, 1975.
5. Melville, B.W. Local Scour at Bridge Sites. Ph.D. Thesis, The University of Auckland, Auckland, New Zealand, 1975.
6. Dey, S.; Bose, S.K.; Sastry, G.L.N. Clear Water Scour at Circular Piers: A Model. *J. Hydraul. Eng.* **1995**, *121*, 869–876. [CrossRef]
7. Johnson, P.A.; Niezgoda, S.L. Risk-Based Method for Selecting Bridge Scour Countermeasures. *J. Hydraul. Eng.* **2004**, *130*, 121–128. [CrossRef]
8. Pagliara, S.; Carnacina, I.; Cigni, F. Sills and gabions as countermeasures at bridge pier in presence of debris accumulations. *J. Hydraul. Res.* **2010**, *48*, 764–774. [CrossRef]
9. Tafarojnoruz, A.; Gaudio, R.; Dey, S. Flow-altering countermeasures against scour at bridge piers: A review. *J. Hydraul. Res.* **2010**, *48*, 441–452. [CrossRef]
10. Dey, S.; Sumer, B.M.; Fredsøe, J. Control of Scour at Vertical Circular Piles under Waves and Current. *J. Hydraul. Eng.* **2006**, *132*, 270–279. [CrossRef]
11. Tafarojnoruz, A.; Gaudio, R.; Calomino, F. Evaluation of Flow-Altering Countermeasures against Bridge Pier Scour. *J. Hydraul. Eng.* **2012**, *138*, 297–305. [CrossRef]
12. Izadinia, E.; Heidarpour, M. Simultaneous use of cable and collar to prevent local scouring around bridge pier. *Int. J. Sediment Res.* **2012**, *27*, 394–401. [CrossRef]

13. Farooq, R.; Ghumman, A.R. Impact Assessment of Pier Shape and Modifications on Scouring around Bridge Pier. *Water* **2019**, *11*, 1761. [CrossRef]
14. Melville, B.W.; Chiew, Y.-M. Time Scale for Local Scour at Bridge Piers. *J. Hydraul. Eng.* **1999**, *125*, 59–65. [CrossRef]
15. Lee, S.O.; Sturm, T.W. Effect of Sediment Size Scaling on Physical Modeling of Bridge Pier Scour. *J. Hydraul. Eng.* **2009**, *135*, 793–802. [CrossRef]
16. Sheppard, D.M.; Melville, B.; Demir, H. Evaluation of Existing Equations for Local Scour at Bridge Piers. *J. Hydraul. Eng.* **2014**, *140*, 14–23. [CrossRef]
17. Dey, S.; Raikar, R. Characteristics of Loose Rough Boundary Streams at Near-Threshold. *J. Hydraul. Eng.* **2007**, *133*, 288–304. [CrossRef]
18. Goring, D.G.; Nikora, V.I. Despiking Acoustic Doppler Velocimeter Data. *J. Hydraul. Eng.* **2002**, *128*, 117–126. [CrossRef]
19. Oliveto, G.; Hager, W.H. Morphological Evolution of Dune-Like Bed Forms Generated by Bridge Scour. *J. Hydraul. Eng.* **2014**, *140*, 06014009. [CrossRef]

fluids

MDPI

Article

Combination of Riprap and Submerged Vane as an Abutment Scour Countermeasure

Abazar Fathi [1], S. M. Ali Zomorodian [1,*], Masih Zolghadr [2], Aaron Chadee [3,*], Yee-Meng Chiew [4], Bimlesh Kumar [5] and Hector Martin [6]

1 Water Engineering Department, Shiraz University, Shiraz 71946-84334, Iran
2 Department of Water Sciences and Engineering, Agricultural College, Jahrom University, Jahrom 74148-46199, Iran
3 Department of Civil and Environmental Engineering, University of the West Indies, St Augustine P.O. Box 331310, Trinidad and Tobago
4 School of Civil and Environmental Engineering, Nanyang Technological University, Singapore 639798, Singapore
5 School of Civil Engineering, Indian Institute of Technology Guwahati, Guwahati 781039, India
6 School of Natural and Built Environment, Queen's University Belfast, Belfast BT7 1NN, UK
* Correspondence: mzomorod@shirazu.ac.ir (S.M.A.Z.); aaron.chadee@sta.uwi.edu (A.C.)

Abstract: Scour is one of the main causes of hydraulic structural failures. The present experimental study examines the use of riprap, submerged vanes, and a combination of these for scour reduction around vertical walls and spill-through abutments under clear-water conditions. Specifically, the influence of placing riprap stones with different apron shapes (geometry) and/or a group of submerged vanes of constant height and length on abutment scour was examined. The main aim is to propose the optimum apron geometry and placement of submerged vanes to (1) reduce edge failure at vertical walls and spill-through abutments; and (2) prevent shear failure at the spill-through abutment (no shear failure is observed around the vertical wall abutment). The results show that using ripraps for scour protection is more effective than submerged vanes. However, the highest reduction in scour depth was achieved when a combination of riprap and submerged vanes was used together. This arrangement can reduce the maximum clear-water scour depth by up to 54% and 39% with vertical walls and spill-through abutments, respectively. Furthermore, selecting appropriate apron scale ratios reduces the required riprap volume by up to 46% and 31% for the vertical wall and spill-through abutment, respectively. In addition, the installation of vanes increased the riprap stability and reduced edge failure in both abutments tested. Finally, using riprap aprons with proper scales ratios at the downstream side of the spill-through abutment also prevents shear failure in this zone.

Keywords: scour countermeasure; vertical wall abutment; spill-through abutment; riprap; submerged vanes

Citation: Fathi, A.; Zomorodian, S.M.A.; Zolghadr, M.; Chadee, A.; Chiew, Y.-M.; Kumar, B.; Martin, H. Combination of Riprap and Submerged Vane as an Abutment Scour Countermeasure. *Fluids* **2023**, *8*, 41. https://doi.org/10.3390/fluids8020041

Academic Editors: Jaan H. Pu, Prashanth Reddy Hanmaiahgari, Manish Pandey, Mohammad Amir Khan and Mehrdad Massoudi

Received: 27 November 2022
Revised: 4 January 2023
Accepted: 12 January 2023
Published: 21 January 2023

1. Introduction

Local scour around bridge foundations has caused bridge failure, leading to financial, time and even human losses worldwide [1–4]. To reduce such occurrences, researchers worldwide have investigated bridge scour using various approaches. To understand the mechanism of bridge scour, it is essential to study the flow pattern around the bridge foundation, and the related scour mechanism, which is highly complex. The main factors influencing the abutment scour mechanism are downflow, primary vortex and wake vortices [5,6]. After the flow hits abutment face, a downward flow or downflow develops because of the formation of a pressure gradient. The downflow first excavates a groove around the abutment wall, which in turn expands to form a helical flow (primary vortex) that can effectively entrain and transport sediment particles around the abutment. The flow separation at the outer edge of the abutment creates wake vortices that can lift the bed material like a cyclone in this zone to be transported downstream [7].

Several recent studies [8–12] have investigated different types of scour countermeasures to reduce local scour around bridge foundations. Local scour countermeasures can generally be divided into two categories [13], namely (a) armoring, and (b) flow-altering countermeasures. The former includes installing heavy elements such as riprap stones, gabions, cable-tied blocks, reno-mattresses, grout bags, etc. Flow-altering countermeasures, however, include the installation of submerged vanes, collars, sacrificial piles, spur dikes, and slots [14–17].

Many researchers [18–23] have studied riprap protection at bridge foundations. It is considered an environmentally friendly protection technique because it provides suitable conditions for aquatic organisms to thrive [24]. The permeable media through the particles allows native vegetation to take root and allows the survival of other organisms, helping to restore the stream's natural conditions. Many rock structures providing possible stream restoration with environmental benefits have been proposed and investigated by researchers such as Bhuiyan et al. [25] and Pagliara et al. [26]. From a scouring perspective, riprap stones should be designed to withstand the local shear stresses that form at the abutment. According to Chiew [9–27]; and Melville et al. [20], riprap layer failures around bridge foundations include shear, winnowing, edge and bed-form undermining failures. Shear failure, which occurs when turbulent flow erodes and moves riprap stones around bridge foundations, can be mitigated by using an appropriate riprap size. Winnowing failure is a consequence of the removal of finer materials (uplifting parent materials) through the riprap stones. Its occurrence can be prevented by using synthetic or granular filters. Edge failure occurs when a depression at the border of the sediments and riprap stones propagates and causes riprap stones to slide and fall into the hole. This can be reduced by providing a suitably designed apron. Bed-form undermining is the result of passing bed features (dunes and anti-dunes) around bridge foundations. This failure will bury riprap stones within sediments and can be prevented by placing the riprap below the level of the dune trough [18].

Flow-altering countermeasures exist in the form of in-stream structures, such as W-weirs, U-weirs, J-hook vanes and submerged vanes, the latter of which is not expensive to build [28]. The appropriate design of these structures prevents bank erosion, local scour, and channel degradation and improves grade control [8,18].

In addition to the above-mentioned structures, submerged vanes also have a broad range of applications for the reason that they can modify the local flow field, diverting the incoming flow from directly impinging on bridge piers or abutments and protecting river bend scour. Baltazar et al. [29] used submerged vanes to change flow pattern in a lateral diversion under live bed conditions. The authors reported that the amount of sediment entering to diversion reduced up to 26% when vanes were in place. Vanes prohibit the diversion vortex induced in the main channel by creating tip vortices and modifying velocity flied. Bahrami Yarahmadi et al. [30] used triangular vanes in a 900-flume bend. They concluded that triangular vane reduced bed shear stress near the outer bank, and at the position of 0.8 times of vane length in the downstream, vanes have the best performance in producing secondary flow. Bahrami Yarahmadi and Shafai Bejestan [31] applied triangular vanes in a flume bend. They reported that a single vane reduces average velocity near the outer bank and increases the number of vanes, resulting in the thalweg being pushed from the outer bank towards the channel midway. They reported that the best vanes performance was achieved when the lateral spacing is five times the vanes' length. With regard to bridge foundation, submerged vanes change the magnitude and direction of shear stress upstream of the pier or abutment. One of the first studies on the use of submerged vanes as a pier-scour countermeasure was carried out by Odgaard and Wang [32] and claimed that submerged vanes push the sediment bed toward pier. Later, Lauchlan [33] used vanes to reduce scour around bridge pier and stated that applying vanes resulted in a 34% scour depth reduction. Ghorbani and Kells [14] used vanes as pier scour countermeasures. Ghorbani and Kells [14] concluded that for vanes height other than zero (vanes height measured from sand bed), scour depth increases and maximum scour depth (87%) was achieved when two vanes with the angle of 18.50 were attached to the pier. By using vanes, Johnson et al. [15] used vanes as an abutment scour countermeasure. They reported that vanes reduce velocities and shear stress in the vicinity of foundation and if abutment is within the area affected by vanes, the scour depth will be decreased greatly. They stated that maximum flow control will be achieved when the angle of attack is between 25 and 300, and two structures instead of one improve vanes' performance which resulted in a 96% scour depth reduction around the bridge abutment. Shafai Bejestan et al. [34] used vanes to reduce local scour around a vertical wall abutment. They concluded that vanes reduce velocity and shear stress at the abutment nose and push this area toward the middle channel. The authors utilized different vanes angles and positions and stated that the best performance was achieved when vanes are attached to an abutment at an angle of 400.

In an attempt to overcome the shortages of each method and enhance scour countermeasure techniques, many researchers have applied a combination of different methods to reduce scour depth significantly. Zarrati et al. [35] experimented with the combination of riprap and collar to mitigate local scour around pier groups. They observed that incorporating tow methods resulted in 50% and 60% scour reduction in front and rear piers, respectively; however, using collar independently reduced scour depth by 25% and 30% for front and rear piers. Garg et al. [36] employed sleeve, collar and submerged vanes to protect a bridge pier against local scour. They reported that the combination of these techniques is more effective than using them individually, so that incorporating vanes with collar and vanes with sleeves reduces scour depth by 86% and 70%, respectively, in comparison with applying vanes, collar and sleeves which resulted in 57%, 78% and 39%. In another laboratory experiment conducted by Biswas and Barbhuiya [37], riprap and submerged vanes were used to mitigate river bend scour. They observed that, in a permanent river, it is not possible to mitigate scour, either with riprap or submerged vanes individually; however, bend protection can be attained if a combination of two techniques is applied. Zolghadr et al. [38] performed a laboratory examination to evaluate the effect of riprap and Six Pillar Concrete (SPC) elements separately and in combination on scour around a bridge abutment in different Froude Numbers under clear water conditions. They stated that average scour reductions using SPC elements and riprap alone were 83% and 30%, respectively; however, the highest reduction was achieved when both scour countermeasures were applied, which was 91%. In addition, they reported that the combination of the two techniques removed edge failure thoroughly.

Many researchers have independently used riprap and submerged vanes to reduce scour at bridge foundations. However, to the best of the authors' knowledge, incorporating these two methods has not been investigated at the vertical wall and spill-though abutments. Incorporating both riprap and vanes together may provide a more effective abutment-scour countermeasure because combining them can overcome the weakness of using each method individually. Some potential advantages include riprap stability enhancement and edge failure protection.

Moreover, very few studies have investigated the volume of riprap needed, although it is an important consideration for engineers to determine the cost-effectiveness of an apron design. An appropriate riprap configuration leads to a reduction in riprap volume and cost [39]. Consequently, the present study aims to examine not only the depth of scouring but also riprap failure and volume. In summary, the main objective of this research is to explore the effect of riprap and submerged vanes (singly or in combination) on scour reduction around vertical walls and spill-through abutments in clear-water conditions. The effect of the different geometry of riprap was evaluated on scour depth, edge failure and volume of riprap. Furthermore, shear failure that occurs at spill-through abutment is discussed as well.

2. Materials and Methods

The present experiments were conducted in a hydraulic laboratory the of water engineering department at Shiraz University, Iran. Tests were performed in a flume with 16 m length, 1.2 m width and 0.4 m depth. The abutment model was placed in a recess that was 2.5 m long, 1.2 m wide, and 0.2 m deep, filled with uniformly distributed sediment. The characteristics of the sediment used in this study were: median grain size, $d_{50} = 0.78$ mm for which 50% by weight is finer [40]; geometric standard deviation, $\sigma_g = \sqrt{\frac{d_{84.1}}{d_{15.9}}} = 1.28$ which satisfies uniformity of sediment [41], in which 84.1% and 15.9% of the particles are finer by weight, and specific gravity = 2.63. Based on Raudkivi [42], median grain size was selected to prevent ripple forming ($d_{50} > 0.7$ mm) and the effect of sediment size on scour depth was omitted from consideration in this study since $\frac{L}{d_{50}} > 50$ [1], where L is the abutment length. A rock-filled box was installed at the flume entrance to eliminate the effect of large circulations induced at the flume entrance. The distance between the flume entrance and the sediment recess (test section) was 8.5 m to ensure the formation of a fully developed flow. Two concrete false floors were placed at the upstream and downstream ends of the sediment recess to prevent the leaching of particles, and a layer of the same sediment was glued on the concrete false floors so that a uniform bed roughness existed along the flume. A circulatory flume system was used to introduce the necessary flow for the tests. Three pumps (each with a capacity of 120 L/s) were used to circulate water from a large underground reservoir to the head tank to ensure a constant head. The flow discharge was adjusted using a butterfly valve, and the resulting flow rate was measured with an electronic flow meter installed along the pipe. The flow depth was regulated by using a hand-operated tailgate located at the downstream end of the flume. The scheme of the flume and related facilities are shown in Figure 1.

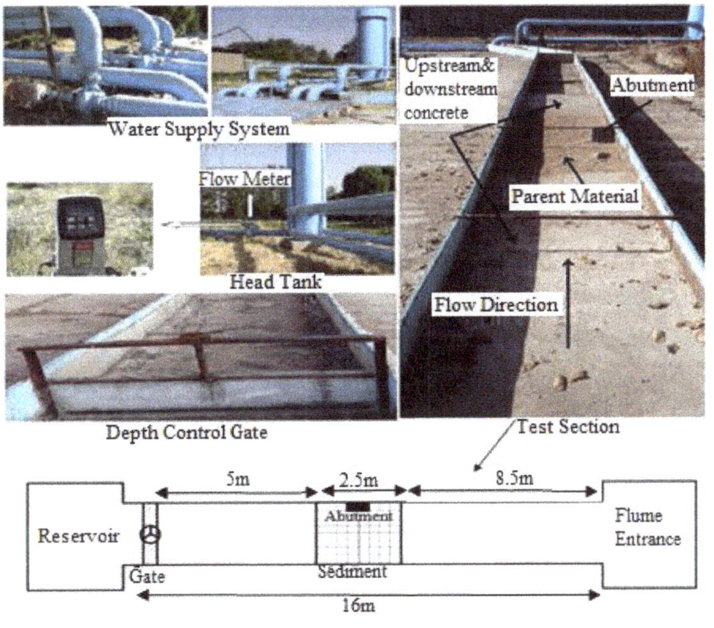

Figure 1. The scheme of test section, instrument and plan view of sediment recess.

Figure 2 shows the two types of abutment examined in this study: (1) vertical wall with three different dimensions; and (2) spill-through abutment. The length of the vertical wall abutments (L) used in the study were 15, 25 and 35 cm with a width of 10 cm. The length of the spill-through abutment (L) was 35 cm; its top width = 10 cm, top length (L') = 20 cm and side slope = 1:1 ($H : V$).

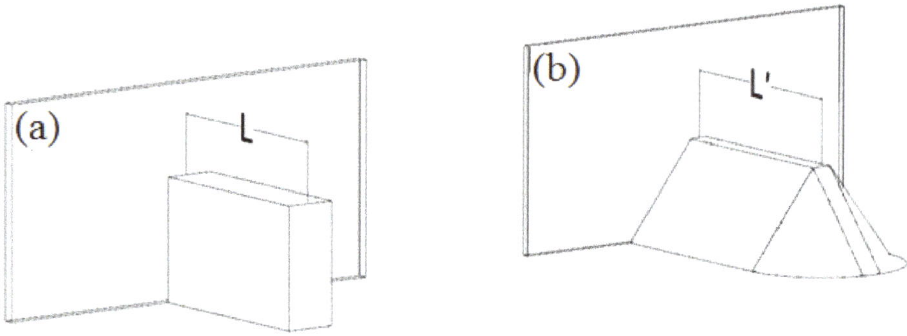

Figure 2. Schematic illustration of the abutments: (**a**) Vertical wall; (**b**) Spill-through.

Riprap stones in the form of fine gravel with median grain size $D_{R50} = 7$ mm were used in this study. The riprap stones were selected based on the study of Chiew [9]. Similar scale ratios of riprap stones also were used by Cardoso et al. [21] for abutment countermeasures ($D_{R50} = 7$ mm and $d_{50} = 0.96$ mm) and Zarrati et al. [35] for pier countermeasure ($D_{R50} = 5$ mm and $d_{50} = 0.95$ mm) in their studies. The scale ratios of the apron (a_R, b_R and t, see Figure 3 for the definition of these scale ratios), which are selected according to Melville and Coleman [1] and Cardoso et al.'s [21] recommendations, are shown in Table 1.

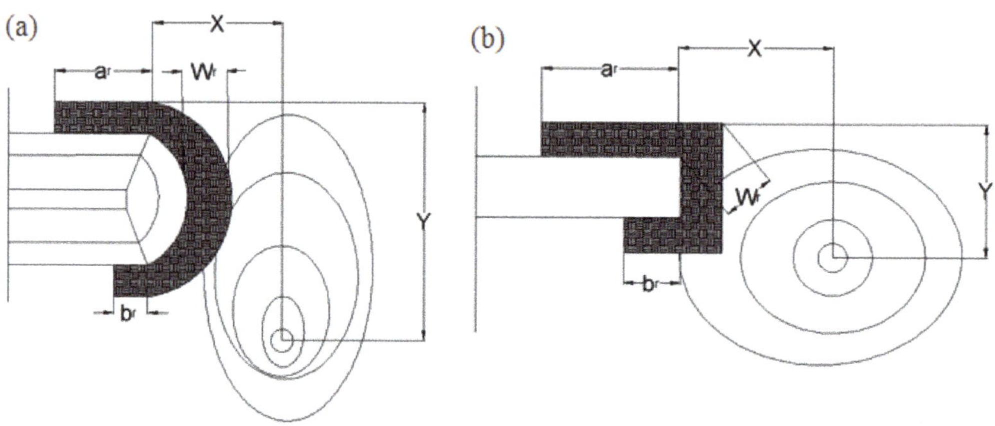

Figure 3. Scale ratios of the riprap apron and position of the maximum scour depth at the (**a**) spill-through abutment; (**b**) vertical-wall abutment.

Table 1. Characteristics of riprap stones.

Type of Abutment	L (cm)	Riprap				
		a_R(cm) =min{L,2y}	b_R(cm) =3D_{R50}	t(cm) =3D_{R50}	w_R(cm) Formula	Value
Vertical wall	15	15	2	2	$w_R = 0.75y(L^+)^{0.55}$	11
	25	25	2	2	$w_R = 0.5y\left[2K(L^+)^{0.5}\right]^{1.35}$	26
	35	30	2	2	$w_R = 0.5y\left[2K(L^+)^{0.5}\right]^{1.35}$	33
Spill-through	35	20	2	2	$w_R = 2y$	28

Note: K is the abutment shape factor = 1 for vertical wall abutment. $L^+ = \frac{L}{y}$ in which y is flow depth. w_R is the apron width. a_R and b_R, respectively, are the upstream and downstream lengths of the apron; and t is the apron thickness.

The apron width (w_R) in the spill-through abutment was selected according to Richardson and Davis's [43] recommendation. For vertical wall abutments with 25 and 35 cm lengths, Melville et al.'s [20] relation was applied; for the length of 15 cm, the Cardoso et al. formula [21] was used (Table 1). A granular filter with median particle size = 1.8 mm and the same scale ratios of the riprap layer was used to prevent winnowing failure. The filter was selected according to Terzaghi's criteria. To install the filter and riprap layers, the volume of the bed material that was the same as that of the apron was first carefully removed, before it was backfilled with the filter and riprap stones to the same level as the initial bed.

The location of the maximum scour depth was also recorded, as shown in Figure 3, in which X and Y are the coordinates of the maximum scour depth locations, respectively. The scale ratios were measured from the abutment toe; these locations are similar to that presented in Cardoso et al. [21].

The parameters that are involved in defining the geometrical layout of submerged vanes are shown in Table 2. The values were selected based on the findings of Johnson et al. [15] and Fathi and Zomorodian [44]. The geometrical layout of the vanes is shown in Figure 4.

Table 2. Parameters of geometrical layout of vanes.

Parameter	Value	
	L_{V15}	L_{V25}, L_{V35}, L_{S35}
Length of vanes (L_S)	$4H_S = 22.5$ cm	22.5 cm
Height of vanes (H_S)	$0.4y = 5.6$ cm	5.6 cm
Angle of vanes relative to flow direction (α)	30°	30°
Distance of vanes from the abutment (P)	$\frac{B}{3.2}$	$\frac{B}{3.2}$
Number of vanes in a row (M)	1	2
Number of rows of vane (N)	1	1
Position of first vane relative to the edge of the abutment (d_V)	$\frac{L}{3}$	$\frac{L}{3}$
Lateral spacing of the vanes (e)	$2H_S < e = 12$ cm $< 3H_S$	$2H_S < e = 12$ cm $< 3H_S$

Note: B is channel width, and y is flow depth. The subscripts V and S represent vertical wall and spill-through, respectively, and 15, 25 and 35 are the abutment lengths in cm.

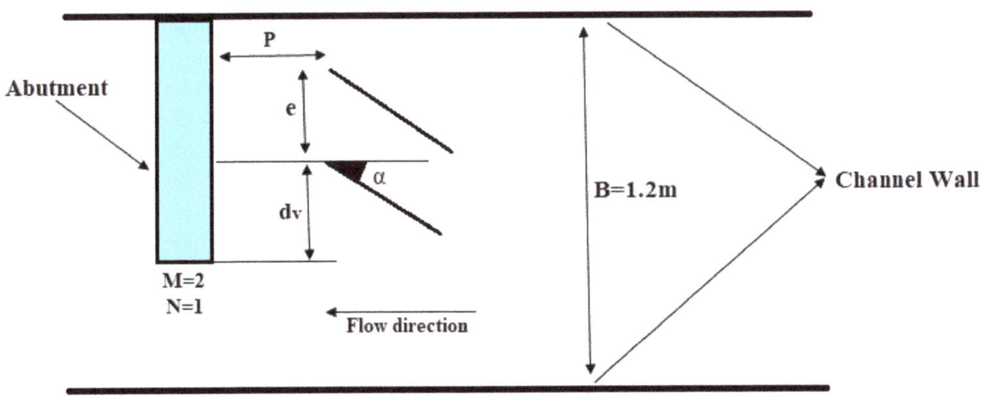

Figure 4. Plan view of channel and arrangement of vanes.

Laboratory Experiments

In Table 2, the angle of attack (α), number of vanes in a row (M), length (L_S) and the height of the vanes (H_S), and the distance of the vane from the abutment (P) are found to have the most dominant effect on the performance of vanes. This is inasmuch as vanes produce horizontal circulations that modify the flow pattern. In submerged vanes, the vortex sheet separation occurs at their top edges, creating a vortex that affects the vertical pressure distribution and, hence, a lift force. The strength of the horizontal circulations depends on the lift force. If the angle of attack is high, for example, the vortex formed and the lift force induced will correspondingly be high [8,45]. Moreover, Odgaard and Wang [8] stated that a vane affects a distance approximately twice its height in the transverse direction. Thus, increasing the number of vanes in a row (M) produces more circulations and a large lateral extent, thereby effecting a superior outcome.

The strength of the vortex circulations induced by vanes decays as they move downstream for the reason of viscous diffusions and bed resistance [8]. Van Zwol [45] reported that the circulations persist within a distance that is twice the vane's length. Accordingly, vanes should be installed at an appropriate distance from the abutment (P) to reduce flow velocity within the area between the vanes and the abutment.

The maximum scour depth occurs at the threshold condition for bed sediment entrainment, ($\frac{V}{V_c} = 1$) where V = mean approach velocity and V_c = critical velocity [46]. Accordingly, all laboratory experiments were conducted near the threshold condition in this study. The critical velocity was computed using the customary mean velocity logarithmic law [9] for a rough bed with the roughness height = $2d_{50}$ as follows:

$$\frac{V_c}{u_{*c}} = 5.75log\frac{y}{2d_{50}} + 6 \quad (1)$$

$$R_e = \frac{V\,y}{\nu} \rightarrow R_e = \frac{0.33 \times 0.14}{10^{-6}} = 46200 \quad (2)$$

In the hydraulic rough turbulent area, the Reynolds Number was 46,200 which confirms using customary mean velocity logarithmic law. In Equation (1), the critical shear velocity, u_{*c}, was calculated from Shield's Diagram. Consequently, the critical shear and mean velocities were determined to be 0.0192 m/s and 0.33 m/s, respectively. The value of flow discharge, flow depth and mean velocity were calculated as 0.045 m^3/s, 0.14 m and 0.27 m/s, respectively.

To determine the test duration needed in the study, a preliminary test that lasted 24 h was first performed, and the temporal evolution of the scour at the location of the eventual maximum scour was plotted. After 6 h, it was found that the rate of change of the scour depth was not significant. In other words, after 6 h, 90 percent of the 24 h scour depth had occurred [47]. Consequently, the test period used in this study is 6 h. In general, reaching equilibrium scour depth is really time-consuming [48]. Long-duration experiments are needed to develop a true equilibrium scour depth [49]. A 96 h duration was reported by Kothyari et al. [50], stating that such a period is required to develop a near-equilibrium scour hole at bridge piers under clear-water conditions [51]. However, since the current study aims to compare how different arrangements of the countermeasures affect abutment scour reduction, the duration used in the study was shortened. Similar studies with shorter experiments dealing with scouring can be found in the literature with 4 h duration or less [16,52].

After the completion of each experiment, the bed topography around the abutments was measured in a 2 cm × 2 cm grid using a laser displacement meter with 1 mm accuracy. The displacement meter, which was capable of moving in both streamwise and lateral directions, was placed on a platform.

Four types of experiments were conducted: (I) abutments without protection (baseline experiments); (II) abutments with riprap apron; (III) abutments with submerged vane; and (IV) abutments with the combination of riprap and submerged vanes. To evaluate the effect of the scour countermeasures, the maximum scour depths measured in the main tests (Types II, III and IV) are compared with that in the baseline test (Type I). Figure 5 shows the location of the maximum scour depth in the baseline experiments for both abutment types.

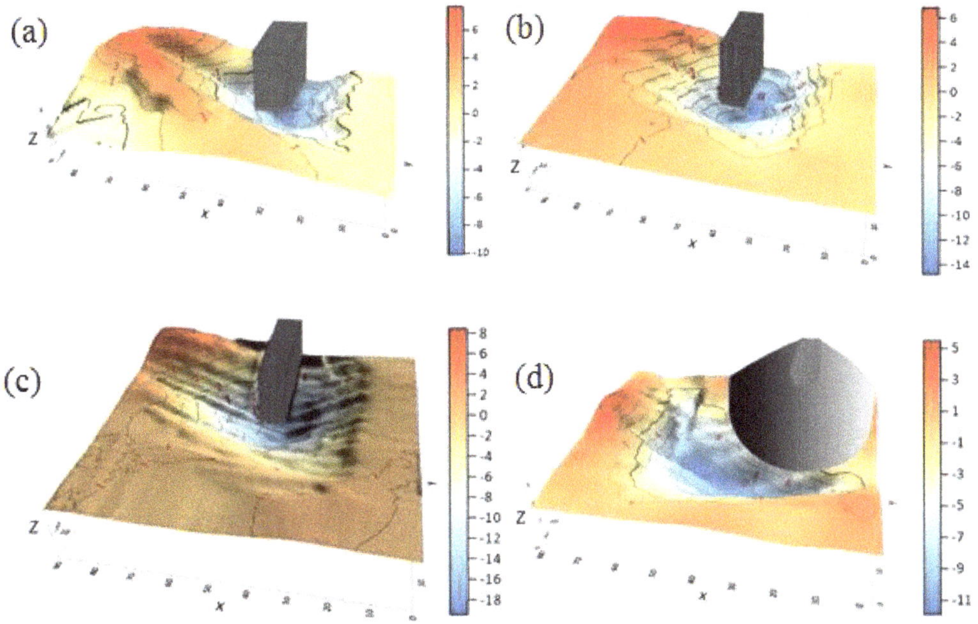

Figure 5. Location of the maximum scour depth at vertical wall and spill-through abutments. (**a**) L_{V15}; (**b**) L_{V25}; (**c**) L_{V35}; (**d**) L_{S35} (Flow from right to left).

In the experiments, two riprap failure modes were observed: edge and shear failure. The former appeared in both abutment shapes; however, it was minor with the vertical wall abutments. Hence, the discussion of edge failure is devoted to spill-through abutments. The latter was witnessed only at the downstream side of the spill-through abutment. To prevent shear failure at the spill-though abutment, larger stones were needed. Consequently, a riprap layer (R_{15}) with a diameter = 15 mm, width = 15 cm and the same thickness of D_{R50} was installed at the downstream side of the spill-through abutment to prevent shear failure.

3. Results and Discussion

3.1. *Application of Submerged Vanes and Riprap Individually*

In essence, vanes act as a kind of flow-altering countermeasure tool, whereas riprap stones create an armor layer to shelter the finer underlying parent materials from erosion (armoring countermeasure). Tests T_2 and T_3 (Table 3) show the effectiveness of using submerged vanes and riprap placement independently on scour reductions, respectively, as an illustration. As shown in Table 3, the use of a riprap layer (T_3) on scour reduction is more effective than that of submerged vanes (T_2), for two reasons. First, in the case of submerged vanes installations, due to the presence of the scour hole around abutments, the edge of the scour hole produces flow separation to create a curved flow that triggers the formation of the primary vortex, which is one of the dominant factors in scour mechanisms. Second, using ripraps increases the distance between the unprotected materials and the abutment such that the highest turbulence intensity no longer affects the scouring process as long as the location with this intensity is now covered with larger rocks [53,54]. As a result, the reduced flow circulations and disturbances farther away from the abutment are already subsumed into the main flow before encroaching onto the bed material. Thus, using riprap resulted in shallower scour depths compared to using submerged vanes alone. In addition, observations in Test T_3 showed that the riprap layer displaced the maximum scour depth from the abutment toe to the middle of the channel. However, in the submerged vane Test, T_2, the maximum scour depth location remained in the vicinity of the abutment toe. In this case, riprap layers not only reduced the scour more effectively, but also pushed the maximum scour depth location away from the abutment. Relocating the scour hole farther away from the abutment is a key factor in preventing bridge failure [20] (see Figure 6a,b).

Table 3. Summary of results.

Tests Number	Abutment Type	Scour Countermeasures	X (cm)	Y (cm)	Q (cm)	d_S(cm)	d_s/L	Scour Reduction Compared to Baseline Tests (%)
T_1	L_{V25}	–	–	–	0.045	15.2	0.608	–
T_2	L_{V25}	S	–	–	0.045	10	0.4	34.21
T_3	L_{V25}	R	65	24	0.045	9.4	0.376	38.75
T_4	L_{V25}	SR	50	30	0.045	8.9	0.356	41.45
T_5	L_{S35}	–	–	–	0.045	11.8	0.337	–
T_6	L_{S35}	S	–	–	0.045	8.2	0.234	30.5
T_7	L_{S35}	SR	95	26	0.045	7.5	0.214	36.44
T_8	L_{S35}	R_{15}, R	80	22	0.045	7.8	0.223	33.89
T_9	L_{S35}	S, R_{15}, R	98	29	0.045	7.2	0.206	38.98
T_{10}	L_{V35}	–	–	–	0.045	17.8	0.508	–
T_{11}	L_{V35}	S	–	–	0.045	16.1	0.460	9.55
T_{12}	L_{V35}	R	60	40	0.045	13.2	0.377	25.84
T_{13}	L_{V35}	SR	62	35	0.045	12.1	0.346	32.02
T_{14}	L_{V15}	–	–	–	0.045	10.1	0.673	–
T_{15}	L_{V15}	S	–	–	0.045	8.7	0.580	13.86
T_{16}	L_{V15}	R	36	24	0.045	8.3	0.553	17.82
T_{17}	L_{V15}	SR	22	21	0.045	5.5	0.360	54.45

Note: S, R, R_{15} and S, R represent submerged vanes, riprap with a median diameter of 7 mm, and riprap with a diameter of 15 mm, a combination of submerged vane and riprap stones, respectively. Scour countermeasures refer to the device used to reduce scour. The subscripts V and S represent vertical wall and spill-through abutments, respectively, and 15, 25 and 35 are abutment lengths in cm.

Figure 6. Maximum scour depth location and erosion of riprap stones around L_{V25} in case of submerged vane and riprap. (**a**) T_2; (**b**) T_3; (**c**) T_3; (**d**) T_4 (flow from right to left).

3.2. Combination of Submerged Vanes and Riprap

Although using submerged vanes and riprap stones as abutment scour countermeasures provides positive results, causing definitive scour depth reductions, if they are used together, one could surmise that their combined effects will be even more significant. Consequently, this study examines whether using these methods together can lead to an improved scour countermeasure method, and if so, by how much? Table 3 (T_4) shows the experimental results obtained by using a combination of submerged vanes and riprap, clearly revealing that fewer riprap stones are displaced compared to T_3 (see Figure 6c,d). This is due to the reduction of the local shear stresses and turbulence around the abutment for the presence of the submerged vanes [15,34]. In summary, the experimental data show that using both these methods together can reduce the maximum scour depth by 41% more than that using individual methods separately (compare T_4 to T_3 and T_2 in Table 3).

For the other abutments (L_{V35}, L_{V15} and L_{S35}), similar results to that of the L_{V25} abutment types were obtained when submerged vanes and riprap were used together, although with different percentages of scour reduction. The reason probably is due to the differences in the length and shape of the abutments. In summary, although the effect of riprap on scour reduction is superior to that of submerged vanes, the latter can enhance riprap stability due to their ability to modify the local flow field. The present data show that the highest reduction in scour depth for all the abutments tested occurs when these two methods are used together (Table 3, Tests T_4, T_9, T_{13} and T_{17}).

3.3. Geometry and Scale Ratios of Riprap Apron

Without the armoring layer, maximum scour depth happens at the downstream end of the spill-through abutment (Figure 7a) and with the presence of the apron layer, the edge failure occurs at this zone. To reduce edge failure, apron geometry plays an important role [13]. Since Richardson and Davis [43] have provided a well-known and widely used formula to determine (w_R), a test was performed based on their design (Test T_{18} in Table 4). Test T_{18} is the baseline test to examine the effectiveness of other riprap scale ratios. Despite their recommendation, Cardoso et al. [21] reported that designs using the method proposed by Richardson and Davis [43] are overly conservative. In order to design an apron to achieve optimum cost-effectiveness, its width and thickness are decreased, according to Table 4. The results of four additional tests based on different scale ratios and apron geometry at the spill-through abutment are shown in Table 4. Furthermore, the riprap volume reduction percentages of all of the tests as compared to the baseline test also are shown.

Tests T_{19} to T_{22}, with different apron widths and thicknesses, were conducted to obtain an apron configuration without compromising its effectiveness in this study. In Test T_{19}, the width of the apron was 25 cm (Figure 7b, Geometry I), 11% shorter than that of the baseline Test, T_{18}. The experimental results revealed the presence of edge failure at the apron. Observation clearly shows how a depression is first formed at the interface of the riprap stones and original bed sediments, which is the onset of "edge failure". With time, the depression propagates, causing the riprap stones to slide into it. If this hole becomes excessively large, a total disintegration of the apron may occur [9]. Edge failure and the extent of riprap coverage are closely related; if there is a sufficient supply of riprap stones, this problem may be mitigated. Under this condition, the riprap stones that slide into the scour hole can re-armor the hole to prevent further erosion and total disintegration [9,55,56]. Accordingly, to overcome the observed edge failure, the apron was changed from a circular to a square shape (see T_{20} in Figure 7c, Geometry II), with the aim of enhancing the riprap stones' stability here (compare T_{20} to T_{19}). This observation, which shows a distinct improvement, is consistent with the result of Simarro et al. [13], who reported that the apron geometry of Figure 7c is superior to that of Figure 7b.

Table 4. Results of different geometry and scale ratios of apron at spill-through abutment.

Tests Number	w_R (cm)	w_R/L	a_r (cm)	b_r(cm)	X (cm)	Y (cm)	Shape of Apron	d_s/L	Changes in Scour Depth (%)	Volume of Riprap (cm³)	Riprap Volume Reduction (%)
T_{18}	28	0.8	20	2	73	25	Geometry I	0.268	–	7393	–
T_{19}	25	0.714	20	2	70	27	Geometry I	0.277	+3	6353	14
T_{20}	25	0.714	20	2	80	25	Geometry II	0.234	−12.7	7228	2.2
T_{21}	20	0.571	20	2	60	23	Geometry II	0.240	−10.6	5439	26.4
T_{22}	25	0.714	20	2	74	30	Geometry II	0.234	−12.7	5006	32.3

Note: The positive and negative signs in Column 9 indicate an increase and decrease in scour depth compared to T_{18}, respectively.

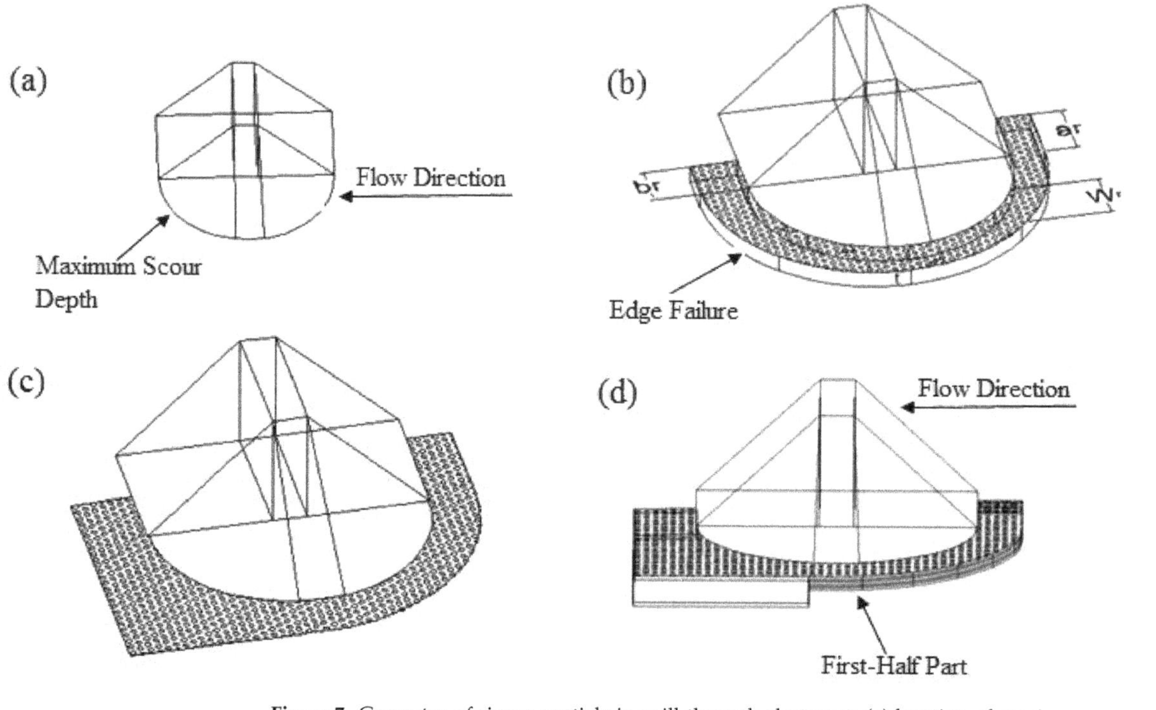

Figure 7. Geometry of riprap particle in spill-through abutment: (**a**) location of maximum scour depth (**b**) Geometry I, T_{19}; (**c**) Geometry II, T_{20}; (**d**) shape of apron with $1.5D_{R50}$, T_{22}. (flow from right to left).

In Test T_{21}, the apron width was further decreased to 20 cm to test the over-conservatism of the method of Richardson and Davies [43]. With this change, the result shows that the location of the maximum scour depth has migrated closer to the abutment toe ($X = 60$ cm and $Y = 23$ cm) because the scour hole position is dependent on apron width [20]. However, edge failure, which occurs at the downstream end of the abutment as shown in Figure 7b, is more prominent when compared to that in Test T_{20}. Hence, the apron width was reverted to 25 cm. However, since scouring and the dislodgement of riprap stones (edge failure) did not occur at the upstream end of the riprap apron, the apron thickness at this location was decreased by $1.5D_{R50}$ in Test T_{22} to reduce the volume of the riprap layer needed (Figure 7d). With this change, the result shows that reducing the thickness of the riprap layer has no effect on edge failure and the resulting scour depth, i.e., it did not exacerbate edge failure. The results also show that edge failure and scour depth were reduced in Test T_{22} compared with Test T_{19} by 12.7%. In addition, the volume of the riprap layer in Test T_{22} is reduced by 32% in comparison to the baseline Test T_{18}.

Similar to the spill-through abutment, the riprap volume is also reduced by decreasing its thickness at the vertical wall abutment so that the apron thickness at the upstream end is reduced by $1.5D_{R50}$ for each vertical wall abutment, and the riprap volume is reduced significantly (see Figure 8). The Figure shows that, despite the significant reduction of the riprap volume in all the abutments (Classes 1 and 2), the scour depth did not increase. Apart from edge failure (Tests T_{18} to T_{22}), which is reflected by the erosion of riprap stones at the downstream end of the spill-through abutment, shear failure is also is observed. To address this problem, an appropriate size of the riprap layer was proposed; an issue that will be discussed in the following section. As mentioned earlier, shear failure was not observed at the vertical wall abutment.

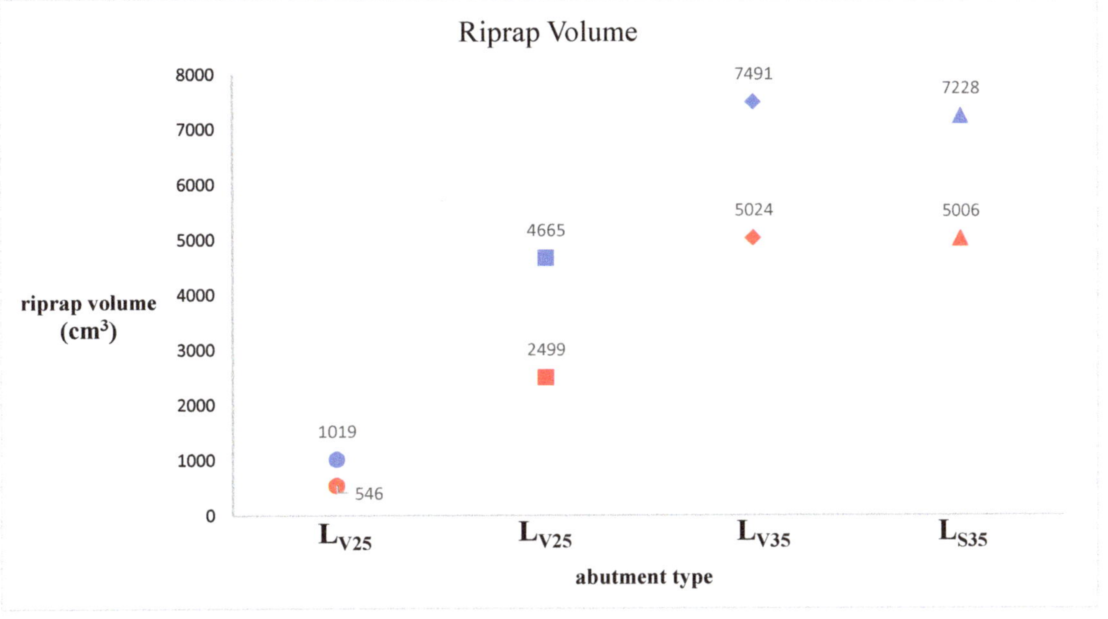

Figure 8. Volume of riprap layer around abutments. Blue symbols represent riprap with $t = 3D_{R50}$ and red symbols represent $t = 1.5D_{R50}$.

3.4. Shear Failure Prevention at Spill-Through Abutment

Shear failure occurs when the riprap stones are not heavy enough to withstand the turbulent flow field [5,13]. In Tests T_{18} to T_{22}, riprap stones at the downstream end of the spill-through abutment were eroded in both geometries I and II (see Figures 7 and 9a) because flow separation has caused an upward flow (wake vortices) that entrains the bed sediments [57]. In addition, the roughness difference between the course riprap and fine bed sediments creates a vulnerable point where the finer materials are eroded. These combine to create a depression (hole) at the interface of the apron and bed sediments. As this scour hole enlarges, riprap stones start sliding and rolling into it (edge failure), as was reported by Chiew [9]. With time, the scour depth continues to grow until the shear stress and turbulence fluctuations around the abutment can no longer erode the parent materials [53]. It should be noted that in the area around the abutment, the shear stress and vortices are higher than those of the approach flow (10 and 1.5 times, respectively, [57]). As a result, reducing the strength of the downflow and primary vortex at the upstream side of the abutment can reduce the strength of wake vortices and edge failure to enhance riprap stability. To achieve this, using submerged vanes is an appropriate alternative. They produce vortices (horizontal circulations) that move downstream with the main flow, augmenting the shear stresses. In addition, vanes reduce velocity within the area between themselves and the abutment, resulting in the formation of a lower pressure gradient and downflow [34,53]. Consequently, the flow separation correspondingly becomes weaker.

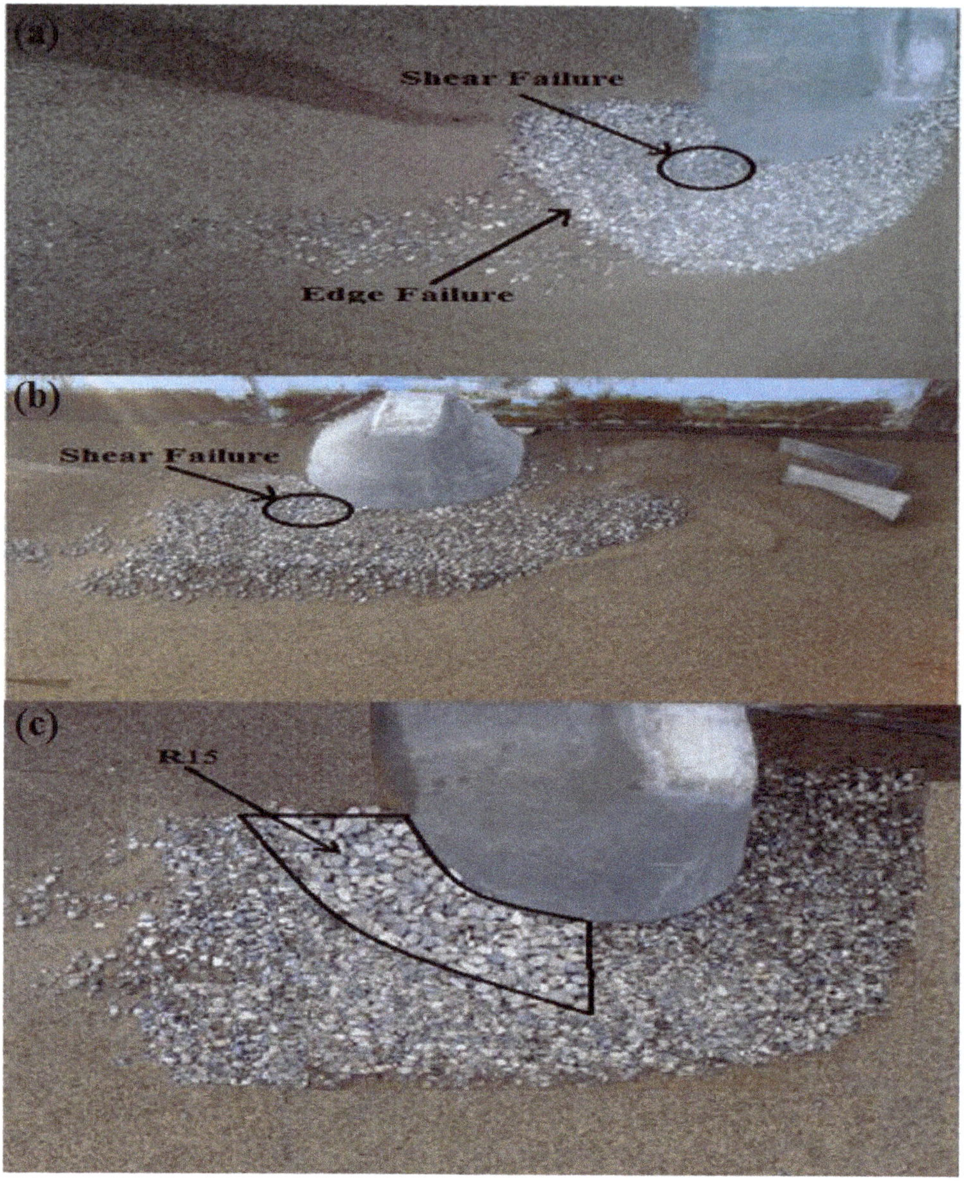

Figure 9. (**a**) T_{19}: shear and edge failure; (**b**) T_7: effect of vane on shear failure; (**c**) T_8: use of R_{15} to arrest shear failure (flow from right to left).

If the placement of vanes still cannot prevent such failures (Figure 9b), installing another riprap layer may be needed. In this study, a riprap layer (R_{15}) with diameter = 15 mm, thickness = $3D_{R50}$ and width = 15 cm was placed around the abutment (Figure 9c). The aim is to use larger riprap stones to arrest shear failure. Test T_8 (Table 3) shows that using R_{15}, the riprap layer is left intact, and shear failure is prevented (Figure 9c). One may infer from these two tests that vanes and riprap size offer different remedial actions in abutment scour countermeasures. The engineer must understand the purpose of their design when using different scour countermeasure approaches.

4. Conclusions

In this study, both armoring (riprap) and flow-altering (submerged vanes) scour countermeasures were investigated individually and together around vertical walls and spill-through abutments. The maximum scour depth and location, armor layer failure, volume and scale rations are discussed in detail.

The following conclusions are drawn from the study:

1. Installing a riprap layer reduces the maximum scour depth by up to 39% and 34% in vertical wall and spill-through abutments, respectively.
2. Submerged vanes reduce the maximum scour depth by up to 34% and 30% in vertical walls and spill-through abutments, respectively. This shows that the effect of riprap on scour reduction is more than that of submerged vanes.
3. The largest decrease in scour depth is achieved through a combination of riprap and submerged vanes. These reductions were 54% and 39% in vertical walls and spill-through abutments, respectively.
4. By installing a riprap layer, the location of the maximum scour depth is relocated away from the abutment toe while it remains close to it by using submerged vanes alone.
5. Applications of submerged vanes enhance riprap stability and reduce edge failure. With vanes, fewer riprap stones are eroded by the flow.
6. A square-shaped riprap layer at the downstream end of spill-through abutments is more effective in promoting riprap stone stability and reducing-edge failure than circular-shaped riprap layers.
7. Using submerged vanes is not an effective way to prevent riprap shear failure at the downstream side of spill-through abutments. Utilizing a larger riprap layer (R_{15}) is needed.
8. By decreasing the thickness of the riprap layer proposed by Cardoso et al. (2010) in the upstream half of the apron, the volume of the riprap layer needed is reduced up to 46% without affecting the riprap efficacy.

Author Contributions: Conceptualization, A.F. and B.K.; methodology, A.F. and M.Z.; validation, A.F. and S.M.A.Z.; formal analysis, S.M.A.Z., M.Z., Y.-M.C., B.K. and H.M.; investigation, A.F.; resources, S.M.A.Z.; data curation, Y.-M.C.; writing—original draft preparation, A.F.; writing—review and editing, M.Z., Y.-M.C. and H.M.; visualization, S.M.A.Z., A.C. and Y.-M.C.; supervision, A.C.; project administration, M.Z. and A.C.; funding acquisition, S.M.A.Z. All authors have read and agreed to the published version of the manuscript.

Funding: The costs for this research were provided from the grants of the second author, Water Engineering Department, Shiraz University (04/02/2015).

Data Availability Statement: The data presented in this study are available on request from the corresponding author.

Conflicts of Interest: The authors declare no conflict of interest.

Nomenclature

The following symbols are used in this paper:

a_R	Upstream apron width
B	Channel width
b_R	Downstream apron width
D_{R50}	Median diameter of riprap apron
d_s	Scour depth
d_V	Distance of first vane relative to the edge of the abutment
d_{50}	Median size of sediment bed
e	Lateral spacing of the vanes
H_S	Vane height
K	Abutment shape factor
L	Length of abutment

L'	Top length of spill-through abutment
L_S	Vane length
M	Number of vanes in a row
N	Number of rows of vane
P	Distance of vanes from the abutment
Q	Flow rate
R	Riprap
R_{15}	Riprap with diameter of 15 mm
S	Submerged vane
t	Thickness of riprap layer
u_{*c}	Critical shear velocity
V	Mean velocity
V_c	Critical velocity
w_R	Width of the apron
X	Position of maximum scour depth along the flow direction
Y	Position of maximum scour depth transverse to the flow direction
y	Flow depth
α	Vane angle (degree) corresponding to flow direction
σ_g	Sediment gradation

References

1. Melville, B.W.; Coleman, S.E. *Bridge Scour*; Water Resources: Highland Ranch, CO, USA, 2000.
2. Hong, J.H.; Chiew, Y.M.; Lum, J.Y.; Lai, J.S.; Lin, Y.B. Houfeng Bridge Failure in Taiwan. *J. Hydraul. Eng.* **2012**, *138*, 186–198. [CrossRef]
3. Proske, D. *Bridge Collapse Frequencies versus Failure Probabilities*; Springer International Publishing: Cham, Switzerland, 2018.
4. Pizarro, A.; Manfreda, S.; Tubaldi, E. The science behind scour at bridge foundations: A review. *Water* **2020**, *12*, 374. [CrossRef]
5. Melville, B.W. Pier and abutment scour: Integrated approach. *J. Hydraul. Eng.* **1997**, *123*, 125–136. [CrossRef]
6. Ettema, R.; Constantinescu, G.; Melville, B.W. Flow-field complexity and design estimation of pier-scour depth: Sixty years since Laursen and Toch. *J. Hydraul. Eng.* **2017**, *143*, 03117006. [CrossRef]
7. Fakhimjoo, M.S.; Ardeshir, A.; Behzadian, K.; Karami, H. Experimental investigation and flow analysis of clear-water scour around pier and abutment in proximity. *Water Sci. Eng.* 2022; *in press*. [CrossRef]
8. Odgaard, A.J.; Wang, Y. Sediment management with submerged vanes. I: Theory. *J. Hydraul. Eng.* **1991**, *117*, 267–283. [CrossRef]
9. Chiew, Y.M. Mechanics of riprap failure at bridge piers. *J. Hydraul. Eng.* **1995**, *121*, 635–643. [CrossRef]
10. Melville, B.W.; Hadfield, A.C. Use of sacrificial piles as pier scour countermeasures. *J. Hydraul. Eng.* **1999**, *125*, 1221–1224. [CrossRef]
11. Morales, R.; Ettema, R.; Barkdoll, B. Large-Scale Flume Tests of Riprap-Apron Performance at a Bridge Abutment on a Floodplain. *J. Hydraul. Eng.* **2008**, *134*, 800–809. [CrossRef]
12. Valela, C.; Nistor, I.; Rennie, C.D.; Lara, J.L.; Maza, M. Hybrid Modeling for Design of a Novel Bridge Pier Collar for Reducing Scour. *J. Hydraul. Eng.* **2021**, *147*, 04021012. [CrossRef]
13. Simarro, G.; Civeira, S.; Cardoso, A.H. Influence of riprap apron shape on spill-through abutments. *J. Hydraul. Res.* **2012**, *50*, 138–141. [CrossRef]
14. Ghorbani, B.; Kells, J.A. Effect of submerged vanes on the scour occurring at a cylindrical pier. *J. Hydraul. Res.* **2008**, *46*, 610–619. [CrossRef]
15. Johnson, P.A.; Hey, R.D.; Tessier, M.; Rosgen, D.L. Use of vanes for control of scour at vertical wall abutment. *J. Hydraul. Engrg.* **2001**, *127*, 772–778. [CrossRef]
16. Zolghadr, M.; Shafai Bejestan, M. Six-legged concrete (SLC) elements as scour countermeasures at wing wall bridge abutments. *Int. J. River Basin Manag.* **2021**, *19*, 319–325. [CrossRef]
17. Osroush, M.; Hosseini, A.; Kamanbedast, A.A.; Khosrojerdi, A. The effects of height and vertical position of slot on the reduction of scour hole depth around bridge abutments. *Ain Shams Eng. J.* **2019**, *10*, 651–659. [CrossRef]
18. Chiew, Y.M.; Lim, F.H. Failure behavior of riprap layer at bridge piers under live-bed conditions. *J. Hydraul. Engrg.* **2000**, *126*, 43–55. [CrossRef]
19. Lauchlan, C.S.; Melville, B.M. Riprap protection at bridge piers. *J. Hydraul. Eng.* **2001**, *127*, 412–418. [CrossRef]
20. Melville, B.W.; Ballegooy, S.; Coleman, S.; Barkdoll, B.D. Countermeasure toe protection at spill-through abutments. *J. Hydraul. Eng.* **2006**, *132*, 235–245. [CrossRef]
21. Cardoso, A.H.; Simarro, G.; Fael, C.; Le Doucen, O.; Schleiss, A. Toe protection for spill-through and vertical abutments. *J. Haul. Res.* **2010**, *48*, 491–498. [CrossRef]
22. Ji, U.; Yeo, W.; Kang, J. Subsidence of riprap protection without filters for different installation types of riprap around a pier in sands. *J. Hydro Environ. Res.* **2013**, *7*, 41–49. [CrossRef]

23. Valela, C.; Whittaker, C.N.; Rennie, C.D.; Nistor, I.; Melville, B.W. Novel Riprap Structure for Improved Bridge Pier Scour Protection. *J. Hydraul. Eng.* **2022**, *148*, 04022002. [CrossRef]
24. Cope, E.D. Environmentally Friendly and Sustainable Stream Stability in the Vicinity of Bridges. Master Thesis, Brigham Young University, Provo, UT, USA, 2022.
25. Bhuiyan, F.; Hey, R.D.; Wormleaton, P.R. Hydraulic evaluation of W-weir for river restoration. *J. Hydraul. Eng.* **2007**, *133*, 596–609. [CrossRef]
26. Pagliara, S.; Kurdistani, S.M.; Cammarata, L. Scour of clear water rock W-weirs in straight rivers. *J. Hydraul. Eng.* **2014**, *140*, 06014002. [CrossRef]
27. Chiew, Y.M. Local Scour and Riprap Stability at Bridge Piers in a Degrading Channel. *J. Hydraul. Eng.* **2004**, *130*, 218–226. [CrossRef]
28. Rodríguez-Amaya, C.; Durán-Ariza, A.; Duarte-Méndez, S. Submerged Vane Technology in Colombia: Five Representative Projects. *Water* **2020**, *12*, 984. [CrossRef]
29. Baltazar, J.; Alves, E.; Bombar, G.; Cardoso, A.H. Effect of a Submerged Vane-Field on the Flow Pattern of a Movable Bed Channel with a 90° Lateral Diversion. *Water* **2021**, *13*, 828. [CrossRef]
30. Bahrami Yarahmadi, M.; Shafai Bejestan, M.; Pagliara, S. An experimental study on the secondary flows and bed shear stress at a 90-degree mild bend with and without triangular vanes. *J. Hydro Environ. Res.* **2020**, *33*, 1–9. [CrossRef]
31. Bahrami Yarahmadi, M.; Shafai Bejestan, M. Sediment management and flow patterns at river bend due to triangular vanes attached to the bank. *J. Hydro Environ. Res.* **2015**, *10*, 64–75. [CrossRef]
32. Odgaard, A.J.; Wang, Y. Scour prevention at bridge piers. In *Hydraulic Engineering: Proceedings on the 1987 National Conference on Hydraulic Engineering*; ASCE: Williamsburg, VA, USA, 1987; pp. 523–527.
33. Lauchlan, C.S. Countermeasures for pier scour. Ph.D. Dissertation, University of Auckland, Auckland, New Zealand, 1999.
34. Shafaei Bejestan, M.; Khademi, K.; Kozeymehnezhad, H. Submerged vane-attached to the abutment as scour countermeasure. *J. Ain Shams. Eng.* **2015**, *6*, 775–783. [CrossRef]
35. Zarrati, A.R.; Nazariha, M.; Mashahir, M.B. Reduction of local scour in the vicinity of bridge pier groups using collars and riprap. *J. Hydraul. Eng.* **2006**, *132*, 154–162. [CrossRef]
36. Garg, S.V.; Setia, B.; Verma, D.V.S. Combination of scour protection devices around oblong bridge pier. *J. Hydraul. Eng.* **2008**, *14*, 56–68. [CrossRef]
37. Biswas, P.; Barbhuiya, A.K. Countermeasure of river bend scour using a combination of submerged vanes and riprap. *Int. J. Sediment Res.* **2018**, *33*, 478–492. [CrossRef]
38. Zolghadr, M.; Shafai Bejestan., M.; Fathi, A.; Hoseinreza, A. Protecting Vertical-Wall Bridge Abutment Using Six-Pillar Concrete Elements. *Arab. J. Geosci.* **2022**, *15*, 1226. [CrossRef]
39. Guan, D.; Liu, J.; Chiew, Y.M.; Zhou, Y. Scour Evolution Downstream of Submerged Weirs in Clear Water Scour Conditions. *Water* **2019**, *11*, 1746. [CrossRef]
40. Raudkivi, J.A. Functional Trends of Scour at Bridge Piers. *J. Hydraul. Eng.* **1986**, *112*, 1–13. [CrossRef]
41. Melville, B. W: Local scour at bridge abutment. *J. Hydraulic. Eng.* **1992**, *118*, 615. [CrossRef]
42. Raudkivi, J.A. *Loose Boundary Hydraulics*; Balkema, A.A., Ed.; CRC Press: Rotterdam, The Netherlands, 1998.
43. Richardson, E.V.; Davis, S.R. Evaluating scour at bridges. In *Hydraulic Engineering Circular 18, HEC-18 Report FHWA-IP-90-017: 204*; Federal Highway Administration: Washington, DC, USA, 1995.
44. Fathi, A.; Zomorodian, S.M.A. Effect of Submerged Vanes on Scour around a Bridge Abutment. *KSCE J. Civ. Eng.* **2018**, *22*, 2281–2289. [CrossRef]
45. van Zwol, J.A. Design Aspect of Submerged Vanes. Master Thesis, Faculty of Civil Engineering and Geosciences, Delft University of Technology, Delft, The Netherlands, 2004.
46. Chiew, Y.M.; Melville, B.W. Local Scour around Bridge Piers. *J. Hydraul. Res.* **1987**, *25*, 15–26. [CrossRef]
47. Melville, B.W.; Chiew, Y.M. Time Scale for Local Scour at Bridge Piers. *J. Hydraul. Eng.* **1999**, *125*, 59–65. [CrossRef]
48. Liang, B.; Du, S.; Pan, X.; Zhang, L. Local scour for vertical piles in steady currents: Review of mechanisms, influencing factors and empirical equations. *J. Mar. Sci. Eng.* **2019**, *8*, 4. [CrossRef]
49. Cheng, N.S.; Chiew, Y.M.; Chen, X.W. Scaling analysis of pier-scouring processes. *J. Eng. Mech.* **2016**, *142*, 06016005. [CrossRef]
50. Kothyari, U.C.; Garde, R.C.J.; Ranga Ranju, K.G. Temporal variation of local scour around circular bridge pier. *J. Hydraul. Eng.* **1992**, *118*, 1091–1106. [CrossRef]
51. Sheppard, D.M.; Demir, H.; Melville, B.W. *Scour at Wide Piers and Long Skewed Piers*; Transportation Research Board: Washington, DC, USA, 2011; Volume 682. [CrossRef]
52. Bozkus, Z.; Osman, Y. Effect of inclination of bridge piers on scouring depth. *J. Hydraul. Eng.* **2004**, *130*, 827–832. [CrossRef]
53. Afzalimehr, H.; Moradian, M.; Singh, V.P. Flow Field around Semielliptical Abutments. *J. Hydrol. Eng.* **2018**, *23*, 04017057. [CrossRef]
54. Barbhuiya, A.K.; Dey, S. Vortex flow field in a scour hole around abutment. *Int. J. Sed. Res.* **2003**, *18*, 310–325. [CrossRef]
55. Cardoso, A.H.; Fael, C.M.S. Protecting vertical-wall abutments with riprap-mattresses. *J. Hydraul. Eng.* **2009**, *135*, 457–465. [CrossRef]

56. Chiew, Y.M.; Lai, J.S.; Link, O. Experimental, Numerical and Field Approaches to Scour Research. *Water* **2020**, *12*, 1749. [CrossRef]
57. Molinas, A.; Kheireldin, K.; Wu, B. Shear stress around vertical wall abutments. *J. Hydraul. Eng.* **1998**, *124*, 822–830. [CrossRef]

fluids

MDPI

Article

Experimental Study on the Optimum Installation Depth and Dimensions of Roughening Elements on Abutment as Scour Countermeasures

Masih Zolghadr [1], Seyed Mohammad Ali Zomorodian [2,*], Abazar Fathi [3], Ravi Prakash Tripathi [4], Neda Jafari [5], Darshan Mehta [6], Parveen Sihag [7] and Hazi Mohammad Azamathulla [8,*]

1 Department of Water Sciences and Engineering, Jahrom University, Jahrom 74148-46199, Iran
2 Water Engineering Department, Shiraz University, Shiraz 71946-84334, Iran
3 Faculty of Agriculture, Shiraz University, Shiraz 71946-84334, Iran
4 Civil Engineering Department, Rajkiya Engineering College, Sonbhadra 231206, India
5 Department of Water Sciences Engineering, Shiraz University, Shiraz 71946-84334, Iran
6 Department of Civil Engineering, Dr. S. & S. S. Ghandhy Government Engineering College, Surat 395008, India
7 Department of Civil Engineering, Chandigarh University, Gharuan 140413, India
8 Professor of Civil Engineering, University of the West Indies, St. Augustine P.O. Box 331310, Trinidad and Tobago
* Correspondence: mzomorod@shirazu.ac.ir (S.M.A.Z.); mdazmath@gmail.com (H.M.A.)

Abstract: The causes of many bridge failures have been reported to be local scour around abutments. This study examines roughening elements as devices with which to intercept the downflow responsible for the formation of the principal vortex, which is what triggers local scour around abutments. Two vertical wall abutments with different widths were examined under four different hydraulic conditions in a clear-water regime. Elements with different thicknesses (t) and protrusions (P) with the same dimensions, (P = t = 0.05 L, 0.1 L, 0.2 L, and 0.3 L, where L is the length of the abutment) and with varying depths of installation (Z) were considered. Elements were installed in two positions: between the sediment surface and water elevation and buried within the sediment. To determine the optimum depth of installation, one element was first installed on the sediment surface, and the number of elements was increased in each subsequent test. The results show that installing elements between water surface elevation and the sediment's initial level did not show any defined trend on scour depth reduction. However, the optimum installation depth of the elements is 0.6–0.8 L below the initial bed level. Moreover, the roughening elements with thickness and protrusion of P = t = 0.2 L resulted in the most effective protection of the foundation. The best arrangement, (P = t = 0.2 L and Z = >0.6–0.8 L) reduced the maximum scour depth by up to 30.4% and 32.8% for the abutment with smaller and larger widths, respectively.

Keywords: roughening elements; bridge foundation; river engineering; depth of installation; flow altering

Citation: Zolghadr, M.; Zomorodian, S.M.A.; Fathi, A.; Tripathi, R.P.; Jafari, N.; Mehta, D.; Sihag, P.; Azamathulla, H.M. Experimental Study on the Optimum Installation Depth and Dimensions of Roughening Elements on Abutment as Scour Countermeasures. *Fluids* **2023**, *8*, 175. https://doi.org/10.3390/fluids8060175

Academic Editor: D. Andrew S. Rees

Received: 30 March 2023
Revised: 26 May 2023
Accepted: 29 May 2023
Published: 5 June 2023

1. Introduction

Ref. [1] have reported that, among 383 bridge failures in the U.S., 72% were due to local scour around the abutments. Vortex formation around bridge foundations imposes excessive stress on the bed, increases erosion rate, degrades the streambed, and triggers a scour hole. To establish a clear outlook toward abutment scour, it is vital to understand the flow field around it (Figure 1). Three main factors trigger scour around abutments, namely downflow, primary vortex, and wake vortices [2]. The pressure gradient at the upstream side of the abutment drives the flow toward the bed, which, in turn, returns up and develops a primary vortex. These two factors excavate the parent materials and create a hole at the abutment tip. At the downstream side of the abutment, flow separation causes upward flow, known as wake vortices, that sweep and entrain sediments particles away from the abutment, as demonstrated in Figure 1. The three-dimensional flow field around a bridge foundation is detailed by [3–6]. Recent numerical studies have also been conducted to predict flow field and scour around inline hydraulic structures. A comprehensive study by [7] can be referred to as an example.

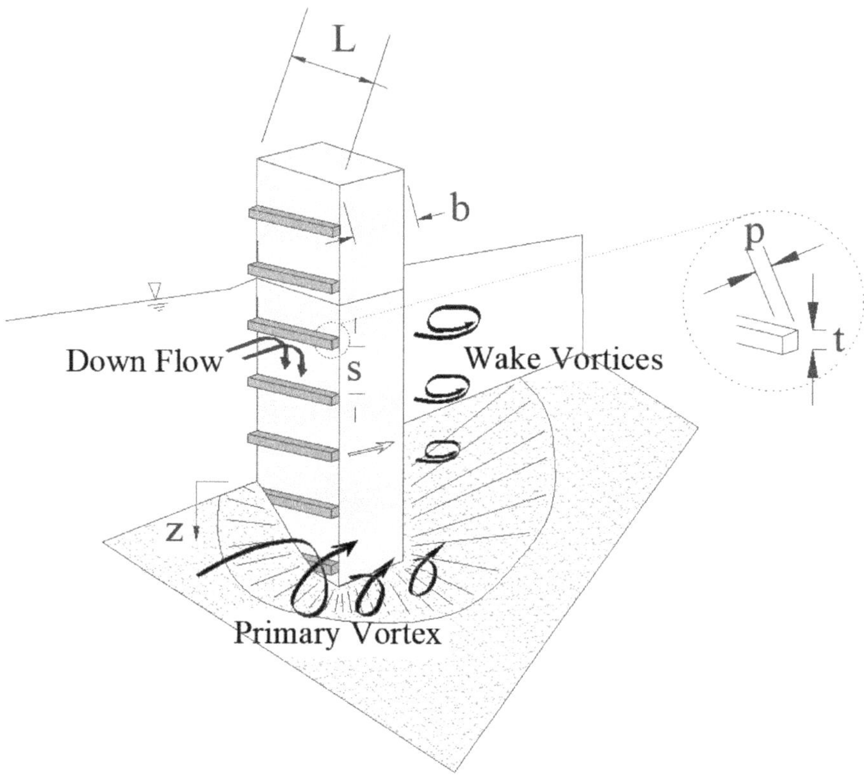

Figure 1. Flow field around an abutment, roughening elements at abutment upstream face.

To cope with local scour around bridge abutments, three approaches are available. One approach is to apply no scour countermeasure, and instead, the bridge foundation is constructed deeper than the equilibrium scour depth. In this approach, accurate estimation of scour depth is crucial, and long-term tests need to be conducted under desired hydraulic conditions. Ref. [8] evaluated local scour in a non-uniform gravel bed and used both laboratory and field data. They proposed new K-factors for the Melville and Coleman relation and developed a new relation to estimate maximum scour depth. Ref. [9] proposed a new equilibrium scour depth relation by estimating the total sediment transport around a bridge pier exposed to local scour. Various researchers have attempted to develop relations to estimate local scour around bridge foundation, and a list of empirical relations can be found in [10], describing the ultimate scour depth as a function of flow, sediment, foundation, and channel properties. The relations are valid under certain conditions, such as live bed or clear water. A second approach involves an increase in the resistance of the bed material by the placement of different coatings. In this method, heavy elements are installed to create a physical barrier that enhances sediment stability against approaching flow, and includes the placement of riprap, concrete blocks, gabions, etc. Various researchers have studied this approach, such as [11–14]. Most of these studies aimed to introduce relations for designing the size and placement extent of the pieces. Finally, for the third approach the flow is altered or deviated from the foundation to reduce erosive factors, followed by installation of collars, slots, submerged vanes, sacrificial piles [6,15–24]. These studies aimed to determine the geometric properties and effectiveness of the proposed methods. Some other researchers have studied the combination of both methods. Ref. [25] investigated the application of riprap alone and a combination of riprap and collar for scour countermeasures around rectangular bridge piers aligned with the flow. They developed relations for predicting stable riprap size and extent with different aspect ratios and skew angles of piers. They concluded that increasing the skew angle and aspect ratio of the piers increase the required riprap size and extent. Ref. [26] conducted similar studies (combination of riprap and collar) for cylindrical piers and tested seven riprap sizes with two different dimensions of collars. Their results show that the application of collar reduces stable riprap size and extension. Ref. [27] conducted 48 laboratory experiments to study the combination of riprap and Six Pillar Concrete (SPC) elements on scour mitigation around a vertical wall bridge abutment. They concluded that SPC elements deviate the flow and scour from the abutment tip. Moreover, they found that the installation of pebbles around pillars enhances the stability of the SPC elements, prevents edge failure, and reduces abutment scour depth.

Among flow altering devices, the use of roughening elements is a newly emerged method that has received less attention. This method requires more investigation in order to explore various aspects of the approach and to improve its performance as a scour countermeasure strategy. Roughening elements can change the flow field and weaken the downward flow, which is the main cause of the formation of horseshoe vortices [28]. The factors affecting the performance of the roughening elements are the element protrusions (P), thicknesses (t), spacing (S), and installation depth of the elements (Z). Figure 1 shows a protected abutment with roughening elements and effective geometric parameters with a flow pattern included. This method resembles that involving the use of threading cables to reduce scour around piers [29].

The primary attempt to use roughening elements as scour countermeasure was undertaken by [30]. They applied these elements as a flow altering method and reduced the maximum scour depth to 7.2% at the abutment wall, concluding that roughening elements can be used as suppressor of downflow to reduce the action of the vortex. Ref. [28] investigated the element's spacing (S) and protrusion (P) at a given thickness (t). They claimed that smaller spacings create a barrier against downflow. In contrast, a large space between two elements enables the downflow to reattach to the abutment wall, which causes the performance of the elements to plummet. Briefly, they claimed that the least scouring occurs for $S/P = 2$.

Based on literature review, few studies have been conducted on the performance of roughening elements as a scour countermeasure method. Moreover, to the best of our knowledge, the optimum placement depth, thickness and protrusion of the elements have not been investigated. Therefore, this research aims to build on previous studies and answer questions about the optimum installation depth and dimensions of the elements.

2. Materials and Methods

The scour depth (ds) is a function of different factors, including the geometry of fluid, flow, channel, abutment and roughening elements, as well as sediment properties. In order to develop dimensionless parameters sixteen variables are considered as follows:

$$d_s = f_1(\rho, \mu, g, y, U, B, L, b, t, P, S, Z, \rho_s, d_{50}, \sigma, T) \tag{1}$$

where, ρ is water density, μ is water dynamic viscosity, g is gravitational acceleration, y is flow depth, U is flow average velocity, B is channel (flume) width, L is abutment length (m) (perpendicular to the flow), b is abutment width (m) (in direction of flow), t and P are roughening element thickness and protrusion, respectively (m), S is the roughening element's spacing (m), Z is the depth of the roughening element installation below initial bed level (m), ρ_s is the sediment density, d_{50} is median sediment size, σ is geometric standard deviation, and T is the time in seconds. Using the Buckingham theory and taking ρ, U, L as repetitive variables, the general form of the equation for predicting scour around and abutment armed with roughening elements can be expressed as follows:

$$\frac{d_s}{y} = f_2\left(R_e,Fr,\frac{L}{y},\frac{L}{B},\frac{b}{L},\frac{t}{L},\frac{p}{L},\frac{S}{L},\frac{Z}{L},\frac{\rho_s}{\rho},\frac{L}{d_{50}},\sigma,\frac{TU}{L}\right) \tag{2}$$

where R_e represents the Reynold number and F_r represents the Froude number defined in the following equations:

$$R_e = \frac{\rho.U.y}{\mu} \tag{3}$$

$$F_r = \frac{U}{\sqrt{g.y}} \tag{4}$$

The length scale "L" can be replaced by water depth "y" (or a combination of "y, L") as suggested in other studies [2,28,31]. The graphs are presented with dimensionless parameters as variables to make it possible to expand the results to real world scale. Some parameters, such as R_e, $\rho s/\rho$, σ, L/y and L/B, are constant in each hydraulic condition. Therefore, desired parameters including P/L, t/L, Z/L, and ds/L are studied and presented in terms of tables and graphs. As a result, Equation (2) can be summarized to the following form:

$$\frac{d_s}{y} = \left(\frac{P}{L},\frac{t}{L},\frac{Z}{L},F_r\right) \tag{5}$$

The experiments were conducted in a laboratory flume with dimensions of 16 m (length) × 0.7 m(width) × 0.6 m (depth) and a longitudinal slope of 0.0027. The walls and bottom of the flume were made of glass (see Figure 2). Water was pumped from an underground reservoir to a head-tank to maintain a constant head water and was then poured to the flume. The flow depth was adjusted by a sharp, crested, rectangular gate located at the end of the flume, and the flow discharge was measured using an electronic flow meter and adjusted with a butterfly valve. To evaluate local scour, cohesionless sediments with a median diameter of 0.71 mm and a thickness of 20 cm was used. According to [32], using sediments with an average diameter of less than 0.7 mm escalates the possibility of bed form creation. In addition, the geometric standard deviation of the sediments $\sigma_g = \sqrt{\frac{d_{84}}{d_{16}}}$) was 1.25, which satisfies the uniformity condition [33,34] and is less than 1.3.

Two vertical wall abutments with same length of 0.14 m and widths (b) of 0.07 and 0.14 m were used. The abutment dimensions were chosen to eliminate the contraction ratio, as the effect of the contraction ratio on scouring is eliminated when the ratio of the abutment length to flume width is less than 0.33 [35]. Two different widths were chosen to study the effect of the abutment width on scouring. Roughening elements with the same length as the abutment and with different protrusions and thicknesses were investigated. The thickness and the protrusion of the elements were equal to P = t = 0.05 L, 0.1 L, 0.2 L, and 0.3 L. Based on the findings of [28], the spacing between the elements was set to S/P = 2.

According to [2], an abutment is classified as short if the ratio of its length to the flow depth is less than one, and where the scouring will be independent of the flow depth. To determine hydrodynamic conditions, a flow discharge of 31 lit/s was used to establish a constant flow depth of 15 cm. The threshold velocity was measured by using a one-dimensional velocimeter at the desired flow depth, and was found to be 0.3 m/s, which is consistent with the value obtained from Melville and Sutherland's [36] formula:

$$u_C/u_{*C} = 5.75\log\left(5.53\frac{y}{d_{50}}\right) \tag{6}$$

where U_C is critical velocity, U_{*C} is the critical shear velocity, y is flow depth, and d_{50} is the median diameter of sediments. The critical shear velocity was calculated using Shield's diagram. In this study, the ratio of flow velocity (U) to critical velocity was 0.94. Because this ratio is less than 1, the clear-water conditions were achieved [37]. To provide a better understanding of experimental results, other hydraulic conditions were tested in addition to U/U_C = 0.94, including U/U_C = 0.9, 0.8 and 0.7 were tested as well.

Figure 2. Schematic illustration of flume setup.

In the literature, test durations range from 2 to 96 h [38,39]. In this study, a primary test was conducted for 24 h, during which two parameters, the scour depth and time at the abutment toe, were continuously monitored in order to determine the temporal evolution of the scour depth.

It was observed that the majority of the scouring occurs at the beginning of experiments. After 5 h, negligible changes in scour depth were observed, which is consistent with the findings of [28,40]. Refs. [27,41,42] have also reported that variations in the final percentage of scour depth versus time become negligible after 4 h for different Froude numbers and abutment geometries. However, it should be mentioned that the goal of the current study was not to propose a relation in order to estimate the maximum scour depth or reach an equilibrium condition. To reach a fully developed condition, the test must be prolonged for several days [43,44]. Some researchers, such as [45], believe that such a condition is not attainable. However, a duration of 5 h was chosen for the tests in this study.

In the initial set of experiments, two unprotected abutments were tested, and it was observed that the highest scour depth occurred at the abutment toe. Therefore, the abutments toe was chosen as the reference point for comparison of the scouring in all subsequent experiments, as it was in similar studies [28,46,47]. The topography around the abutment was recorded in a 2 × 2 cm grid using a laser distance meter that could move in longitudinal and cross-sectional directions with 1 mm accuracy. To evaluate the effectiveness of the scour countermeasure in subsequent experiments, the maximum scour depth at the abutment toe was used as the reference point for comparing results. Figure 3 shows the maximum scour depth at two abutment tips in the baseline experiments.

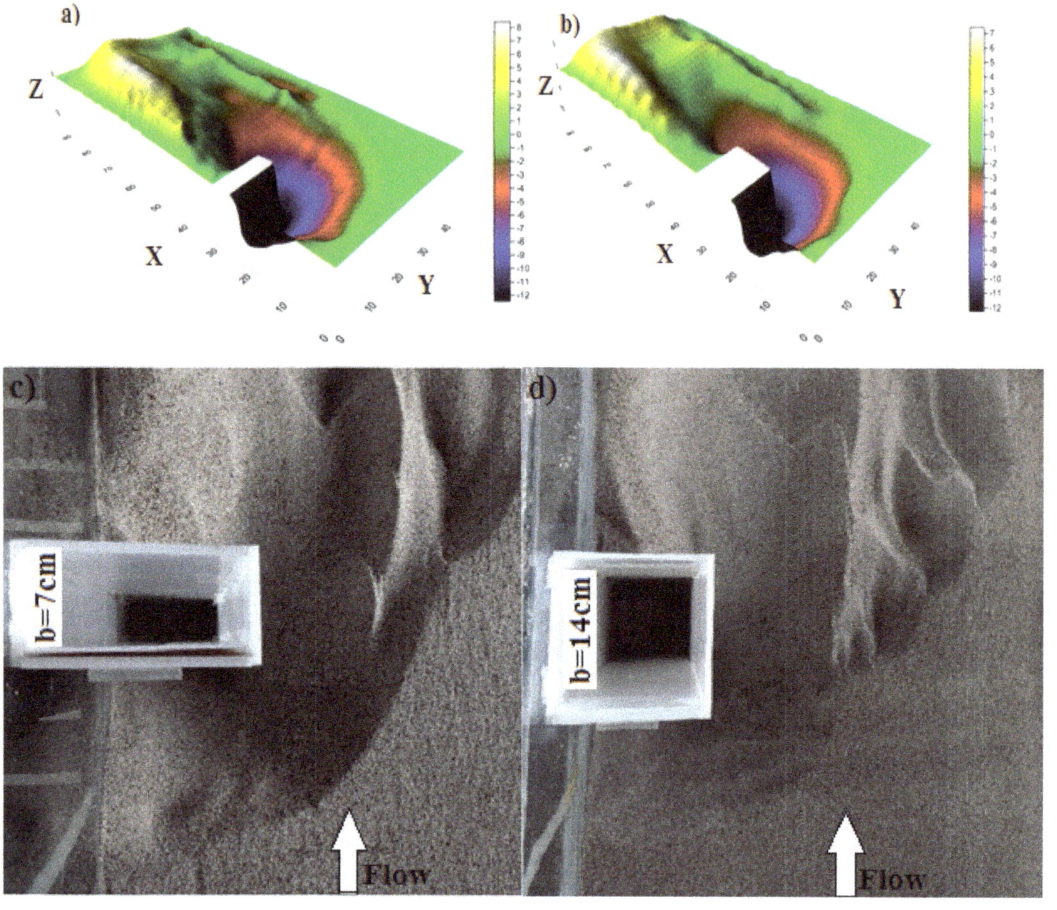

Figure 3. Maximum scour depth at abutment toe. (**a**) Width = 7 cm (b/L = 0.5), (**b**) width = 14 cm (b/L = 1) (Flow Direction is from right to left), and (**c**,**d**) runs of baseline experiments.

The assessment experiments were designed to investigate the optimum installation depth and dimensions of the scour countermeasure elements. In all of the experiments, the space between two successive elements was set to $S/P = 2$, which was selected based on the findings of [28]. The experiments were conducted in the following order: (1) the effect of installing elements between sediment surface and water level, with the dimension of t = P = 0.05 L, was evaluated at the abutment face (b = 7 cm), In this experiment, an element was first installed, and subsequent tests were conducted by adding one element to investigate the effect on scour depth of placing this element above the sediment (Figure 4a). (2) The optimum depth of installation was determined by placing elements under the initial level of the sediment (buried in sediments, t = P = 0.05 L) (Figure 4b). (3) After determining optimum installation depth, tests were conducted with the presence of elements both under and above the bed level (Figure 4c). In these tests, four elements were installed under the bed level at a depth of $Z/L = 0.6$ (the optimum depth), while the number of elements installed above the bed was increased from 1 to 8 to reach the water surface. This step was undertaken in order to evaluate the installation of elements buried in sediment and between the water level and sediment surface. Steps 1–3 were repeated for elements with different dimensions (thickness and protrusion). (4) The effect of different dimensions was evaluated under the optimum depth of installation (Figure 4d). All roughening elements were tested with a square cross-section, so the values of P (protrusion) and t (thickness) were equal. Steps 1–4 were conducted under the hydraulic condition of $u/u_c = 0.94$. (5) In the final experiments, all four steps were repeated under different hydraulic conditions, of u/u_c = 0.9, 0.8, and 0.7, to obtain more authentic results. It should be noted that the best results obtained from each step were applied in the subsequent step. For example, the optimum depth of installation (Z/L) determined in step 2 was used in step 4 to determine the optimum dimension.

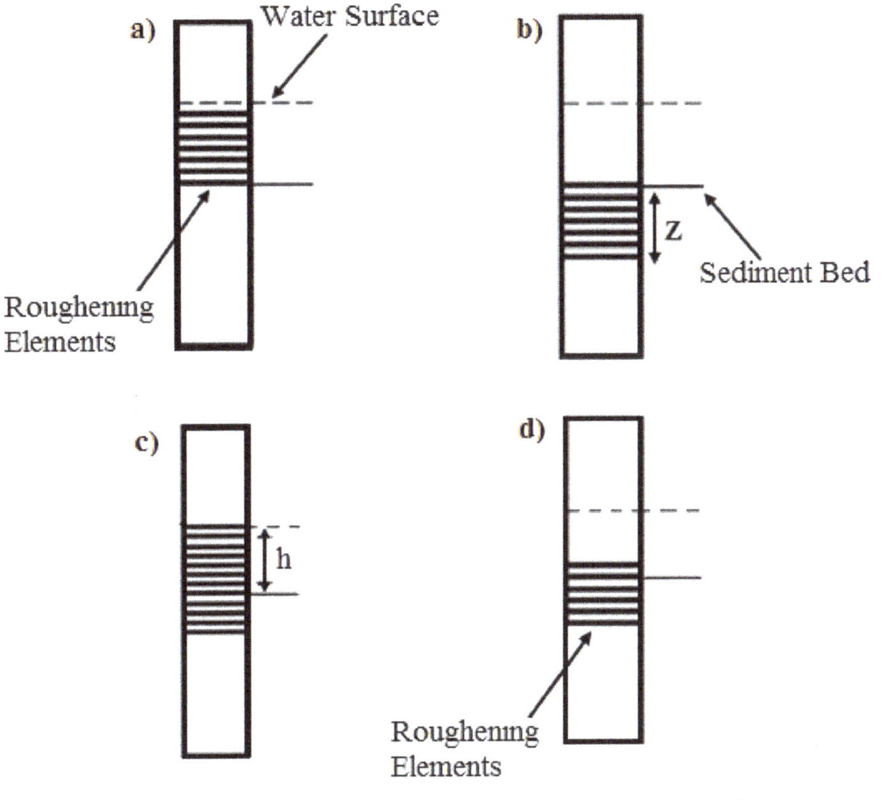

Figure 4. Sketch of the installation of roughening elements on the upstream face of the abutment, under and above the bed (**a**): T_1, (**b**) T_7, (**c**) T_8, (**d**) T_{14} (The roughening elements face the flow direction).

In the assessment experiments, some of the results obtained for the small abutment (*b* = 7 cm) were repeated for the larger abutment. For example, the optimum depth of installation determined for the small abutment was repeated for the larger one. Therefore, the number of experiments conducted for the abutment with width of 14 cm is less than for the abutment with the width of *b* = 7 cm (Table 1).

Table 1. Overall view of experiments.

Test Conditions		Baseline Experiments	Assessment Experiments				Total Experiments
			Elements under Bed Level	Elements above and under Bed Level	Optimum Thickness and Protrusion	Different Hydraulic Conditions	
Number of Tests	b = 7 cm	1	7	7	4	16	35
	b = 14 cm	1	4	4	4	16	29

3. Results and Discussion

The results of the baseline experiments indicate that the widths of the abutment do not have significant effect on the maximum scour depth at the abutment toe, which can be attributed to the role of contraction in the scouring process. A larger abutment width leads to greater volume of scour depth, which can result in more severe changes in bed topography. This is because the longer the extension of the abutment, the further the progression of scouring downstream. Ref. [48] also found that the maximum scour depth is independent of the contraction length (abutment length) which is consistent with the current study's findings. They observed that the variation of scour depth at the abutment toe was less than 10% for abutments with widths of 5 and 110 cm. This suggests that the abutment width does not have a significant effect on the maximum scour depth at the abutment toe. Additionally, they have reported that the velocity vectors were very similar around the abutment toe, where the maximum scour depth occurs, which indicates that the flow pattern is not significantly affected by the abutment width.

4. Installing Elements between the Bed and Water Surface (Step 1 in Assessment Experiments)

The findings indicate that installing more than one element on the sediment bed, regardless of the desired thickness and protrusion, does not reduce scour depth and in some cases, may even increase it. There is no clear trend in the results.

5. Determining the Optimum Installation Depth of the Elements

In the second step of the assessment experiments, (Table 2, Tests T_1 to T_7) (Figure 4b), it was found that installing elements at a depth of more than 8.4 cm ($Z/L = 0.6$) from the bed level has no effect on scour reduction. This is evident in tests T_5, T_6 and T_7 in Table 2. The reason for this may be that, as a scour hole deepens, the strength of the primary vortex and three-dimensional flow field inside the scour hole decreases, and therefore the deeper placement of elements has no effect on scour depth. Consequently, the number of elements under the bed was set at four to satisfy the $Z/L = 0.6$ criterion (Test T_5). Based on Table 2 (Test T_8 to T_{14}), installing elements between the sediment bed and water surface has no consistent pattern regarding scour reduction (compare T_9, T_{10}, T_{11} and T_{12}).

In the previous section, the effect of installing elements between the sediment's initial bed and water surface was investigated (in this case, no elements were installed under the bed level), and it was concluded that installing elements above the bed level did not have a specific effect on scour depth and varied randomly. In this section (Figure 4c), elements were installed both under and above the bed level, and the same results were obtained. Comparing the results with the previous section (in which elements were installed between the bed and water surface), it can be inferred that installing elements above the bed level does not impose a specific trend on scour reduction, even if some elements are installed within the sediments. Therefore, it is concluded that elements must be installed within a specific range to affect the primary vortex action that develops inside the scour hole [4]. Thus, installing elements under the sediment surface resulted in better protection of the abutment.

Table 2. Summary of results of scour depth around both abutments (b = 7 and b = 14 cm) using roughening elements with different protrusion and thickness.

Test Number	Number of Elements under the Bed	Number of Elements above the Bed	Abutment Width	t = P	U/U_C	ds/y	Scour Depth Reduction %	Z (cm)	Z/L
T_1	0	1	b/L = 0.5	0.05 L (0.7 cm)	0.94	0.77	7.2	0	0
T_2	1	1	b/L = 0.5	0.05 L (0.7 cm)	0.94	0.76	8.8	2.1	0.15
T_3	2	1	b/L = 0.5	0.05 L (0.7 cm)	0.94	0.75	9.6	4.2	0.3
T_4	3	1	b/L = 0.5	0.05 L (0.7 cm)	0.94	0.74	11.2	6.3	0.45
T_5	4	1	b/L = 0.5	0.05 L (0.7 cm)	0.94	0.69	17.6	8.4	0.6
T_6	5	1	b/L = 0.5	0.05 L (0.7 cm)	0.94	0.69	17.6	10.5	0.75
T_7	6	1	b/L = 0.5	0.05 L (0.7 cm)	0.94	0.69	17.6	12.6	0.9
T_8	4	2	b/L = 0.5	0.05 L (0.7 cm)	0.94	0.7	16	8.4	0.6
T_9	4	3	b/L = 0.5	0.05 L (0.7 cm)	0.94	0.7	16	8.4	0.6
T_{10}	4	4	b/L = 0.5	0.05 L (0.7 cm)	0.94	0.69	17.6	8.4	0.6
T_{11}	4	5	b/L = 0.5	0.05 L (0.7 cm)	0.94	0.73	12	8.4	0.6
T_{12}	4	6	b/L = 0.5	0.05 L (0.7 cm)	0.94	0.71	14.4	8.4	0.6
T_{13}	4	7	b/L = 0.5	0.05 L (0.7 cm)	0.94	0.72	13.6	8.4	0.6
T_{14}	4	8	b/L = 0.5	0.05 L (0.7 cm)	0.94	0.7	15.2	8.4	0.6
T_{15}	4	1	b/L = 0.5	0.05 L (0.7 cm)	0.94	0.69	17.6	8.4	0.6
T_{16}	2	1	b/L = 0.5	0.1 L (1.4 cm)	0.94	0.65	22.4	9.8	0.7
T_{17}	1	1	b/L = 0.5	0.2 L (2.8 cm)	0.94	0.58	30.4	10.5	0.75
T_{18}	1	1	b/L = 0.5	0.3 L (4.2 cm)	0.94	0.6	28	11.2	0.8
T_{19}	4	1	b/L = 1	0.05 L (0.7 cm)	0.94	0.73	12	8.4	0.6
T_{20}	2	1	b/L = 1	0.1 L (1.4 cm)	0.94	0.6	27.2	9.8	0.7
T_{21}	1	1	b/L = 1	0.2 L (2.8 cm)	0.94	0.56	32.8	10.5	0.75
T_{22}	1	1	b/L = 1	0.3 L (4.2 cm)	0.94	0.59	29.6	11.2	0.8

Note: the 7th column is obtained by comparing with baseline tests (ds = 12.5 cm).

The results for other dimensions of roughening elements were similar. For t = P = 0.1 L, 0.2 L and 0.3 L, placing elements under the bed was found to be more effective than placing them above the sediment, and the best depth of element installation was at *Z/L* = >0.6–0.8. The results of the experiments, i.e., the dimensionless scour depths at the abutment toe for different installation depth of elements, are shown in Figure 5.

Based on Figure 5, it can be observed that for all element dimensions, the ratio of *Z/L* = >0.6–0.8 has the lowest scour depth. Therefore, the optimum installation depth of the elements under the sediments is proposed to be 0.6–0.8 times the abutment length. Similar results were obtained for the other abutment (*b* = 14 cm).

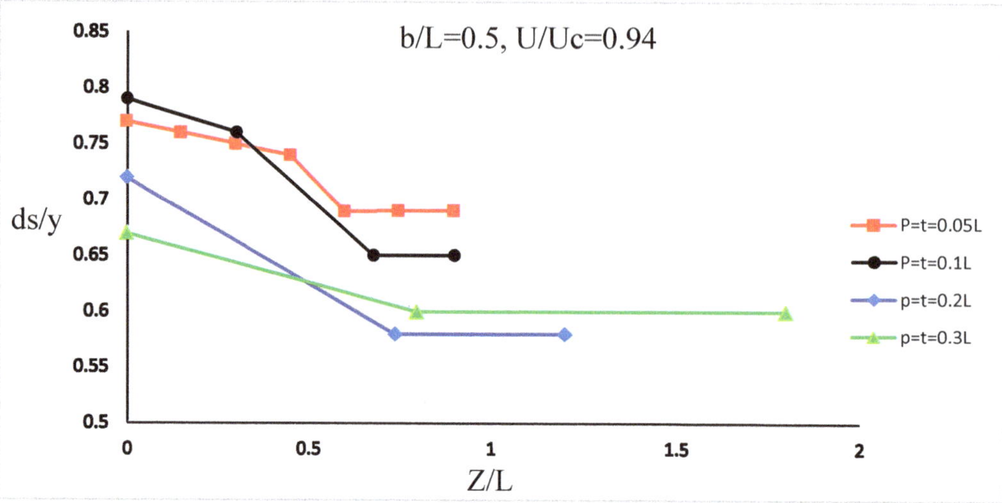

Figure 5. Effect of the installation depth of elements with different dimensions at scour depth.

6. Determining Optimum Thickness and Protrusion of the Elements

From a hydraulic perspective, after the flow collides with the upstream abutment face, a downward flow is formed that turns into the bed and creates the principal vortex and abutment scour. The presence of roughening elements creates disturbance in the downward flow, leading to energy dissipation. As a result, the principal vortex is weakened, and less intense scouring occurs compared with baseline experiments.

The results indicate that, as the dimensions of roughening elements increase (Table 2, T_{15} to T_{17} for b = 7 cm, T_{19} to T_{21} for b = 14 cm), the scour depth decreases to a certain extent and then increases (compare T_{17} to T_{18} and T_{21} to T_{22}). The results for both abutments are similar (Figure 6), indicating that abutment width does not affect the maximum scour depth. However, it can be concluded that, after increasing the protrusion of the elements beyond a certain value, they act as a new wall, reducing the effectiveness of the elements and leaving the abutment unprotected. This condition is similar to moving the abutment upstream by the same amount as the element's protrusion, resulting in an increase in scour depth. The results suggest the best results can be obtained when the thickness and protrusion of the elements are equal to 20% of the abutment length. The scour depth at the abutment toe with the width of 7 and 14 cm was reduced by up to 30.4 and 32.8%, respectively.

Figure 7 shows the effect of different hydraulic conditions on scour depth in different element dimensions.

Figure 7 demonstrates that the best performance of roughening elements in reducing scour was achieved at t = P = 0.2 L (2.8 cm) even under different hydraulic conditions. The figure also shows that increasing the dimension of the elements to a certain degree reduces scour depth. It is clear that, under different hydraulic conditions and when increasing the dimension of the elements to a certain amount, the scour depth reduces. After P/L = t/L = 0.2, larger dimension has a reverse effect on scour depth.

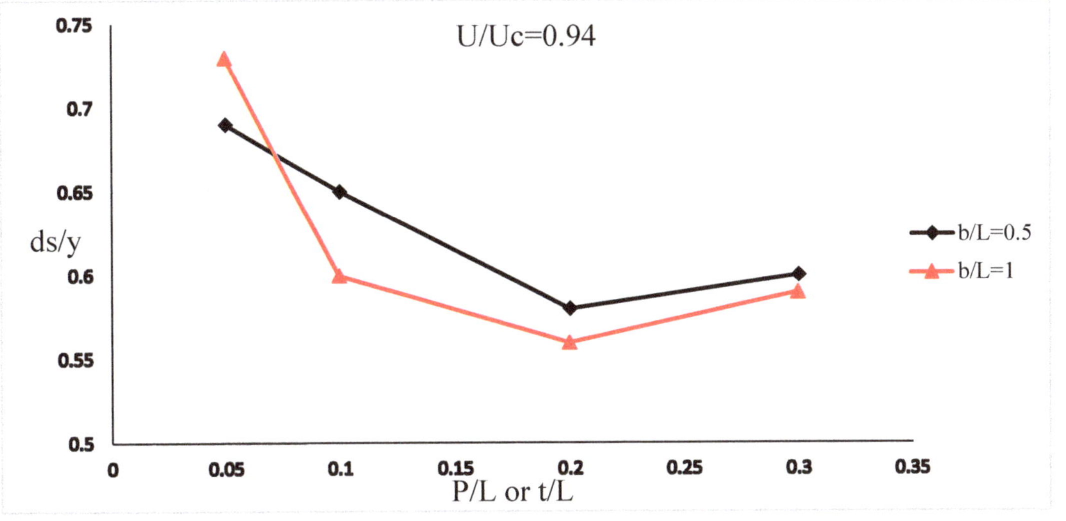

Figure 6. Results of the effects of the elements protrusion size on scour depth for both abutments. All elements' dimensions were tested under the best depth of installation (optimum value of Z/L).

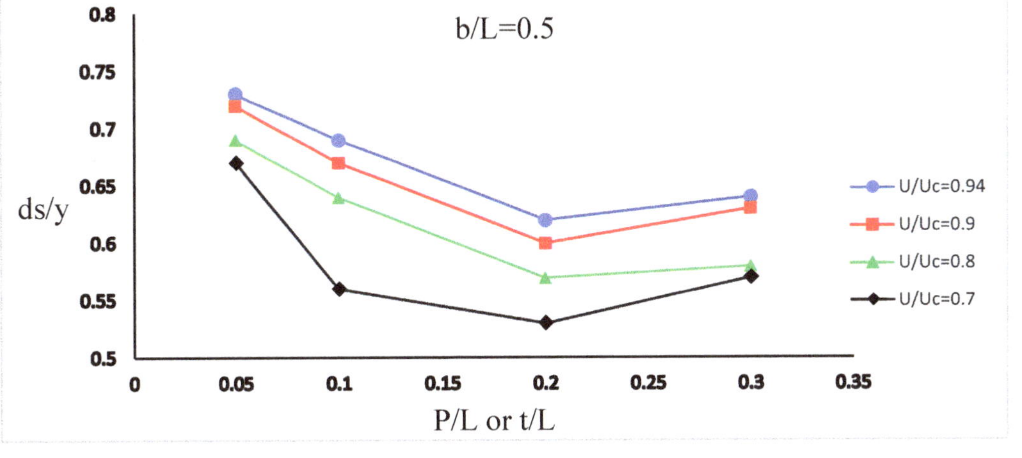

Figure 7. Evaluation of scour depth at different hydraulic conditions.

7. Conclusions

This study investigated the effectiveness of roughening elements as a countermeasure for bridge abutment scour. Different arrangements and dimensions of roughening elements were tested on two rectangular abutments, and the local scour at the abutment toe was measured as the baseline point. The optimal depth of installation was found to be at Z/L = >0.6–0.8 under different hydraulic conditions. The best countermeasure performance was achieved by increasing the dimensions of the elements (protrusion (P) and thickness (t)) up to 0.2 L (abutment length), However, increasing the element dimensions beyond this point (t = p = 0.3 L), led to an increase in scour depth. Overall, the best performance was achieved at t = P = 0.2 L and Z/L = >0.6–0.8, resulting in a reduction of scour depth by up to 30.4% (b = 7 cm) and 32.8% (b = 14 cm), respectively. Further investigation is needed to investigate the effectiveness of these elements on cohesive soils and to expand their application.

Author Contributions: M.Z.: Methodology, Writing—original draft, Data curation, validation; S.M.A.Z.: validation, investigation, supervision; A.F.: Writing—review & editing, Formal analysis, Conceptualization; R.P.T.: Writing—review & editing, visualization; N.J.: Methodology, investigation; D.M.: resources, formal analysis; P.S.: Formal analysis, visualization; H.M.A.: Writing—review & editing, Formal analysis. All authors have read and agreed to the published version of the manuscript.

Funding: The costs for this research were provided from the grants of the second author, Water Engineering Department, Shiraz University (08, 05, 2016).

Data Availability Statement: Some or all data, models, or code that support the findings of this study are available from the corresponding author upon reasonable request. (List items).

Conflicts of Interest: The authors declare no conflict of interest.

Notations

Following symbols are used in this study:

B = channel (flume) width
b = Abutment width
d_{50} = Median diameter of sediment
Fr = Froud number
g = gravitational acceleration,
L = abutment length (m) (perpendicular to the flow),
P = Elements' protrusion
Re = Reynold number
S = roughening element's spacing t = Elements thickness
t = roughening element's thickness
T = time
U = Flow velocity
U_C = Critical velocity
U_{*C} = Critical shear velocity
y = Flow depth
Z = Installation depth of elements
σ_g = Geometric standard deviation
ρ_s = the sediment density,
ρ = water density
μ = water dynamic viscosity

References

1. Richardson, E.V.; Harrison, L.J.; Richardson, J.R.; Davies, S.R. *Evaluating Scour at Bridges*; US Department of Transportation, Federal Highway Administration: Washington, DC, USA, 1993.
2. Melville, B.W. Local scour at bridge abutments. *J. Hydraul. Eng.* **1992**, *118*, 615–631. [CrossRef]
3. Afzalimehr, H.; Moradian, M.; Singh, V.P. Flow Field around Semielliptical Abutments. *J. Hydrol. Eng.* **2018**, *23*, 04017057. [CrossRef]
4. Barbhuiya, A.K.; Dey, S. Vortex flow field in a scour hole around abutment. *Int. J. Sed. Res.* **2003**, *18*, 310–325. [CrossRef]
5. Molinas, A.; Kheireldin, K.; Wu, B. Shear Stress around Vertical Wall Abutments. *J. Hydraul. Eng.* **1998**, *124*, 822–830. [CrossRef]
6. Tripathi, R.P.; Pandey, K.K. Numerical investigation of flow field around T-shaped spur dyke in a reverse-meandering channel. *Water Supply* **2022**, *22*, 574–588. [CrossRef]
7. Lai, Y.G.; Liu, X.; Bombardelli, F.A.; Song, Y. Three-Dimensional Numerical Modeling of Local Scour: A State-of-the-Art Review and Perspective. *J. Hydraul. Eng.* **2022**, *148*, 03122002. [CrossRef]
8. Pandey, M.; Oliveto, G.; Pu, J.H.; Sharma, P.K.; Ojha, C.S.P. Pier Scour Prediction in Non-Uniform Gravel Beds. *Water* **2020**, *12*, 1696. [CrossRef]
9. Gazi, A.H.; Afzal, M.S. A new mathematical model to calculate the equilibrium scour depth around a pier. *Acta Geophys.* **2019**, *68*, 181–187. [CrossRef]
10. Melville, B.; Coleman, S. *Bridge Scour*; Water Resources Publications: Littleton, CO, USA, 2000; 550p.
11. Melville, B.; Van Ballegooy, S.; Coleman, S.; Barkdoll, B. Scour Countermeasures for Wing-Wall Abutments. *J. Hydraul. Eng.* **2006**, *132*, 563–574. [CrossRef]
12. Korkut, R.; Martinez, E.J.; Morales, R.; Ettema, R.; Barkdoll, B. Geobag Performance as Scour Countermeasure for Bridge Abutments. *J. Hydraul. Eng.* **2007**, *133*, 431–439. [CrossRef]

13. Cardoso, A.H.; Fael, C.M. Protecting Vertical-Wall Abutments with Riprap Mattresses. *J. Hydraul. Eng.* **2009**, *135*, 457–465. [CrossRef]
14. Sui, J.; Afzalimehr, H.; Samani, A.K.; Maherani, M. Clear-water scour around semi-elliptical abutments with armored beds. *Int. J. Sediment Res.* **2010**, *25*, 233–245. [CrossRef]
15. Dargahi, B. Controlling Mechanism of Local Scouring. *J. Hydraul. Eng.* **1990**, *116*, 1197–1214. [CrossRef]
16. Chiew, Y.M. Scour protection at bridge piers. *J. Hydraul. Eng.* **1992**, *118*, 1260–1269. [CrossRef]
17. Kumar, V.; Raju, K.G.R.; Vittal, N. Reduction of Local Scour around Bridge Piers Using Slots and Collars. *J. Hydraul. Eng.* **1999**, *125*, 1302–1305. [CrossRef]
18. Johnson, P.A.; Hey, R.D.; Tessier, M.; Rosgen, D. Use of Vanes for Control of Scour at Vertical Wall Abutments. *J. Hydraul. Eng.* **2001**, *127*, 772–778. [CrossRef]
19. Zarrati, A.R.; Gholami, H.; Mashahir, M. Application of collar to control scouring around rectangular bridge piers. *J. Hydraul. Res.* **2004**, *42*, 97–103. [CrossRef]
20. Li, H.; Barkdoll, B.D.; Kuhnle, R.; Alonso, C. Parallel Walls as an Abutment Scour Countermeasure. *J. Hydraul. Eng.* **2006**, *132*, 510–520. [CrossRef]
21. Heidarpour, M.; Afzalimehr, H.; Izadinia, E. Reduction of local scour around bridge pier groups using collars. *Int. J. Sediment Res.* **2010**, *25*, 411–422. [CrossRef]
22. Fathi, A.; Zomorodian, S.M.A. Effect of Submerged Vanes on Scour Around a Bridge Abutment. *KSCE J. Civ. Eng.* **2018**, *22*, 2281–2289. [CrossRef]
23. Valela, C.; Rennie, C.D.; Nistor, I. Improved bridge pier collar for reducing scour. *Int. J. Sediment Res.* **2021**, *37*, 37–46. [CrossRef]
24. Tripathi, R.P.; Pandey, K.K. Scour around spur dike in curved channel: A review. *Acta Geophys.* **2022**, *70*, 2469–2485. [CrossRef]
25. Mashahir, M.B.; Zarrati, A.R.; Mokallaf, E. Application of Riprap and Collar to Prevent Scouring around Rectangular Bridge Piers. *J. Hydraul. Eng.* **2010**, *136*, 183–187. [CrossRef]
26. Zarrati, A.R.; Chamani, M.R.; Shafaie, A.; Latifi, M. Scour countermeasures for cylindrical piers using riprap and combination of collar and riprap. *Int. J. Sediment Res.* **2010**, *25*, 313–322. [CrossRef]
27. Zolghadr, M.; Bejestan, M.S.; Fathi, A.; Hoseinreza, A. Protecting vertical-wall bridge abutment using six-pillar concrete elements. *Arab. J. Geosci.* **2022**, *15*, 1226. [CrossRef]
28. Radice, A.; Davari, V. Roughening Elements as Abutment Scour Countermeasures. *J. Hydraul. Eng.* **2014**, *140*, 06014014. [CrossRef]
29. Dey, S.; Sumer, B.M.; Fredsøe, J. Control of Scour at Vertical Circular Piles under Waves and Current. *J. Hydraul. Eng.* **2006**, *132*, 270–279. [CrossRef]
30. Radice, A.; Lauva, O. On flow-altering countermeasures for scour at vertical-wall abutment. *Arch. Hydro Eng. Environ. Mech.* **2012**, *59*, 137–153.
31. Oliveto, G.; Hager, W.H. Temporal evolution of clear-water pier and abutment scour. *ASCE J. Hydraulic. Eng.* **2002**, *128*, 811–820. [CrossRef]
32. Breusers, H.N.C.; Raudkivi, A.J. *Scouring. Hydraulic Structures Design Manual2*; IAHR: Rotterdam, The Netherlands, 1991.
33. Raudkivi, A.J.; Ettema, R. Clear-water scour at cylindrical piers. *J. Hydraul. Eng.* **1983**, *109*, 338–350. [CrossRef]
34. Haghnazar, H.; Sangsefidi, Y.; Mehraein, M.; Tavakol-Davani, H. Evaluation of infilling and replenishment of river sand mining pits. *Environ. Earth Sci.* **2020**, *79*, 362. [CrossRef]
35. Ballio, F.; Teruzzi, A.; Radice, A. Constriction Effects in Clear-Water Scour at Abutments. *J. Hydraul. Eng.* **2009**, *135*, 140–145. [CrossRef]
36. Melville, B.W.; Sutherland, A.J. Design Method for Local Scour at Bridge Piers. *J. Hydraul. Eng.* **1988**, *114*, 1210–1226. [CrossRef]
37. Raudkivi, A.J. *Loose Boundary Hydraulics*, 3rd ed.; Technical Report; Pergamon Press: Oxford, UK, 1990.
38. Bozkus, Z.; Yildiz, O. Effects of Inclination of Bridge Piers on Scouring Depth. *J. Hydraul. Eng.* **2004**, *130*, 827–832. [CrossRef]
39. Vittal, N.; Kothyari, U.C.; Haghighat, M. Clear-water scour around bridge pier group. *J. Hydraul. Eng.* **1994**, *120*, 1309–1318. [CrossRef]
40. Mohammadpour, R.; Ghani, A.A.; Azamathulla, H.M. Estimation of dimension and time variation of local scour at short abutment. *Int. J. River Basin Manag.* **2013**, *11*, 121–135. [CrossRef]
41. Zolghadr, M.; Bejestan, M.S.; Rezaeianzadeh, M. Investigating the Effect of Six-Legged Element Placement Density on Local Scour at Wing-Wall Bridge Abutments. In Proceedings of the World Environmental and Water Resources Congress, West Palm Beach, FL, USA, 22–26 May 2016. [CrossRef]
42. Zolghadr, M.; Bejestan, M.S. Six legged concrete (SLC) elements as scour countermeasures at wing wall bridge abutments. *Int. J. River Basin Manag.* **2020**, *19*, 319–325. [CrossRef]
43. Coleman, S.E.; Lauchlan, C.S.; Melville, B. Clear-water scour development at bridge abutments. *J. Hydraul. Res.* **2003**, *41*, 521–531. [CrossRef]
44. Cheng, N.S.; Chiew, Y.M.; Chen, X. Scaling Analysis of Pier-Scouring Processes. *J. Eng. Mech. ASCE* **2016**, *142*, 06016005. [CrossRef]
45. Lanca, R.; Fael, C.; Cardoso, A. Assessing equilibrium clear water scour around single cylindrical piers. *River Flow* **2010**, *2010*, 1207–1213.
46. Khazimenejad, H.; Ghomeishi, M.; Shafai Bajestan, M. Comparison of Symmetrical and Unsymmetrical Rectangular Collars on Reduction of Local Scour at Bridge Abutment. *J. Irrig. Sci. Eng.* **2014**, *37*, 1–12.

47. Mahdi, M.; Mehdi, F.; Venkat, A. Using data mining methods to improve discharge coefficient prediction in Piano Key and Labyrinth weirs. *Water Supply* **2021**, 22, 1964–1982. [CrossRef]
48. Xiong, X.; Melville, B.W.; Feriedrich, H. Effect of contraction length on abutment scour. In Proceedings of the International IAHR World Congress, Athens, Greece, 5–7 September 2013.

MDPI
St. Alban-Anlage 66
4052 Basel
Switzerland
www.mdpi.com

Fluids Editorial Office
E-mail: fluids@mdpi.com
www.mdpi.com/journal/fluids

www.ingramcontent.com/pod-product-compliance
Lightning Source LLC
Chambersburg PA
CBHW061447210926
51850CB00008B/1643
* 9 7 8 3 7 2 5 8 1 4 2 7 5 *